# Springer Tracts in Mechanical Engineering

Springer Tracts in Mechanical Engineering (STME) publishes the latest developments in Mechanical Engineering - quickly, informally and with high quality. The intent is to cover all the main branches of mechanical engineering, both theoretical and applied, including:

- Engineering Design
- Machinery and Machine Elements
- Mechanical Structures and Stress Analysis
- Automotive Engineering
- Engine Technology
- Aerospace Technology and Astronautics
- Nanotechnology and Microengineering
- Control, Robotics, Mechatronics
- MEMS
- Theoretical and Applied Mechanics
- Dynamical Systems, Control
- Fluids Mechanics
- Engineering Thermodynamics, Heat and Mass Transfer
- Manufacturing
- Precision Engineering, Instrumentation, Measurement
- Materials Engineering
- Tribology and Surface Technology

Within the scope of the series are monographs, professional books or graduate textbooks, edited volumes as well as outstanding PhD theses and books purposely devoted to support education in mechanical engineering at graduate and post-graduate levels.

**Indexed by SCOPUS, zbMATH, SCImago.**

Please check our Lecture Notes in Mechanical Engineering at https://link.springer.com/bookseries/11236 if you are interested in conference proceedings.

To submit a proposal or for further inquiries, please contact the Springer Editor **in your region** :

Ms. Ella Zhang (China)
Email: ella.zhang@springernature.com
Swati Meherishi (India)
Email: swati.meherishi@springer.com
Dr. Leontina Di Cecco (All other countries)
Email: leontina.dicecco@springer.com

All books published in the series are submitted for consideration in Web of Science.

Guoqing Zhang · Jianpeng Wang

# Material-Oriented Cutting Processes in Precision Machining

Springer

Guoqing Zhang
College of Mechatronics and Control Engineering
Shenzhen University
Shenzhen, Guangdong, China

Jianpeng Wang
College of Mechatronics and Control Engineering
Shenzhen University
Shenzhen, Guangdong, China

ISSN 2195-9862 ISSN 2195-9870 (electronic)
Springer Tracts in Mechanical Engineering
ISBN 978-981-96-2503-1 ISBN 978-981-96-2504-8 (eBook)
https://doi.org/10.1007/978-981-96-2504-8



This Springer imprint is published by the registered company Springer Nature Singapore Pte Ltd.
The registered company address is: 152 Beach Road, #21-01/04 Gateway East, Singapore 189721, Singapore

# Preface

The influence of material properties on the cutting process in precision machining is a critical aspect of modern manufacturing technologies. This book delves into this subject from a microscopic perspective, exploring how the unique characteristics of single crystal, polycrystalline, amorphous, and ferrous materials affect the cutting mechanisms and surface generation during machining. Additionally, the role of diamond tool materials and the size effects on cutting performance are thoroughly examined, with the goal of optimizing cutting processes to enhance efficiency and precision.

About this book, Chap. 1 investigates the cutting mechanisms in machining single crystal materials, focusing on the brittle-ductile transition and using molecular dynamic simulation to analyze these effects. Chapter 2 continues by addressing surface generation in single crystal materials, particularly in the context of nanosurface creation and surface morphology control. The discussion shifts to polycrystalline materials in Chap. 3, where the properties of irregular microstructures and their impact on surface generation mechanisms are explored. Chapter 4 takes a closer look at the machining of amorphous materials, examining their machining performance and the intricacies of surface generation. Chapter 5 addresses the challenges of tool wear in machining ferrous materials, offering insights into the electrochemical tool wear mechanisms and proposing various methods for suppressing wear. This theme continues in Chap. 6, which presents advanced cutting techniques such as intermittent cutting technology and the effects of lubricants on machining processes. Chapter 7 focuses on the size effects in precision machining, analyzing how the size of the cutting edge and cutting depth influence the overall cutting process. Chapter 8 concludes with a discussion on synthetic diamond, detailing the diamond synthesis process and evaluating the cutting performance of diamond tools in precision machining.

This book aims to provide a comprehensive overview of feasible machining technologies and advanced cutting processes tailored to the specific properties of various materials. It bridges the disciplines of materials science and precision machining, making it an invaluable resource for graduate students, researchers, and engineering technicians in these fields. Additionally, professionals in related areas such as material engineering, optics, semiconductor manufacturing, and advanced manufacturing technologies may find this book insightful and inspiring.

Last, we thank the publisher for their guidance and support in bringing this work to fruition.

Shenzhen, China Guoqing Zhang
Hong Kong, China Jianpeng Wang

**Competing Interests** The authors have no competing interests to declare that are relevant to the content of this manuscript.

# Contents

# About the Authors

**Guoqing Zhang** born in June 1982, received his bachelor's degree in mechanical engineering from Northeast Petroleum University, China, in 2005. In 2009, he received his master's degree in mechanical engineering from Harbin Institute of Technology, China and he received his Ph.D. degree from the Hong Kong Polytechnic University in 2014. Now, he is a full professor at Shenzhen University and selected as a Shenzhen Peacock Program for Overseas/Professional High-Level Talents. He serves as a Deputy Director of Guangdong Key Laboratory of Electromagnetic Control and Intelligent Robots, Deputy Director of Shenzhen Key Laboratory of High Performance Nontraditional Manufacturing, Committee member of Chinese Mechanical Engineering Society Design Branch, Member of Advanced Optical Manufacturing Young Expert Committee of Chinese Optical Engineering Society.

His research interests include ultra-precision machining technology and equipment, robots and intelligent equipment, etc. Until now, he has hosted more than 10 research projects including 4 NSFC (National Natural Science Foundation of China) projects, published more than 110 peer-reviewed papers and has been granted 14 US patents and 42 CN patents and co-edited 2 books published by Springer. He is now served as an academic editor for the international journal *Shock and Vibration* and reviewers for more than 10 international journals. He has won more than

5 prizes including the China Industry-University-Research Cooperation Promotion Award (individual) and the second prize for innovation achievement.

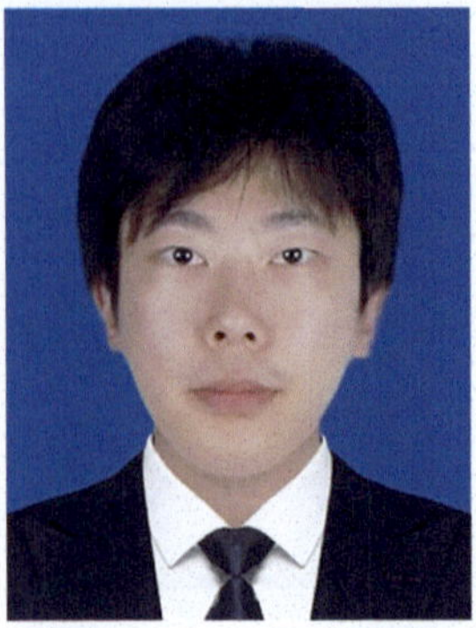

**Jianpeng Wang** born in March 1996, received his bachelor's degree in mechanical engineering from Yantai University, China, in 2018. In 2022, he received his master's degree in mechanical engineering from Shenzhen University, China. He is currently a Ph.D research student at the Hong Kong Polytechnic University.

His primary research interest includes ultra-precision machining technology and equipment. Until now, he has published more than 10 peer-reviewed papers and has been granted 5 CN patents. Co-edited 2 books published by Springer. He has been honored with multiple scholarship awards, including the National Scholarship for Postgraduates and the First-class Academic Scholarship. His thesis was awarded as one of the "Top 100 Outstanding Master Degree Thesis" of Shenzhen University.

# Abbreviations

| | |
|---|---|
| AA | Amorphous alloy |
| AM | Amorphous materials |
| BDT | Brittle-ductile transition |
| BSCM | Brittle single crystal materials |
| CBN | Cubic boron nitride |
| CHPA | Cubic anvil high-pressure apparatus |
| CUCT | Critical undeformed chip thickness |
| CVD | Chemical vapor deposition |
| DBAP | Ductile-brittle alternating pattern |
| DMS | Down milling side |
| DTW | Diamond tool wear |
| EAM | Embedded atom method |
| EDS | Energy dispersive spectrometer |
| EVC | Elliptical ultrasonic vibration cutting |
| FCC | Face-centered cubic |
| FE | Finite element |
| FTIR | Fourier-transform infrared |
| GCT | Gradient cooling technology |
| HCJC | Hybrid constitutive Johnson-Cook |
| HPGM | High pressure gas machining |
| HPHT | High-pressure high-temperature |
| IMS | Irregular micro-structures |
| MD | Molecular dynamics |
| MF | Magnetic field |
| MQLM | Minimum quantity lubrication machining |
| MSB | Micro-slot bottom |
| ODM | Ordinary dry machining |
| OM | Optical microscopy |
| PBC | Periodic boundary conditions |
| PCD | Polycrystalline diamond |
| PSZ | Primary shear zone |

| | |
|---|---|
| PVD | Physical vapor deposition |
| RDF | Radial distribution function |
| SCC | Single-crystal copper |
| SCD | Single-crystal diamond |
| SCG | Single-crystal germanium |
| SEM | Scanning electron microscope |
| SHPB | Split Hopkinson pressure bar |
| SPDT | Single-point diamond turning |
| SRP | Subsurface residual stresses |
| TGGM | Temperature gradient growth method |
| TIS | Tool-in side |
| TOS | Tool-out side |
| UCT | Undeformed chip thickness |
| UMS | Up milling side |
| UPFC | Ultra-precision fly cutting |
| UPM | Ultra-precision machining |
| UVAM | Ultrasonic vibration-assisted machining |
| UVC | Ultrasonic vibration cutting |
| XRD | X-ray diffraction |

# Chapter 1
# Cutting Mechanism in Machining Single Crystal Materials

**Abstract** Single-crystal materials machining involves complex interactions between the cutting tool and the material, significantly influenced by the material's unique anisotropic properties and the transition between brittle and ductile cutting modes. Single-point diamond turning (SPDT) is a crucial machining method used for shaping single-crystal materials to achieve desired surface morphologies. However form accuracy and performance are substantially affected by the machined surface of single-crystal materials' presence of a ductile–brittle alternating pattern (DBAP). This chapter provides a comprehensive overview of these cutting mechanisms, focusing on the critical phenomena of brittle-ductile transition (BDT) and the insights gained from molecular dynamics (MD) simulations.

**Keywords** Single-point diamond turning · Cutting mechanism · Brittle-ductile transition · Single crystal materials

## 1.1 Brittle-Ductile Transition

The advancements in semiconductors, integrated circuits, and optical technology have led to an increased focus on ultra-precision machining technology for fabricating optics, micro/nano sensors, and micro/nano electromechanical systems [1, 2]. However, conventional polycrystalline materials are proving insufficient in meeting the high-quality and high-fidelity requirements of optical-electronic components [3]. Single-crystal materials, characterized by their absence of lateral grain boundaries, offer excellent fidelity and high-speed signal propagation properties, making them ideal candidates for such applications [4–6].

Brittle single-crystal materials (BSCMs) are an important subset of single-crystal materials, and SPDT serves as a prominent ultra-precision machining technique due to its efficiency and high-quality output [7–10]. However, BSCM is susceptible to crack extension and defects on the machined surface due to the brittle removal mechanism during machining [11]. The anisotropic nature of BSCM further complicates controlling the consistency of surface morphology [12, 13]. Consequently,

 
G. Zhang and J. Wang, *Material-Oriented Cutting Processes in Precision Machining*, Springer Tracts in Mechanical Engineering,
https://doi.org/10.1007/978-981-96-2504-8_1

brittle removal combined with anisotropy results in the occurrence of a DBAP on the machined surface [14, 15]. The above phenomenon influences BSCM's machining quality under SPDT greatly. Therefore, it is essential to study the mechanism behind ductile/brittle removal and anisotropy to mitigate the negative effects of DBAP.

This section aims to provide insight into the mechanism behind the BDT. Future research can apply these findings to use or explore similar methods for other BSCMs machining.

### *1.1.1 Brittle-Ductile Transition in SPDT*

The phenomenon of BDT in SPDT of brittle materials mainly occurs in two common machining methods. One is fixed depth cutting, and the other is variable depth cutting. The mechanisms of BDT generated by these two machining methods are different, where the former is induced by the anisotropy of single-crystal materials' surface, manifested as fan-shaped patterns on the machined surface, while the latter is the BDT within the same crystal plane/direction.

In this section, experiments were performed using an ultra-precision lathe (Moore, Nanotech 450UPL) to show the two machining method. The experimental setup and schematic are shown in Fig. 1.1a. The lathe features two translational axes ($X$- and $Z$-axis) and an optional rotary axis ($C$-axis). The workpiece, fixed on a vacuum chuck, can be adjusted to have its crystallographic direction parallel to the $Z$-axis.

Face-turning (Fig. 1.1b) represents the fixed cutting depth method and plunge-cutting (Fig. 1.1c) represents the variable cutting depth method. During face-turning, the workpiece undergoes rotation on the turning spindle as the tool moves along the $X$-axis with a consistent feed in the $X$-direction and a depth of cut in the $Z$-direction. The relative motion between the workpiece and tool facilitates the process of removing material. Plunge-cutting is a machining technique where the workpiece does not rotate. The tool moves along the $X$-axis and cuts at different depths in the $Z$-direction, while maintaining a constant feed in the $X$-direction. The tool's rake face is aligned parallel to the horizontal plane, causing the feed direction with the $X$-axis to be inclined. Material removal is accomplished through the relative motion between the workpiece and tool.

Germanium (Ge) is a representative example of a BDT mechanism, commonly employed in experiments to investigate this phenomenon. It belongs to the category of BSCM. The surface of Ge following face-turning displays alternating patches with foggy and clear morphologies (Fig. 1.1d). The coming sections will discuss the anisotropic features of Ge surfaces under various cutting conditions. Experiments were conducted to investigate the anisotropy mechanism and the BDT in Ge by performing plunge-cutting on several crystal surfaces. Figure 1.1e depicts the distribution diagrams of grooves on various surfaces, guaranteeing the existence of at least one groove in each anisotropic region.

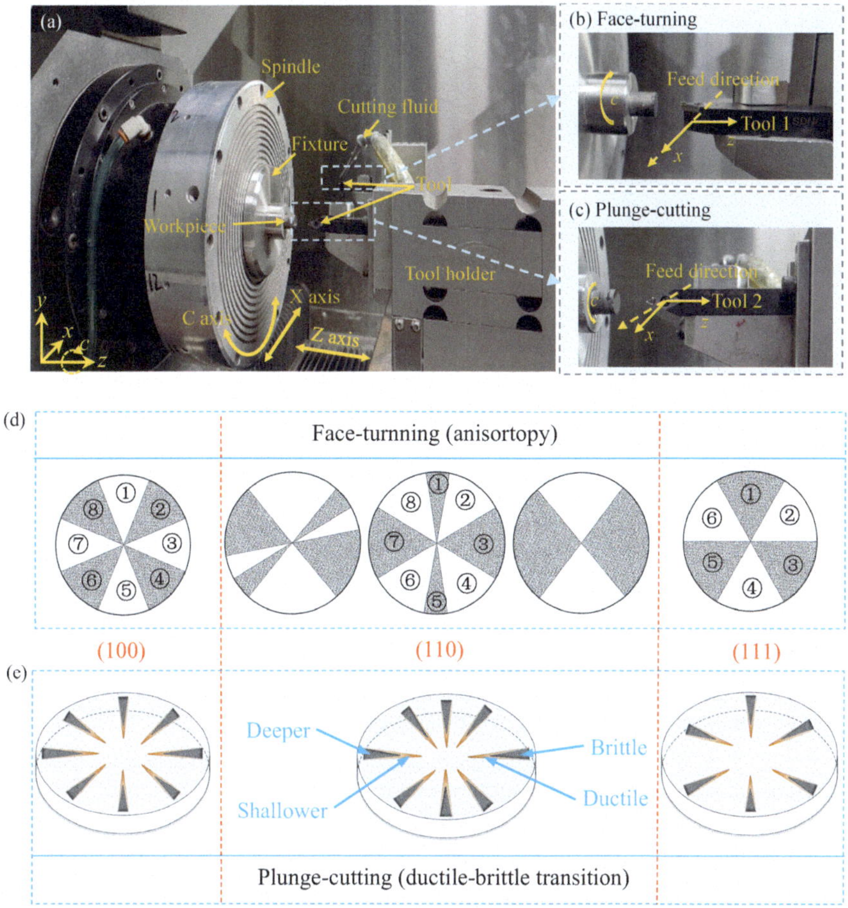


**Fig. 1.1** **a** Photographs depicting the setup and movement of the tool and workpiece in the turning system; **b** an illustration of the face-turning experiments; **c** an illustration of the plunge-cutting experiments; **d** potential surface structures of Ge (100), (110), and (111) crystal planes after face-turning, with alternating cloudy-clear regions labeled to represent different anisotropic areas; **e** the design of groove distribution at different intervals on the surfaces after plunge-cutting, ensuring the presence of at least one groove in one anisotropic region as shown in Fig. 1.1d

Following the completion of the cutting experiments, the surface roughness and morphology were evaluated using a white light interferometer (Bruker, Contour GT-X). The tool edge morphology was analyzed using an scanning electron microscope (SEM) manufactured by FEI, model Quanta450FEG. In addition, the crystal structures of both the surface and subsurface were analyzed and described using a laser confocal Raman spectrometer (Lab Ram Xplora).

Table 1.1 presents the specific characteristics of the workpiece and tool utilized in this section, whereas Table 1.2 lists the cutting parameters.

**Table 1.1** Parameters of diamond tools and workpieces used in the experiments

| Parameters | Values |
|---|---|
| Tool material | Single-crystal diamond |
| Tool rake angle (°) | 0 |
| Tool clearance angle (°) | 10 |
| Tool radius (mm) | 1.041 |
| Workpiece material | Single-crystal germanium with (100) (110) (111) crystal plane |
| Workpiece size (mm * mm) | Ø10 * 5 |

**Table 1.2** Cutting parameters used in the experiments

| Experiments | Spindle speed ($n$) | Feed rate ($f$) | Cutting depth ($a_p$) | Inclination angle |
|---|---|---|---|---|
| Face-turning | 2000 r/min | 5 and 10 mm/min | 1 and 2 μm | 0° |
| Plunge-cutting | 0 | 0.3, 3, and 30 mm/min | 3 μm | arctan (1/1000) |

## 1.1.2 *Mechanical Representation*

The presence of anisotropy can affect the stress distribution and stress concentration in materials, thereby influencing the failure mechanisms and failure modes of the material, which in turn affects the BDT behavior of the material.

During the loading process, anisotropy can lead to stress concentration in certain directions, making these regions more prone to crack initiation or failure, resulting in brittle failure of the material. Conversely, certain crystal orientations can lead to a more uniform stress distribution, reducing stress concentration and making the material more prone to plastic deformation, exhibiting ductile failure. Therefore, a thorough analysis of the BDT process in Ge requires mechanical analysis of its anisotropy.

The anisotropy of Ge can be characterized and visualized through its mechanical properties.

(1) Elastic constants

Table 1.3 shows the elastic constants of Ge, including $c_{11}$, $c_{12}$, and $c_{44}$.

**Table 1.3** The elastic constants and compliance coefficients of Ge

| | | | |
|---|---|---|---|
| Elastic constants (Gpa) | $c_{11}$ | $c_{12}$ | $c_{44}$ |
| | 128.9 | 48.3 | 67.1 |
| Compliance coefficients (Mpa$^{-1}$) | $s_{11}$ | $s_{12}$ | $s_{44}$ |
| | 9.749 | – 2.657 | 14.903 |

(2) Compliance coefficients

Table 1.3 also shows the compliance coefficients of Ge, including $s_{11}$, $s_{12}$, and $s_{44}$. The derivations of these coefficients are shown as follows [16]:

$$\begin{cases} s_{11} = \frac{c_{11}+c_{12}}{(c_{11}-c_{12})(c_{11}+2c_{12})} \\ s_{12} = \frac{-c_{12}}{(c_{11}-c_{12})(c_{11}+2c_{12})} \\ s_{44} = \frac{1}{c_{44}} \end{cases} \quad (1.1)$$

(3) Anisotropic characteristic parameter

When it comes to the anisotropic properties, the evaluation parameter, $d$, is defined below:

$$d = \frac{2c_{44}}{c_{11} - c_{12}} \quad (1.2)$$

When $d \neq 1$, the crystal material exhibits anisotropic, whereas when $d = 1$, it is isotropic. For Ge, the evaluation parameter $d$, calculated using Eq. (1.2), is found to be 1.665. This value indicates that Ge displays crystal anisotropy.

(4) Shear modulus and elastic modulus

The anisotropy of Ge's mechanical properties, such as shear modulus ($G$) and elastic modulus ($E$), can be evaluated through the anisotropy of its shear and elastic modulus. Based on the theory of metal elastic anisotropy, the following formula can be employed to deduce the $G$ and $E$ of Ge [17, 18]:

$$\begin{aligned} G^{-1} &= s_{44} + 4\left[(s_{11} - s_{12}) - \frac{1}{2}s_{44}\right]\left(u^2v^2 + w^2 + w^2u^2\right) \\ E^{-1} &= s_{11} - 2\left[(s_{11} - s_{12}) - \frac{1}{2}s_{44}\right]\left(u^2v^2 + u^2w^2 + v^2w^2\right) \end{aligned} \quad (1.3)$$

where $s_{ij}$ represents the compliance coefficients of Ge, and $u$, $v$, and $w$ represent the direction cosine of crystal planes.

By substituting the corresponding parameters into Eq. (1.3) and solving using MATLAB® software, the values of $G$ and $E$ in different directions of space can be calculated. To visualize the anisotropy of $G$ and $E$, cloud images are generated as shown in Fig. 1.2a–b. It should be noted that the change of G and E with direction is dependent on $(u^2v^2 + u^2w^2 + v^2w^2)$. This quantity is 0 for the directions of (100) and reaches its maximum value of 1/3 in the directions of (111). In addition, it can be observed that the extremum and minimum values of $G$ occur in the (100) and (111) directions of Ge, respectively, while for $E$, it is the opposite. Since this is the case, the cloud picture of $G$ displays protrusions along the edges of the cube, whereas the cloud image of $E$ appears as a cube with rounded corners and indentations in the centre of each face.

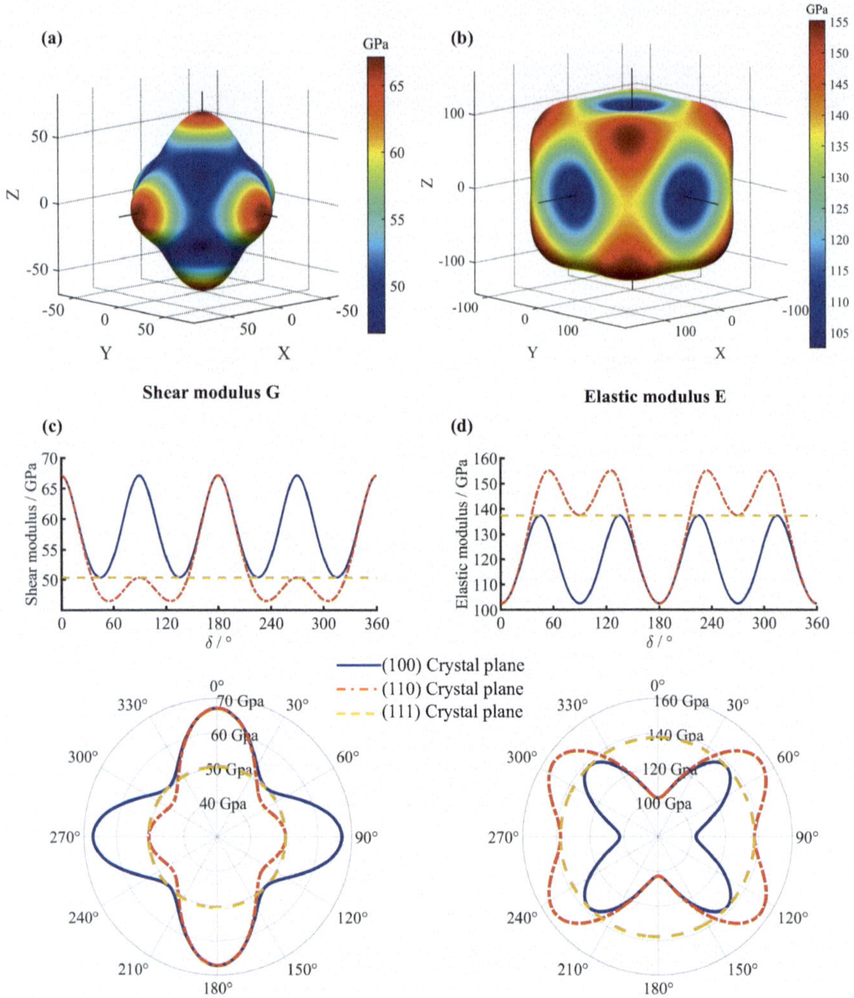


**Fig. 1.2** **a** Anisotropic cloud graph of the shear modulus and **c** theoretical values of the shear modulus; **b** Anisotropic cloud graph of the elastic modulus and **d** theoretical values of the elastic modulus of Ge

By transforming the coordinate system, MATLAB® can generate the continuous curves of G and E for the (111), (110), and (100) crystal planes of Ge, as depicted in Fig. 1.2c–d. Based on the observations, it can be observed that *G* and *E* on the (110) and (100) planes display periodic oscillations, however on the (111) plane, they stay constant. Additionally, the extreme points and values of *G* and *E* vary among these three planes, indicating anisotropic mechanical properties in Ge across different crystal planes. Even within the same plane, except for the elastically isotropic (111) plane, Ge's mechanical properties display directional anisotropy with periodic variations.

### 1.1.3 Critical Cutting Depth

The critical cutting depth refers to the transition depth between brittle cutting and ductile cutting during the SPDT of Ge. The cutting depth is directly related to the surface stress exerted by the diamond tool on Ge. Additionally, due to the surface anisotropy of Ge, the critical cutting depth also varies with different crystal orientations and planes. Therefore, a thorough understanding of the critical cutting depth is crucial for achieving high-quality Ge surfaces during the SPDT process.

Group 14 (IVa) of the periodic table is where the chemical element Ge may be found. It has an atomic weight of 72.64 amu, which places it in the middle of the group that contains tin and silicon. In the same way that silicon is classified as a semiconductor, Ge is classified as a semiconductor because it possesses characteristics that are midway between those of metals and nonmetals. Due to the fact that it possesses four valence electrons and an electron configuration of $[Ar]3d^{10}4s^{2}4p^{2}$, it is capable of effectively producing stable $Ge^{4+}$ and $Ge^{2+}$ ions. Ge has the appearance of a near-metal that is grayish-white in colour, hard and brittle, and has a brilliant, shining, and silvery look. There is a face-centered cubic (FCC) lattice in the crystal structure of germanium, which is similar to the structure of a diamond. In Fig. 1.3a, the unit cell is shown, with the spheres standing in for germanium atoms and the lines standing in for covalent links.

The coordinates of the germanium atoms are $(0, 0, 0)$, $(\frac{1}{2}, \frac{1}{2}, 0)$, $(\frac{1}{2}, 0, \frac{1}{2})$, $(0, \frac{1}{2}, \frac{1}{2})$, $(\frac{1}{4}, \frac{1}{4}, \frac{1}{4})$, $(\frac{3}{4}, \frac{3}{4}, \frac{1}{4})$, $(\frac{3}{4}, \frac{1}{4}, \frac{3}{4})$, $(\frac{1}{4}, \frac{3}{4}, \frac{3}{4})$, respectively. This section focuses on the characteristics of SPDT on the (111), (110), and (100) crystal planes of Ge, according to the emergence mechanism of DBAP. The arrangement and number of atoms differ across various crystal planes. It is clear from looking at Fig. 1.3b–d that the cutting direction is constantly shifting in relation to the crystal orientation as the cutting operation is being carried out.

In order to simplify calculations, the initial crystal direction is typically chosen as [0 1 0], [−11 1 0], [−1 1 0] for the three typical crystal planes, respectively. The determination of the crystal orientation of Ge on certain crystal planes may be accomplished by analysing the angle $\delta$ that exists between the direction of machining and the beginning direction, where $\delta$ is a value that lies within the range of 0 to $2\pi$.

$$
\begin{aligned}
\left[u\ v\ w\right]_{100} &= \left[0\ \cos\delta_{100}\ \sin\delta_{100}\right] \\
\left[u\ v\ w\right]_{110} &= \left[-\tfrac{\sqrt{2}}{2}\cos\delta_{110}\ \tfrac{\sqrt{2}}{2}\cos\delta_{110}\ \sin\delta_{110}\right] \\
\left[u\ v\ w\right]_{111} &= \left[-\tfrac{\sqrt{2}}{2}\cos\delta_{111} - \tfrac{\sqrt{6}}{6}\sin\delta_{111}\ \tfrac{\sqrt{2}}{2}\cos\delta_{111} - \tfrac{\sqrt{6}}{6}\sin\delta_{111}\ \tfrac{\sqrt{6}}{3}\sin\delta_{111}\right]
\end{aligned}
\tag{1.4}
$$

In this equation, $u$, $v$, and $w$ represent the direction cosine of the various crystal planes.

The equation used to determine if a crystal material exhibits anisotropy is expressed as [19]:

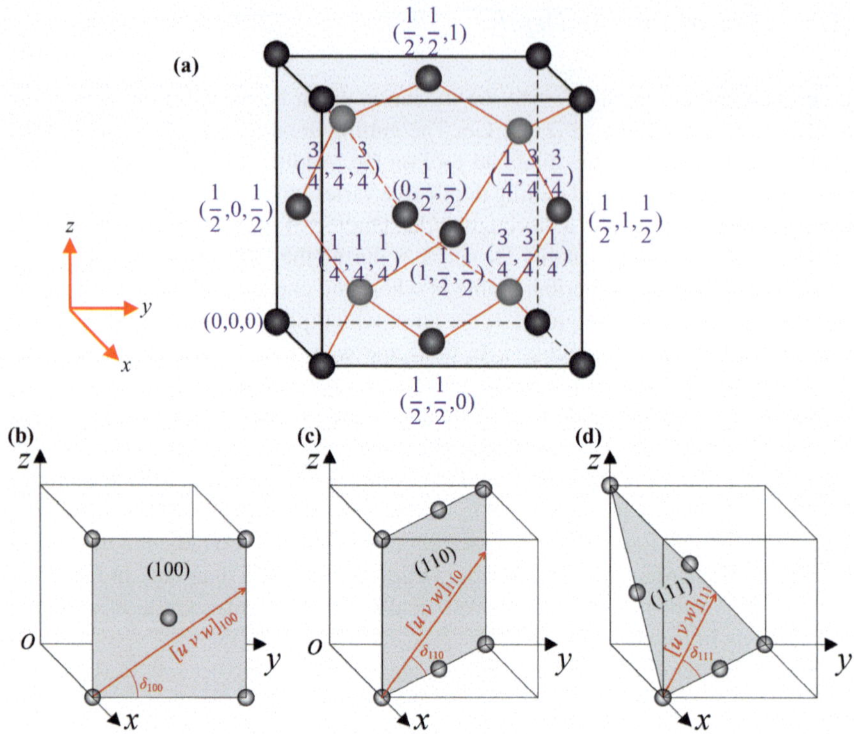


**Fig. 1.3** **a** Schematic diagram of Ge's cell structure and the relative position concerning the cutting direction; **b** (100) crystal planes; **c** (110) crystal planes; **d** (100) crystal planes

$$d = \frac{2C_{44}}{C_{11} - C_{12}} \tag{1.5}$$

The anisotropic characteristic parameter of the material is denoted by $d$, and the elastic constants of the material, $C_{11}$, $C_{12}$ and $C_{44}$, are provided in Table 1.4. By substituting the data into Eq. (1.5), the anisotropic characteristic parameter $d$ of Ge can be calculated as 1.665, indicating that Ge theoretically exhibits anisotropy.

The previous calculations have confirmed that Ge exhibits anisotropic properties. The critical depth of the BDT in Ge is therefore theoretically anisotropic in nature. This is because of the nature of the material. The particular equation that is utilised in order to compute the critical depth of the BDT that occurs during the cutting of Ge is as follows:

**Table 1.4** Values of elastic constants $C$ of Ge

| $C_{11}$ (GPa) | $C_{12}$ (GPa) | $C_{44}$ (GPa) |
|---|---|---|
| 128.9 | 48.3 | 67.1 |

$$D_c = 4\sin^2(\delta - 90°)m_0 \cot\left(\frac{0.5\pi - \alpha - \gamma}{2}\right)\sqrt{\frac{2\lambda}{1.8544}\frac{9\psi EJ}{50H^2}} \tag{1.6}$$

The symbols used in the Eq. (1.6) are defined as follows: $D_c$ represents the critical depth of the BDT, $\delta$ denotes the cleavage angle, $\alpha$ is the clearance angle, $\gamma$ indicates the rake angle, $H$ represents the microhardness, $H \in (7.546, 8.624\ \text{GPa})$, generally $H^{(100)} < H^{(110)} < H^{(111)}$, $m_0$ is the correction coefficient for the deviation angle, $\lambda$ indicates the colligation coefficient, $\lambda = (1 \sim 1.6) \times 10^4$, $\psi$ is the correlation coefficient of surface roughness, $\psi \in (1, 2)$, $E$ denotes the elastic modulus, $E = 1.3 \times 10^2$ GPa, and $J$ is the surface energy required for crack generation. The values of various parameters in Eq. (1.6) can be found in Table 1.5 [19].

During the cleavage process, the covalent bonds between atoms need to be broken. Therefore, the presence of periodic boundary conditions (PBC) strong bonds plays a role in determining the difficulty of material removal in Ge. Moreover, the difficulty of material removal is also influenced by the crystallographic cutting direction, denoted as $\boldsymbol{F}$, and the direction of the PBC vector, denoted as $\boldsymbol{P}$. The angle between $\boldsymbol{P}$ and $\boldsymbol{F}$ is inversely proportional to the material removal difficulty. The coefficient of material removal difficulty, $\varepsilon$ is used to describe the degree of difficulty in processing the material. A lower value of $\varepsilon$ indicates easier material removal. The expression for $\varepsilon$ in Ge can be given by [20]:

$$\varepsilon = \boldsymbol{F} \cdot \boldsymbol{P} \tag{1.7}$$

The following is an expression of the modification equations for the three typical crystal planes, which may be derived from Eq. (1.7):

$$\begin{cases} \varepsilon_{(100)} = 0.8|\sin 2\theta| + 0.2 \\ \varepsilon_{(110)} = 0.7356|\cos 2\theta| + 0.1226 \\ \varepsilon_{(111)} = 0.9966|\sin(3\theta/2 - \pi/4)| + 0.3322 \end{cases} \quad \theta \in [0,\ 2\pi] \tag{1.8}$$

where $\varepsilon_{(hlk)}$ represents the cutting difficulty coefficient for the crystal plane $(hlk)$ of the material, $\boldsymbol{F}$ denotes the composed cutting force, and $\boldsymbol{P}$ strands for the direction of the PBC vector.

By using Eqs. (1.6) and (1.8), it is possible to determine the change in the critical depth of the BDT and the cutting difficulty coefficient during a rotation period. In

**Table 1.5** The lattice structural characteristics of crystal planes (100), (110), and (111)

| Crystal plane | (100) | (110) | (111) |
|---|---|---|---|
| Correction coefficient $m_0$ | $3/2 + \cos 4\omega$ | $1 + 3/4*\cos 2\omega$ | $5/2 + 3/2*\cos 3\omega$ |
| Cleavage angle $\delta$ | 125.27° | 144.73° | 109.5° |
| Surface energy $J$ | 1.6652445 J m$^2$ | 1.1775055 J m$^2$ | 0.9614295 J m$^2$ |

where $\omega$ is the rotation angle during cutting

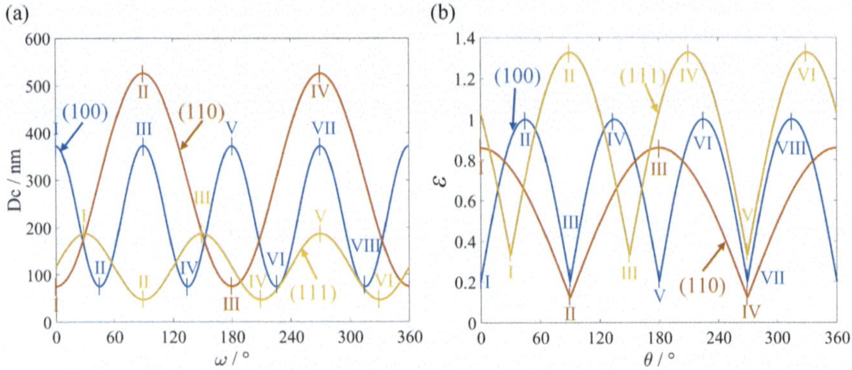


**Fig. 1.4** The fluctuations in both **a** the critical depth of cut ($D_c$) and **b** the cutting difficulty coefficient ($\varepsilon$) throughout the course of a cutting session

general, a coefficient ($\varepsilon$) above 0.9 signifies a difficult removal, while a coefficient below 0.6 indicates easy removal. Figure 1.4a–b depict the regular variation of the critical depth at which the transition from brittle to ductile behaviour occurs, as well as the coefficient that represents the difficulty of cutting Ge in various crystal orientations within a plane. It is worth noting that the maximum and minimum values of $\varepsilon$ differ from plane to plane: for the (100) plane, it is 45° and 0°, for the (110) plane, it is 0° and 90°, and for the (111) plane, it is 90° and 30°. The crystal planes (100), (110), and (111) are subdivided into eight, four, and six distinct areas, respectively. The critical depth of the BDT exhibits an inverse relationship with the cutting difficulty coefficient. This implies a direct correlation between the anisotropy of Ge and its lattice structure. When deciding the critical depth of the BDT, theoretical computation of Ge may be deduced to follow the same rule of anisotropy.

### *1.1.4 Modelling of Critical Cutting Depth*

Plunge-cutting tests were done with different cut depths to provide a thorough knowledge of the cutting process and the crucial transition depth between brittle and ductile areas while utilizing diverse fan forms. These studies were conducted to investigate the anisotropy of Ge following the completion of face-turning. The surface features of the workpiece, which had varying crystal orientations, showed comparable patterns of change.

Figure 1.5a represents a typical surface morphology of the workpiece after plunge-cutting with a fresh tool edge. Meanwhile, Fig. 1.5b presents the longitudinal section diagram of the same surface. The surface morphologies resulting from the plunge-cutting operation may be categorized into ductile and brittle areas. It is crucial to acknowledge that deformation occurs in three separate stages: elastic, ductile, and brittle. It is important to note that the elastic stage has a rather small scale and

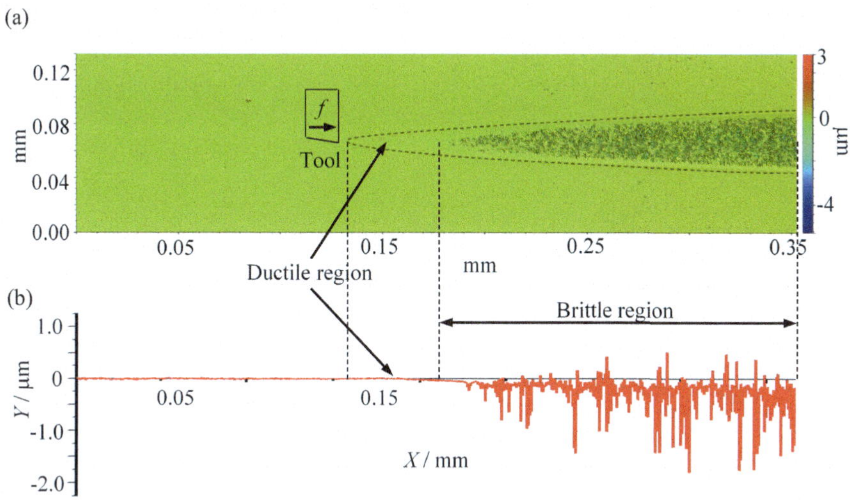


**Fig. 1.5** BDT after plunge-cutting with a fresh tool edge: **a** captured morphology of machined surface and **b** cross-section curve

cannot be precisely identified with ordinary equipment. Hence, additional investigation methods are required to examine the dynamic transformation of the elastic phase.

The model for SPDT of brittle materials is depicted in Fig. 1.6. In this model, $R$ represents the radius of the tool, $a_p$ denotes the cutting depth, $f$ stands for the feed rate, $D_{min}$ refers to the minimum undeformed chip thickness (UCT), $D_{max}$ is the maximum UCT, and $D_\xi$ indicates the instantaneous chip thickness. Furthermore, the critical UCT, $D_c$, is incorporated into the model as a parameter for the BSCM.

When $D_\xi$ does not exceed $D_{min}$, the material is removed through plowing, and the surface that is being machined is produced by the flow of material to the surface of the clearing tool. As soon as $D_\xi$ is greater than $D_{min}$, the material is extracted by means of ductile shear, which ultimately results in the production of chips by the movement of the material towards the rake face. The material is eliminated by brittle fracture that occurs when $D_\xi$ is greater than $D_c$. Because of this, the machined regions of brittle materials are composed of three separate areas: the ploughing area, which is Region I, where $D_\xi$ is less than $D_{min}$; the ductile deformation zone, which is Region II, where $D_{min}$ is less than $D_\xi$ and less than $D_c$; and the brittle fracture zone, which is Region III, where $D_c$ is less than $D_\xi$.

The dynamic interaction that exists between the workpiece and the tool, in addition to the radius of the tool nose, is taken into consideration when determining $D_{min}$ in the context of the SPDT process. In order to get the least UCT, Son et al. suggested a technique that involves computing the tool nose radius and the friction angle. This approach can be represented as Eq. (1.9) [21]:

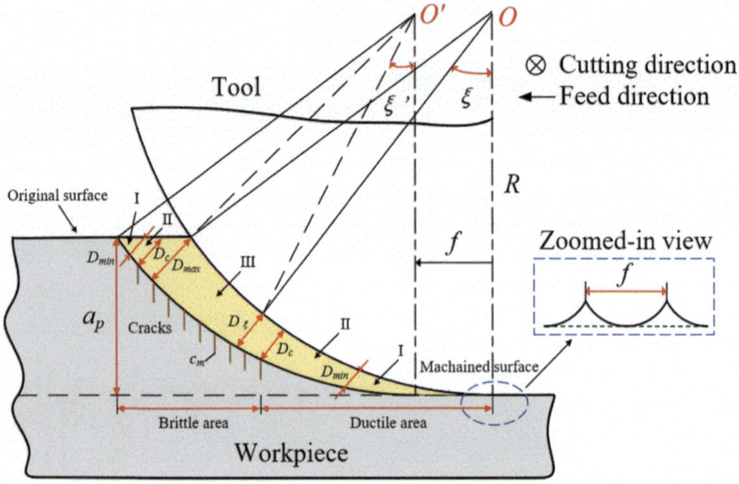


**Fig. 1.6** Model for ductile–brittle transition of brittle materials under SPDT

$$D_{min} = R\left[1 - \cos\left(\frac{\pi}{4} - \frac{\beta}{2}\right)\right] \tag{1.9}$$

where $R$ is the tool nose's radius, and $\beta$ indicates the friction angle.

In accordance with the geometric relationship that exists between the tool and the workpiece, as seen in Fig. 1.6, the $D_{\max}$ can be modelled analytically and obtained by the following equations:

$$D_{\max} = R - \sqrt{R^2 + f^2 - 2f\sqrt{2Ra_p - a_p^2}} \tag{1.10}$$

where $f$ is the feed rate, and $a_p$ indicates the cutting depth.

A prediction model for the $D_c$ of Ge in SPDT that is based on the Griffith fracture expansion criteria claims that the following is true [22]:

$$D_c = \psi \frac{E}{H^2} \frac{K_c^2}{H} \tag{1.11}$$

where $E$ is Young's modulus of the workpiece, $H$ indicates the hardness of the workpiece (approximately 8 GPa), $\psi$ represents the coefficient of ductile–brittle transition, and $K_c$ represents the workpiece's fracture toughness.

The value of $K_c$ can be obtained using Eq. (1.12) [23]:

$$K_c = \left(\frac{G_c E}{1 - \nu^2}\right)^{\frac{1}{2}} \tag{1.12}$$

**Table 1.6** Fracture toughness values ($G_c$) of Ge for different crystal planes

| Crystal planes | (111) | (110) | (100) |
|---|---|---|---|
| $G_c$ (J m$^{-2}$) | 2.1 | 2.6 | 3.7 |

where $G_c$ is the force of critical crack extension, $v$ represents the Poisson's ratio (approximately 0.25).

Additionally, $G_c$ is regarded as the critical strain energy release rate value for energy dissipation during the inelastic process of crack propagation. However, when crack propagation occurs without inelastic changes, $G_c$ represents the surface energy. The values of $G_c$ for Ge on different crystal planes (111), (110), and (100) are provided in Table 1.6 [24].

Therefore, Eq. (1.11) can be rewritten as:

$$D_c = \psi \frac{E^2 G_c}{H^3(1 - v^2)} \tag{1.13}$$

The utilisation of MATLAB® allows for the computation of the theoretical values of $D_c$. Figure 1.7 depicts the fluctuation of $D_c$ across different cutting directions on various crystal planes. A consistent symmetric pattern is observed in the direction of direct current ($D_c$) in each of the distinct crystal planes, as demonstrated by the results. Both the (100) crystal plane and the (110) crystal plane display a total of four maxima and four minima in their respective domains. However, the latter exhibits a larger fluctuation amplitude. Additionally, the $D_c$ of the (111) crystal plane shows isotropic behavior and remains constant.

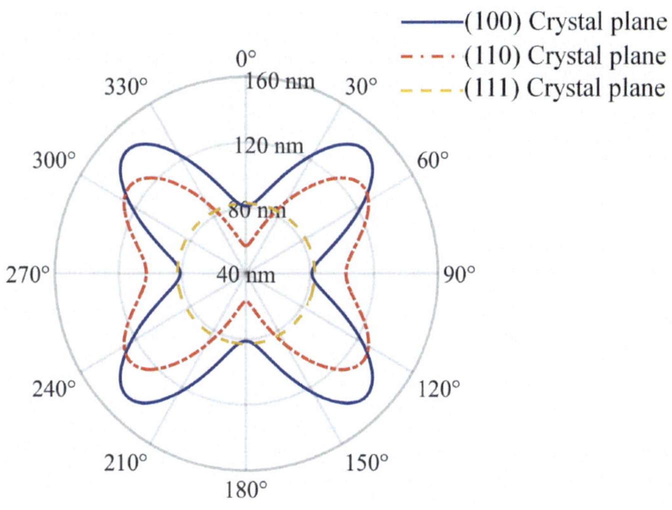


**Fig. 1.7** The theoretical $D_c$ curve in SPDT for different crystal planes of Ge

### 1.1.5 Cutting Force and Chips Characterization

The analysis of chip morphology plays a crucial role in characterizing the material removal mode during cutting. SEM examination of chips obtained from various surfaces and crystal planes was conducted and the results are presented in Fig. 1.8. Three distinct types of chips were identified: brittle chips, ductile–brittle chips, and ductile chips.

Brittle chips, as depicted in Fig. 1.8a, d, g, manifest as discontinuous and fine chips scattered without forming a continuous section. This occurrence can be attributed to the uneven distribution of strain throughout the cutting process. Conversely, ductile chips, showcased in Fig. 1.8c, f, i, display elongated continuous segments that intertwine with each other and tend to cluster together. Ductile–brittle chips portray characteristics of both brittle and ductile chips. They exhibit a needle-shaped morphology resembling brittle chips while also showing chip clusters similar to ductile chips, as evident in Fig. 1.8b, e, h. During the phase in which ductile–brittle chips are produced, this indicates that brittle and ductile removal modes are present simultaneously.

From these observations, it can be concluded that among the three material removal modes (total ductile removal, total brittle removal, and ductile–brittle alternating removal), achieving total ductile removal results in a more stable process.

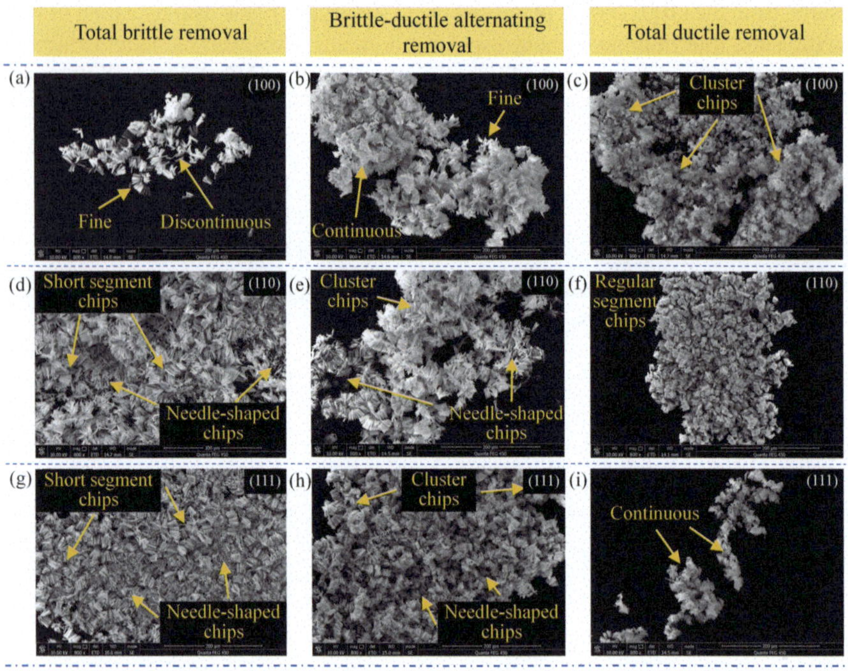


**Fig. 1.8** Chip morphology of total brittle removal, ductile–brittle alternating removal and total ductile removal: **a–c** (100), **d–f** (110), and **g–i** (111) crystal planes

The analysis of force signals plays a crucial role in characterizing the anisotropy of BSCM and differentiating between its brittle and ductile regions. Additionally, it facilitates comprehension of the influence of DBAP on the machining system. Figure 1.9 displays the force signals and spectra that are linked to the occurrence of DBAP on various crystal surfaces.

During the turning process, when the spindle speed is 2000 RPM, the rotation time ($T$) is 0.03s, which corresponds to a frequency ($f$) of 33.33 Hz. The cutting force exerted during the machining of brittle sections is greater in comparison to the cutting force experienced in ductile regions.

A representation of the force signal and spectrum for the (100) crystal plane can be found in Fig. 1.9a. The force peaks that are numbered odd correspond to four brittle zones, each of which has a value of $-1.906 \times 10^{-4}$, $-2.377 \times 10^{-4}$, $-2.765 \times 10^{-4}$, and $-2.831 \times 10^{-4}$ N, respectively. Although force valleys with even numbers correspond to four ductile zones, each of these regions has a value of $2.059 \times 10^{-4}$, $2.461 \times 10^{-4}$, and $2.957 \times 10^{-4}$ N, respectively. Compared to brittle regions, cutting forces in larger ductile regions are slightly larger but more stable, as the ductile regions are larger. A frequency of 133.3 Hz is displayed by the peak spectrum, which is in accordance with the fourfold DBAP of the (100) crystal plane.

Additionally, the force signal and spectrum of the (110) crystal plane are depicted in Fig. 1.9b, which also includes illustrations of force peaks and valleys. Specifically, the force maxima, which are represented by values with odd numbers, are associated with four brittle locations. These regions have respective measurements of $6.730 \times 10^{-4}$, $-1.251 \times 10^{-4}$, $6.852 \times 10^{-4}$, and $-5.73 \times 10^{-5}$ N. Conversely, the force valleys, with even-numbered values, correspond to four ductile regions, measuring $-3.161 \times 10^{-4}$, $-3.35 \times 10^{-4}$, $-4.312 \times 10^{-4}$, and $-3.067 \times 10^{-4}$ N, respectively. It is worth noting that the cutting forces appear to be more stable during the machining of ductile regions compared to brittle regions, and the absolute value of cutting forces may possibly be influenced by the region's size. Additionally, the peak spectrum demonstrates frequencies of 66.67 and 133.3 Hz, which correspond to the double brittle regions and quadruple ductile regions present in the (110) crystal plane's DBAP mechanism.

Similar to Fig. 1.9b, c exhibits these data of (111) crystal plane. The three force peaks, featuring odd-numbered values, align with three brittle regions denoted by force values of $1.3427 \times 10^{-3}$, $1.4318 \times 10^{-3}$, and $1.3534 \times 10^{-3}$ N, respectively. Correspondingly, the three force valleys, displaying even-numbered values, correspond to three ductile regions, with force values of $-1.5065 \times 10^{-3}$, $-1.0742 \times 10^{-3}$, and $-1.7749 \times 10^{-3}$ N, respectively. In addition, the peak spectrum exhibits a frequency of 99.9 Hz, representing the triple DBAP mechanism found within the (111) crystal plane. Furthermore, the force signal within a turning period displayed in Fig. 1.9 correlates well with the shear modulus curves depicted in Fig. 1.2. Notably, during the cutting process of (111) crystal plane, the cutting force exhibits the highest level of stability and concordance with the isotropic properties of the shear modulus. The force signal of the (100) and (110) crystal planes exhibits anisotropy and fluctuation characteristics that align with the corresponding shear modulus properties.

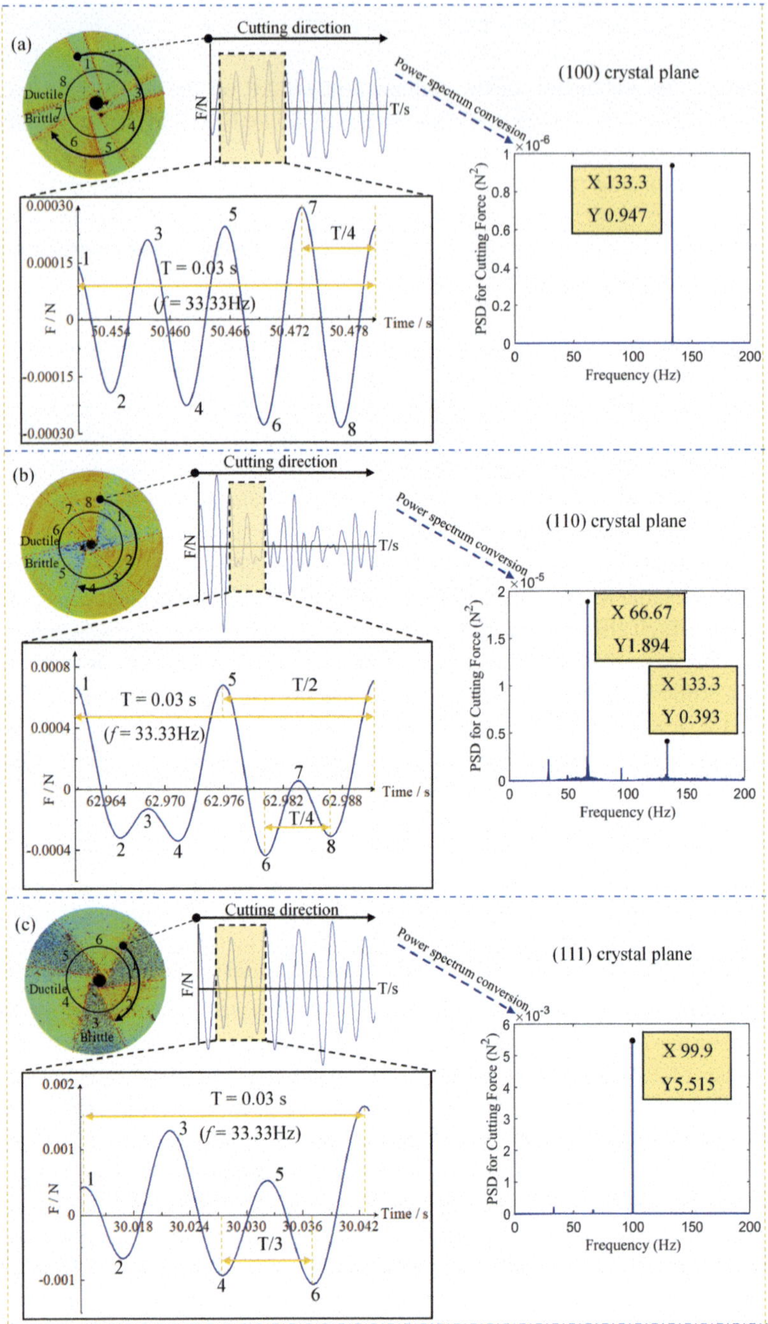


**Fig. 1.9** Cutting forces and their spectrum when machining the surface with DBAP: **a** (100), **b** (110) and **c** (111) crystal planes

Based on the observations, it can be deduced that the DBAP mechanism in BSCM leads to fluctuations in the corresponding frequencies during the machining process. This fluctuation in vibration frequencies can have a negative impact on the ultra-precision machining of BSCM. As a result, it becomes crucial to control the transition from ductile to brittle behavior in BSCM and avoid the occurrence of DBAP in order to enhance the quality of SPDT of BSCM.

On the other hand, it is noteworthy that BSCM can be manipulated to obtain ductile–brittle alternating surfaces at fixed frequencies. This characteristic of BSCM opens up possibilities for its application in specific usage scenarios where such alternating surfaces are desired or advantageous.

## 1.2 Monocular Dynamic Simulation

MD simulation, a nanoscale computational method, is a valuable tool for studying the cutting mechanisms of single-crystal materials during SPDT. For the purpose of determining the connection between the cohesive energy and each atom in the simulation, the LAMMPS software [25] was employed. This section will use MD simulation visualization convenience to showcase the machining characteristics of various single crystal materials from different perspectives.

### *1.2.1 Atomic Arrangement in Nanocutting*

The MD simulation model, depicted in Fig. 1.10, consists of a workpiece with dimensions 36.15 nm in length, 10.845 nm in height, and 3.615 nm in width. Additionally, the model includes a diamond tool that has a radius of 3.56 nm. The purpose of this setup is to explore the process that generates fan-shaped patterns on surfaces that have been cut with single-crystal copper (SCC). There are three different kinds of atoms that make up the workpiece: boundary, thermostat, and Newtonian molecules.

To establish the feasible atomic motion range during cutting and prevent tool and workpiece translation under cutting forces, fixed layers are employed at the bottom and left of the workpiece, representing the first region in the model. This approach, widely used in MD simulations [26], provides a simplified boundary representation. Additionally, thermostat atoms near the boundary are employed to simulate heat dissipation with the temperature adjusted and kept constant at 293 K. This simulation method simplifies the analysis of the cutting mechanism by preventing excessive temperature increase in the workpiece during cutting [27].

The Newtonian atoms represent the main region for studying the cutting mechanism. In the course of the cutting, they represent a simulation of the motion modes and trajectories of atoms. In this part, periodic boundary conditions, also known as PBC, are implemented in the *Z*-direction in order to reduce the size effects that are brought about by the simulation scale. The *X*- and *Y*-directions employ fixed

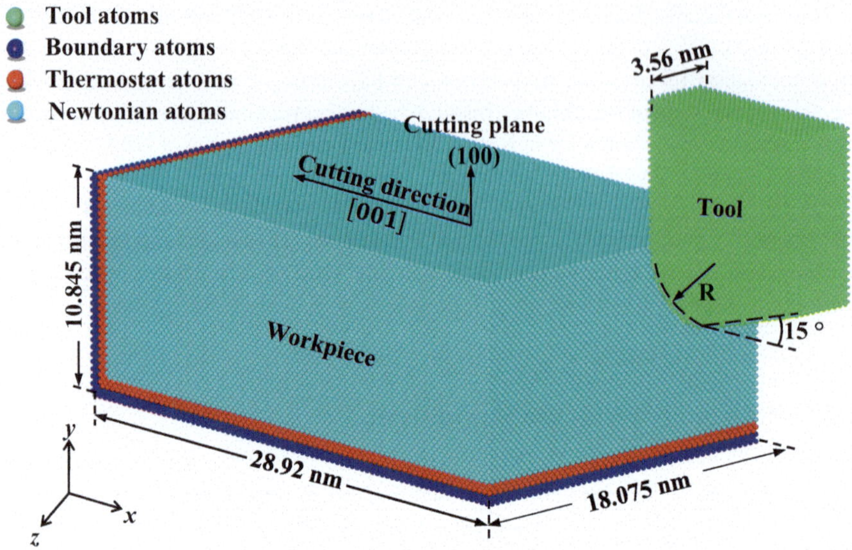


**Fig. 1.10** Schematic description of nanometric cutting model

boundary conditions. Considering computational demands, a high cutting speed of 100 m/s is employed [28]. It should be noted that the cutting speed significantly impacts the temperature of the cutting region, potentially affecting the critical undeformed chip thickness (CUCT) due to thermal effects [29]. However, the simulated cutting distance is short and the generated heat has minimal influence on the CUCT.

Table 1.7 presents the parameters employed in the MD simulations, while Table 1.8 lists the potential function parameters.

In the turning, as shown in Fig. 1.11a, the workpiece remains motionless while the tool cycles through four sequential positions identified as A, B, C, and D. Owing to the fluctuating angles between the tool and the workpiece's crystal orientation, coupled with the inherent anisotropy of single-crystal materials, a range of different turning mechanisms and outcomes emerge in varied SCC orientations. Consequently, these disparities lead to the creation of unique fan-shaped patterns on the workpiece's surface. As a consequence of these differences, the surface of the workpiece is generated with one-of-a-kind designs that resemble fans.

Creating an anisotropic model for SCC necessitates the identification of how the rotation direction correlates with the crystal planes and orientations. Figure 1.12 depicts the spatial relationship among the direction of rotation and the unit cell, particularly when rotating around the three key crystal planes of SCC, which are the (100), (110), and (111) planes.

Here, the symbols $a$, $b$, and $c$ represent the distances from point $Q'$ to the respective axes [100], [010], and [001]. The variable $L$ designates the length of segment O′Q′. The angle $\theta$ is measured between the lines $O'Q$ and $O'Q'$. Figure 1.12a illustrates the Cu (100) plane. The values of $a$, $b$, and $c$ are determined in the following formula:

**Table 1.7** Specified conditions and parameters for simulation

| Items | Parameters |
|---|---|
| Workpiece dimensions | $28.92 \times 10.845 \times 18.075$ nm$^3$ |
| Workpiece crystal structure | FCC |
| Workpiece lattice constant | 3.6149 Å |
| Tool material | Single-crystal diamond |
| Cutting edge radius (tip radius) | 3.56 nm |
| Tool rake angle | 0° |
| Tool clearance angle | 15° |
| Tool crystal structure | Diamond |
| Tool lattice constant | 3.56683 Å |
| Equilibration temperature | 293 K |
| Cutting speed | 100 m/s |
| Timestep | 1 fs |

**Table 1.8** Parameters of potential function employed in the simulation

| Atoms interaction | Potential | Parameters | | | |
|---|---|---|---|---|---|
| C–C | Tersoff potential [30] | $A$ (eV) | 1393.6 | $B$ (eV) | 346.74 |
| | | $\lambda$ (Å$^{-1}$) | 3.4879 | $\mu$ (Å$^{-1}$) | 2.2119 |
| | | $\beta$ | $1.5724 \times 10^{-7}$ | $\eta$ | 0.72751 |
| | | $c$ | $3.8049 \times 10^{4}$ | $d$ | 4.3484 |
| | | $h$ | – 0.57058 | $R$ (Å) | 1.8 |
| | | $S$ (Å) | 2.1 | $\chi$ | 1 |
| C–Cu | Morse potential [31] | $D$ (eV) | 0.1 | $\alpha$ (Å$^{-1}$) | 1.7 |
| | | $r_0$ (Å) | 2.2 | / | |
| Cu–Cu | EAM potential [32] | / | | | |

$$\begin{cases} a = 0 \\ b = L\cos\theta, \ (0^\circ \le \theta \le 360^\circ) \\ c = L\sin\theta \end{cases} \tag{1.14}$$

The direction in which Cu (100) is rotating can be determined by:

$$[uvw] = \left[0 \ \cos\theta \ \sin\theta\right] \tag{1.15}$$

In Fig. 1.12b, the Cu (110) is depicted, with dimensions $a$, $b$, and $c$ determined through the subsequent formula:

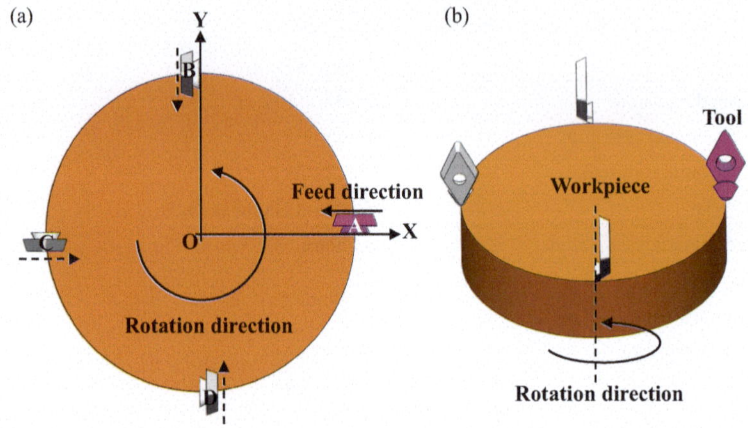


**Fig. 1.11** Schematic diagram of turning: **a** top view and **b** side view

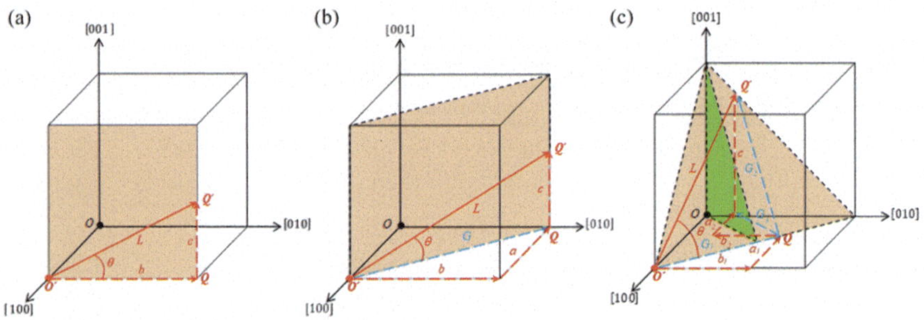


**Fig. 1.12** Turning direction (O′Q′) and its specific location within the unit cell on the following crystal planes: **a** (100), **b** (110), and **c** (111)

$$\begin{cases} a = -\frac{\sqrt{2}}{2} L\cos\theta \\ b = \frac{\sqrt{2}}{2} L\cos\theta, (0^\circ \leq \theta \leq 360^\circ) \\ c = L\sin\theta \end{cases} \tag{1.16}$$

Similarly, the direction of Cu (110) is given by:

$$[uvw] = \left[ -\frac{\sqrt{2}}{2}\cos\theta \;\; \frac{\sqrt{2}}{2}\cos\theta \;\; \sin\theta \right] \tag{1.17}$$

In Fig. 1.12c, the Cu (111) is depicted, with dimensions *a*, *b*, and *c* determined through the subsequent formula:

$$\begin{cases} a = -\frac{\sqrt{2}}{2} L\cos\theta - \frac{\sqrt{6}}{6} L\sin\theta \\ b = \frac{\sqrt{2}}{2} L\cos\theta - \frac{\sqrt{6}}{6} L\sin\theta, (0^\circ \leq \theta \leq 360^\circ) \\ c = \frac{\sqrt{6}}{3} L\sin\theta \end{cases} \tag{1.18}$$

For Cu (111), the turning direction can be derived by:

$$[uvw] = \left[ -\tfrac{\sqrt{2}}{2}\cos\theta - \tfrac{\sqrt{6}}{6}\sin\theta \ \ \tfrac{\sqrt{2}}{2}\cos\theta - \tfrac{\sqrt{6}}{6}\sin\theta \ \ \tfrac{\sqrt{6}}{3}\sin\theta \right] \tag{1.19}$$

The atomic arrangement and turning performance can differ significantly depending on the specific planes and orientation of the crystal. This segment presents the MD simulation outcomes for the nano-scale cutting process of copper with (100), (110), and (111) planes, as shown in Figs. 1.13, 1.14, and 1.15, accordingly. The analysis focuses on material deformation through dislocations and sliding propagation. To enhance visual interpretation, atoms with the same arrangement are represented by the same color. The dislocation area predominantly exhibits the HCP structure resulting from defects in the workpiece during turning [33].

Figure 1.13 presents the atomic arrangement in various crystal orientations of Cu (100) and the ensuing material deformation during the turning process. In Fig. 1.13a, the atomic layout exhibits a recurring pattern of atoms on the Cu (100) surface, creating four symmetrical zones that are in line with the machined surface. During the turning operations oriented along [011], the cloudy regions shown in Fig. 1.13c, e, g, and i exhibit shear slip in the material, which results in dislocations moving both parallel to the turning direction and vertically. These dislocations propagate in front

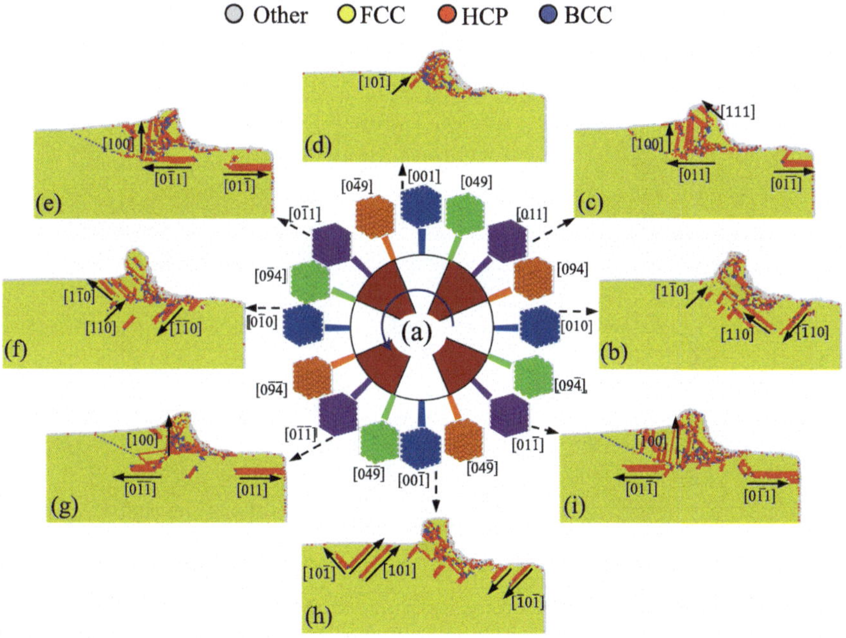


**Fig. 1.13** Different turning directions for the atomic arrangement on the crystal plane (100) of Cu: simulation diagram

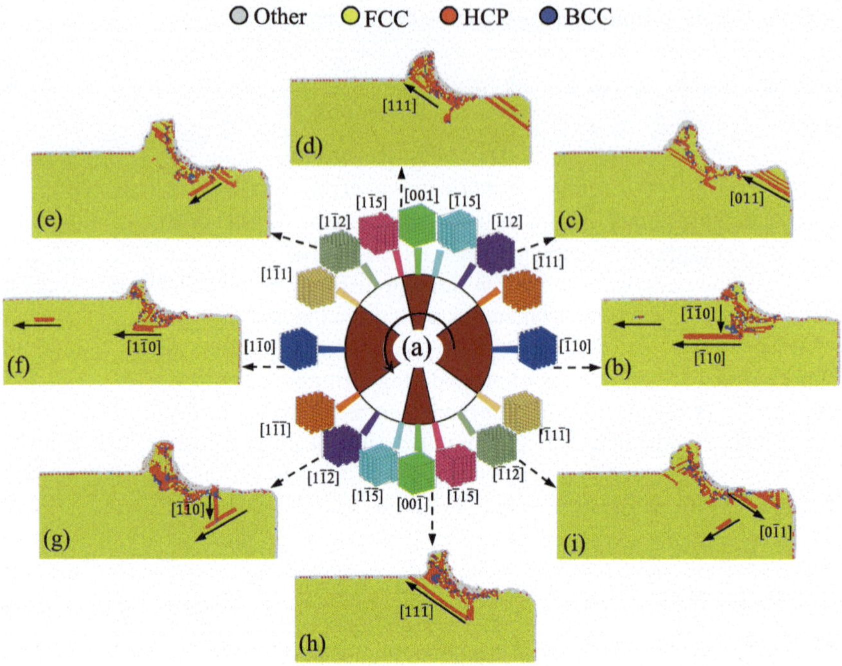


**Fig. 1.14** Different turning directions for the atomic arrangement on the crystal plane (110) of Cu: simulation diagram

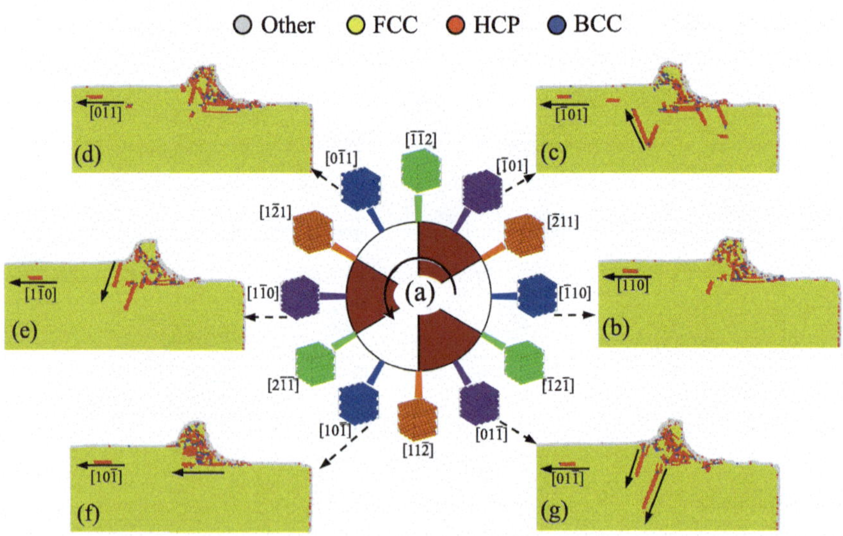


**Fig. 1.15** Different turning directions for the atomic arrangement on the crystal plane (111) of Cu: simulation diagram

of and beneath the cutting tool, inflicting considerable damage on the workpiece's subsurface.

On the contrary, when the cutting is performed along [010], the clean areas depicted in Fig. 1.13b, d, f, and h do not display any vertical slip, but rather show slip at a diagonal to the turning path. It is evident that numerous slips originating from misalignments interact within the turning zone directly in front of the tool, thereby diminishing the propagation of dislocations and reducing the amount of subsurface damage. This occurs because the cutting action is aligned with the direction of slip, thereby minimizing the impact on the workpiece's subsurface integrity.

Upon analysing Fig. 1.14a, it is evident that the atoms on the Cu (110) surface are arranged in a periodic manner, with changes occurring in a circular orientation. This arrangement displays two symmetrical areas that coincide with the four two-fold symmetrical axes. When the turning process occurs along $[\bar{1}10]$ and $[1\bar{1}0]$ (cloudy region), the simulation results displayed in Fig. 1.14b, f demonstrate the propagation of dislocations in both tangential and vertical directions due to material deformation caused by sliding. Thus, these dislocations can readily propagate downwards within the material, leading to significant subsurface damage in the workpiece.

On the other hand, as shown in Fig. 1.14d, h, there is no noticeable vertical slip that takes place while turning along [001] and $[00\bar{1}]$ (cloudy region). Instead, there are some slips that take place at an angle in relation to the direction in which the cutting is taking place. It is important to notice that the turning region in front of the tool is characterised by a number of slips that are induced by misalignment and that interfere with each other. Nevertheless, since turning is accomplished in the sliding direction, the extent of material deformation penetration appears to be less, suggesting reduced subsurface damage.

As depicted in Fig. 1.14c, e, g, i, the slip predominantly transpires at an angle against the turning direction when turning along clean regions ($[\bar{1}12]$, $[1\bar{1}2]$, $[1\bar{1}\bar{2}]$ and $[\bar{1}1\bar{2}]$). This observation reveals that a multitude of slips resulting from misalignments can produce interference within the turning area that precedes the tool. Since the turning process is executed in alignment with the sliding direction, it contributes to a reduction in subsurface damage.

The Cu (111) surface exhibits periodic changes in atom arrangement in a counter-clockwise direction, as depicted in Fig. 1.15a. It displays three symmetric regions that correspond to the surface patterns of the workpiece. Analyzing the simulation results when turning along $[\bar{1}01]$, $[1\bar{1}0]$ and $[01\bar{1}]$ (cloudy region), as shown in Fig. 1.15c, e, g, it is observed that shear slip occurs in the material, with dislocations propagating in both the turning and vertical directions. Consequently, these dislocations propagate both in front of and below the tool, resulting in significant subsurface damage to the workpiece.

In contrast, as seen in Fig. 1.15b, d, f, when the tool cuts the clean area (for example, along $[\bar{1}10]$, $[0\bar{1}1]$ and $[10\bar{1}]$), no vertical slippage is observed. However, there is some slippage detected at a certain angle to the direction of turning. It is crucial to observe that multiple slips, which result from misalignments, interact with one another in the area directly ahead of the tool during the turning process. Despite this, because the turning action is performed in the direction of sliding, it contributes to a reduction in subsurface damage.

The MD simulation results reveal that vertical slippages and dislocations can lead to significant subsurface damage in the workpiece. Turning directions that are prone to vertical slip tend to result in greater subsurface damage, while those less prone to vertical slip exhibit reduced subsurface damage. Consequently, the different deformation modes associated with various turning directions contribute to the formation of fan-shaped patterns on the machined surface of SCC.

The relationship between the friction coefficient and surface morphology can be described by the following equation:

$$\mu = \frac{F_t + F_c \tan\alpha}{F_c - F_t \tan\alpha} \tag{1.20}$$

where $\mu$ represents the friction coefficient, $F_t$ denotes the cutting force in axial direction, $F_c$ stands for the cutting force in tangential direction, and $\alpha$ indicates the tool rake angle.

Understanding the friction coefficient and its connection to atomic density is essential for integrating both micro and macro dynamic models. Table 1.9 illustrates the friction coefficient determined via experimental methods, simulations, and the atomic density across various crystal planes. Atomic density, defined as the number of atoms per unit volume, can be computed using the following formula:

In order to successfully integrate both micro and macro dynamic models, it is vital to have a solid understanding of the friction coefficient and its direct relationship to atomic density. The friction coefficient that was found using experimental methods, simulations, and the atomic density is presented in Table 1.9, which demonstrates the results over a variety of crystal planes. Using the following formula, one may get the atomic density, which is defined as the number of atoms that are contained inside a given volume:

$$N = \frac{n}{V} \tag{1.21}$$

where $N$ represents the atomic density, $n$ denotes the number of atoms in given volume, and $V$ represents the specified volume.

The experimental results in Table 1.9 confirm the accuracy of the theoretical equation by showing that the friction coefficient matches between experiments and simulations. Similar trends in atomic density variations and friction coefficient changes support the connection between them. There is a correlation between the anisotropy of atomic density and the formation of fan-shaped patterns on the surfaces of Cu. For SCC, the formation of fan-shaped patterns on the machined surface is a consequence of fluctuations in the friction coefficient that occur throughout the SPDT process. These fluctuations are caused by the different arrangements of atomic density in the various crystal orientations of the material.

**Table 1.9** Comparison of measured friction coefficient with simulated one and its correlation with atomic density

| Cu | (100) | (110) | (111) |
|---|---|---|---|
| Friction coefficient from experiments | 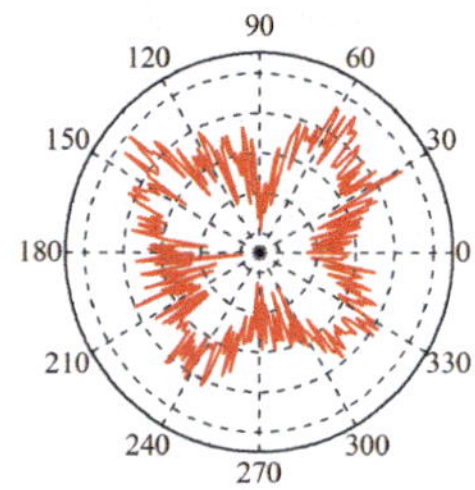 | 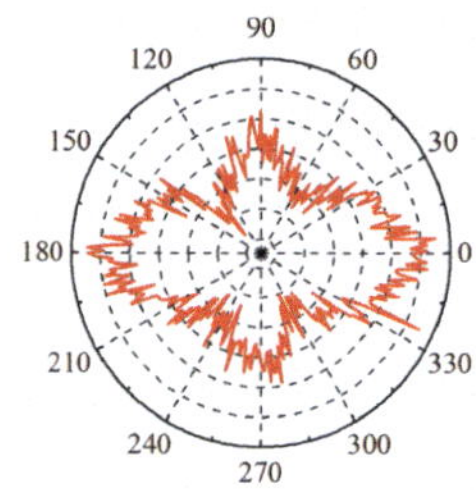 | 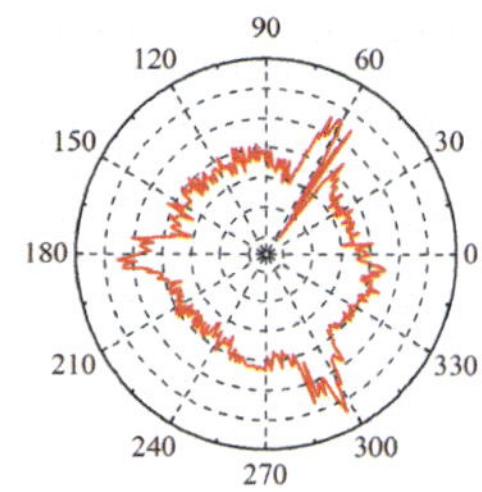 |
| Friction coefficient from simulations | 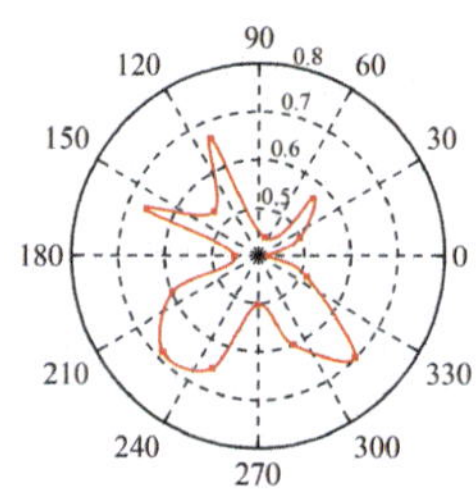 | 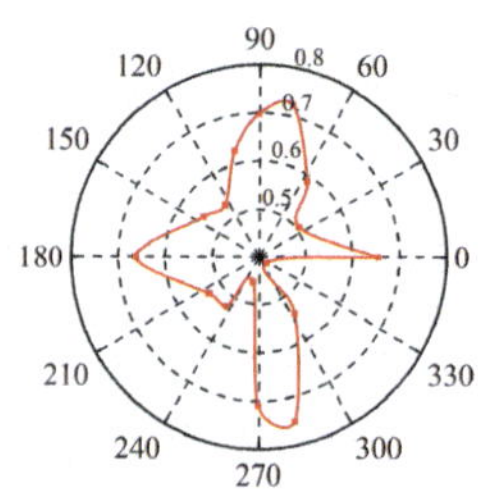 | 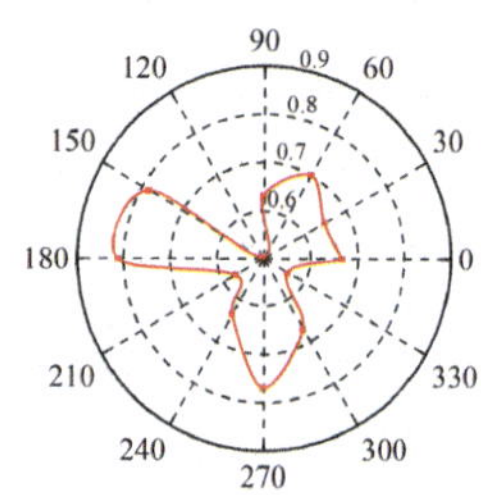 |
| Atomic density | 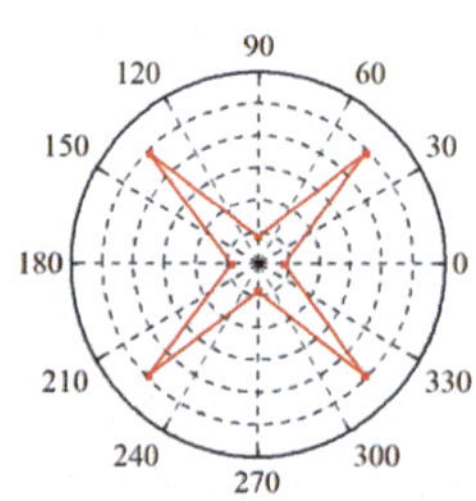 | 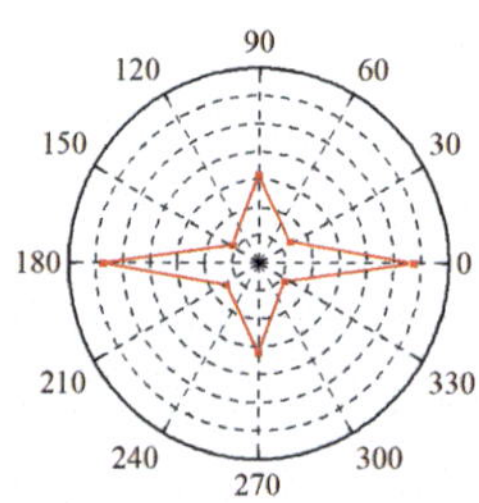 | 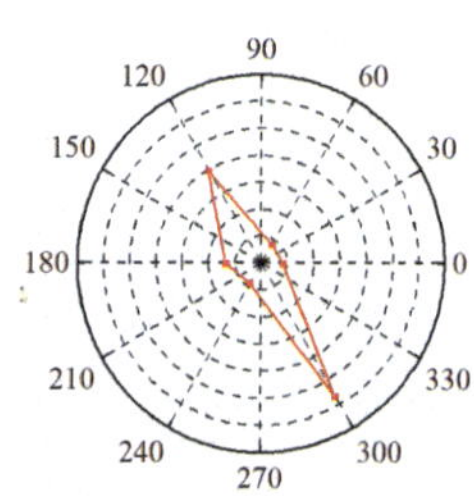 |

## 1.2.2 *Lattice Variation in Nanocutting*

In this section, MD simulation was employed to systematically investigate the lattice variation in BDT and its impact on different crystal planes of Ge, specifically (100), (110), and (111).

Figure 1.16 demonstrates the stable lattice constants of carbon (C) and germanium (Ge) where the cohesive energy reaches its minimum. The calculated lattice constants for C ($a_1$) and Ge ($a_2$) in the simulation align closely with values reported in reference [34], as shown in Table 1.10. In order to replicate the face- and plunge-cutting procedures on the three crystal surfaces (100), (110), and (111), which each have a distinct crystal orientation, MD models were developed.

Table 1.11 provides the simulation parameters. The tool model employed in the MD simulations mirrors the geometry of the tool used in the experiments. This means that the geometry of the tool that was utilised in the tests is reflected in the tool model that was used in the MD simulations. It is necessary to divide the workpiece and the tool into boundary, thermostatic, and Newtonian layers in order to simulate the real execution of the cutting process. Free boundary conditions are applied to the system in the *X*- and *Y*-directions, whereas PBC are imposed in the *Z*-direction [35, 36]. Prior to beginning the simulation of nano-cutting, a period is carried out until the system reaches a state of equilibrium. The entire simulation is conducted at room temperature (298 K).

The interactions between atoms in the system were modeled using different potential functions, as shown in Tables 1.12 and 1.13. The Tersoff potential is utilized to describe the covalent bonds and the effects of bond angles, providing a precise description of covalent systems. However, it tends to overestimate the force required for bond-breaking, which may result in an inaccurate representation of brittle behavior [37–39]. Furthermore, there are currently no known parameters for the interaction between germanium atoms (Ge–Ge) in the available potentials.

In the absence of appropriate parameters, using the Tersoff potential to model the interaction Ge–Ge (workpiece) and C–C (diamond) atoms is considered both reliable and reasonable. The Morse potential also describes the interaction between Ge and

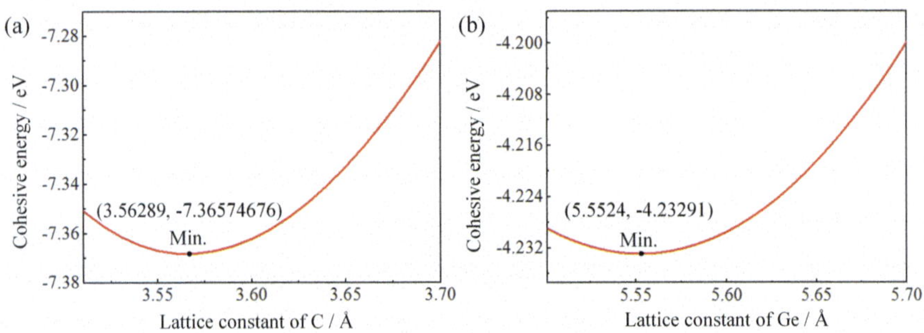


**Fig. 1.16** Variation in cohesive energy versus lattice constants for **a** C and **b** Ge. The lowest point on the curve, which represents the atom's most stable state, was used to select the lattice constant for the simulation parameters

**Table 1.10** Comparison of the lattice constants of C ($a_1$) and Ge ($a_2$) in reference [34] and used in simulation

| Element | Lattice constant in book (Å) | Lattice constant in simulation (Å) |
|---|---|---|
| C ($a_1$) | 3.567 | 3.56289 |
| Ge ($a_2$) | 5.658 | 5.5524 |

**Table 1.11** Detailed parameters in MD simulation

| Items | Parameters |
|---|---|
| Workpiece material | Single-crystal germanium |
| Size of workpiece | $50\ a_2 \times 20\ a_2 \times 30\ a_2$ |
| Number of germanium atoms in workpiece | 249,000 |
| Number of carbon atoms in tool | 22,300 |
| Crystal planes | (100), (110), (111) |
| Tool material | Single-crystal diamond |
| Rake angle (°) | 0 |
| Clearance angle (°) | 10 |
| Tool cutting edge radius (Å) | 5 |
| Cut depth (Å) | 15 |
| Cutting velocity (Å/ps) | 0.3 |
| Relaxation temperature (K) | 298 |
| Time step (ps) | 0.001 |
| Ensemble | NVT (Relaxation)<br>NVT (Thermostat layers)<br>NVE (Newtonian layers) |

*Note* $a_2$ presents the lattice constants of Ge, $a_2 = 5.5524$ Å

**Table 1.12** Tersoff potential parameters for C–C and Ge–Ge [30]

| Parameters | C–C | Ge–Ge |
|---|---|---|
| $A$ (ev) | 1391.6 | 1760.1 |
| $B$ (ev) | 346.7 | 430.0 |
| $\lambda_1$ (Å$^{-1}$) | 3.4879 | 2.4451 |
| $\lambda_2$ (Å$^{-1}$) | 2.2119 | 1.71 |
| $\lambda_3$ (Å$^{-1}$) | 0.0 | 0.0 |
| $R$ (Å) | 1.95 | 2.95 |
| $D$ (Å) | 0.15 | 0.15 |
| $m$ | 3.0 | 3.0 |
| $\gamma$ | 1.0 | 1.0 |
| $c$ | 38,049 | 106,430 |
| $d$ | 4.3484 | 15.2 |

**Table 1.13** Morse potential parameters for C-Ge [41]

| Element | $D$ (eV) | $\alpha$ (Å$^{-1}$) | $r_0$ (Å) | $r_0$ (Å) |
|---|---|---|---|---|
| C-Ge | 0.125778 | 2.58219 | 2.2324 | 10 |

C atoms. This potential is often used in nano-cutting models to show how the tool's atoms interact with the workpiece [40].

The atomic arrangement of Ge varies across different crystal planes, as well as within different crystal orientations of the same plane. This variation is the primary cause of the anisotropy observed in Ge. In order to study this anisotropy, MD simulation was performed based on the crystal orientation relationship depicted in Fig. 1.1 [42]. Figure 1.18 shows the modelling results, which show that crystal orientations with more space between the lattices show a BDT on the (100) plane. In contrast, the (110) plane demonstrates a difference in the order of BDT between crystal orientations with maximum and minimum lattice spacing. On the other hand, the (111) plane requires additional referencing to account for its identical lattice spacing and stable property. Hence, the anisotropy and brittle ductility of Ge directly depend on the lattice spacing and bond classification of its atoms. The variation in lattice spacing consequently leads to anisotropic effects during crystal surface machining.

During the plunge-cutting process, Ge undergoes three stages of deformation: elastic, ductile, and brittle deformation. The sequence of BDTs within the same crystal plane may depend on the sequence of elastic-ductile transitions. However, determining the critical depth for elastic-ductile transition becomes challenging due to its small scale. To overcome this limitation, MD plunge-cutting models are employed.

The simulation results in Fig. 1.19a–f show that the critical depth for elastic-ductile transition on the (100) crystal plane is smaller in brittle regions with crystal orientation <011> compared to ductile regions with crystal orientation <001> . Moreover, the critical thrust forces for the transition in ductile regions are higher than in brittle regions, consistent with previous findings indicating lower forces in brittle regions than in ductile regions within a specific timeframe [43]. The agreement between these results and MD simulations confirms their reliability.

In the (110) crystal plane, the elastic-ductile transitio''s critical depth and thrust force suggest an unstable anisotropy during cutting. The crystal orientation group <001> generally shows higher thrust forces, whereas the critical depth for elastic-ductile transition is greater in the crystal orientation group <011>.

The critical depth required for the elastic-ductile transition on the (111) crystal plane is directly proportional to the critical thrust force needed for the transition. In ductile regions, such as crystal orientations $[0\bar{1}1]$, $[10\bar{1}]$ and $[\bar{1}10]$, the critical depth and thrust force for the elastic-ductile transition are higher compared to brittle regions with crystal orientations $[1\bar{1}0]$, $[01\bar{1}]$ and $[\bar{1}01]$.

The MD simulation results suggest a correlation between dislocations and the elastic-ductile transition. Therefore, examining the correlation between lattice spacing, critical depth, and thrust force is crucial for obtaining useful guidelines in Ge machining during the transition from elastic to ductile behavior.

Ge is not the same in all crystal planes. This includes the critical depths of the elastic-ductile transition and the critical depths and lengths of BDT seen during nano-cutting. By simulating and analysing the cutting process of the Ge subsurface, MD models are used to learn more about this phenomenon. The main difference between amorphous and solid forms is how the atoms are arranged. Amorphous germanium has short-range order but long-range disorder, while single crystal germanium has order at both short and long lengths.

Figure 1.20a–b illustrate the radial distribution function (RDF) of the MD model before and after being cut. The RDF serves to explore the degree of order in a material and describe the correlation among particles. In the given figure, it can be observed that Ge possesses a highly ordered structure prior to cutting, while a disordered structure, possibly indicating an amorphous state, emerges after cutting. In MD simulation, the differentiation between single crystal germanium and amorphous germanium can be determined based on the order of atomic arrangement.

This section identifies the crystal evolution process on the surface and subsurface during nano-cutting of Ge by calculating the coordination number [44]. Prior studies suggest that the coordination numbers for single crystal, liquid, and amorphous Ge mainly fall within the range of (0, 8). For experimental purposes single crystal Ge, surface atoms have a coordination number of 3 or less, whereas that of diamond-like structured atoms is 4. In addition, *β–tin* structured atoms have a coordination number of 6, while that of atoms is 5, which represent a transitional state between the two. In contrast, amorphous atoms have a coordination number of 7 or higher [40].

After face-turning and plunge-cutting, the coordination numbers were found, and OVITO, was used to color Fig. 1.17b–c [45]. As expected from previous experiments, amorphous atoms (shown as red atoms) can be found on the surface and below the surface of the Ge after nano-cutting. These changes happen because of the cutting tool's hydraulic pressure on the workpiece. This pressure causes Ge to change into amorphous germanium while the tool is cutting.

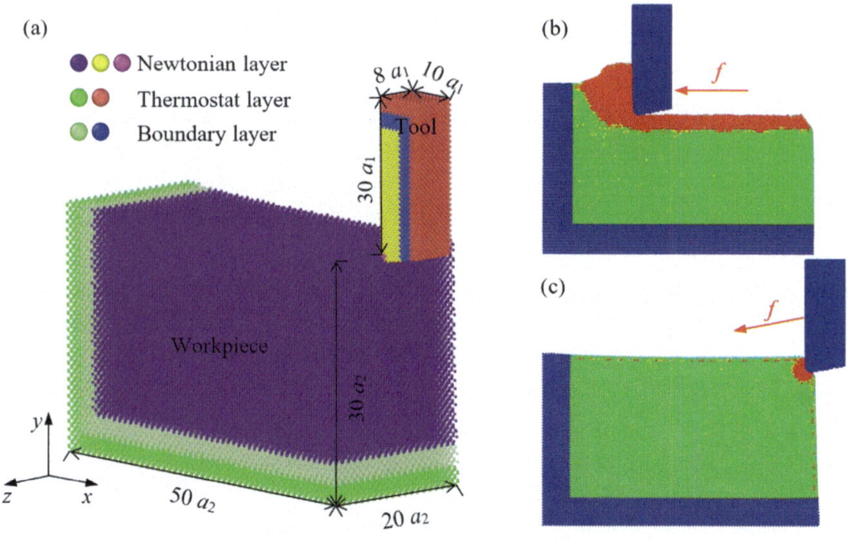


**Fig. 1.17** The MD simulation model for **a** nanocutting of Ge, **b** face-turning, and **c** plunge-cutting

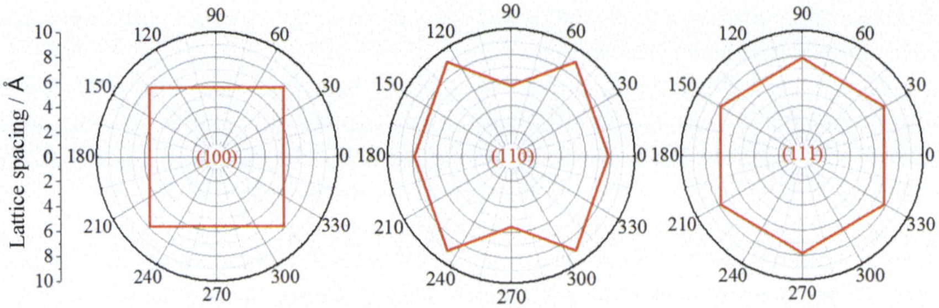


**Fig. 1.18** Determination of lattice spacing (*X* direction) in various crystal orientations of Ge through MD simulation for (100), (110), and (111) crystal planes

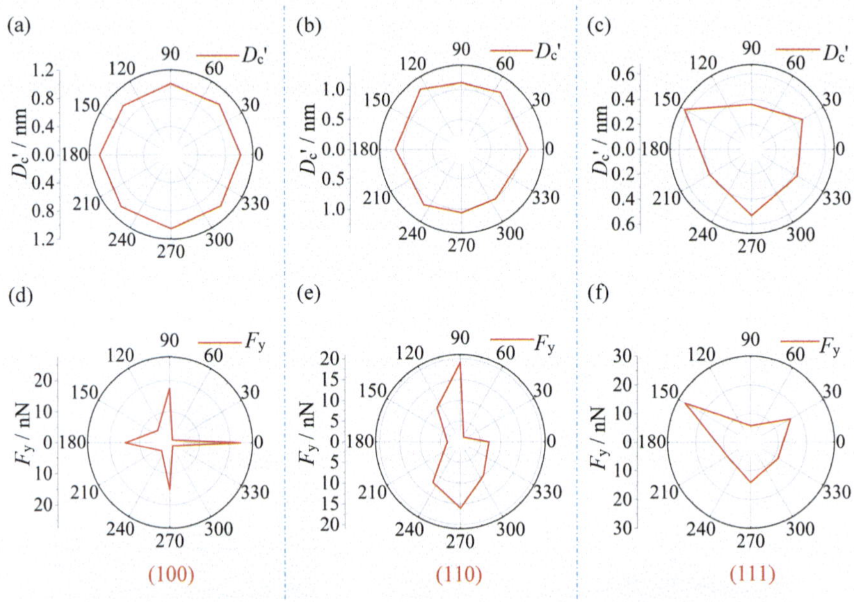


**Fig. 1.19** In the MD simulation of Ge's plunging cutting, the correlation among critical depths for elastic-ductile transition, amplitude of captured forces, and crystal orientations on crystal planes, **a**, **d** (100), **b**, **e** (110), and **c**, **f** (111)

From a micro-atomic point of view, the bending process is based on the diamond tool cutting the diamond-like structure, which is made up of ordered Ge atoms. During this process, kinetic and potential energies change, which causes Ge–Ge bonds to become excited and break.

As a result, the structure quickly loses its order, leading to the creation of a shapeless arrangement. It has been found that when the ratio of the cut depth to the tool edge radius is around 0.6, there are no chips created on the workpiece. Additionally, the depth of the amorphous layer under the surface becomes uneven

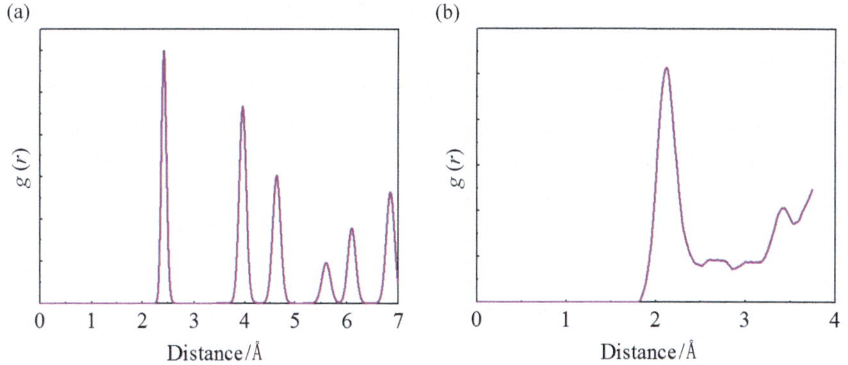


**Fig. 1.20** The RDF of Ge **a** before and **b** after cutting in MD simulation. The *x*-axis indicates the distance *r* from the reference particle, while the *y*-axis indicates the likelihood of other particles being spatially distributed

after the cutting process. However, with a cut depth to tool edge radius ratio increased to 3 or 6, the depth of the amorphous layer in the workpiec"s subsurface area becomes more consistent.

As portrayed in Fig. 1.21a–c, with an increasing cut depth, the thickness of the amorphous layer on each crystal plane gradually augments. When the ratio of cut depth to the tool radius is set at 3, the average amorphous layer thickness on the (100), (110), and (111) crystal planes measures approximately 1.3, 1.26, and 1.22 nm, respectively. The minimum thickness of the amorphous layer on the three crystal planes corresponds to crystal orientations $[00\bar{1}]$, $[\bar{1}10]$ and $[\bar{1}10]$. However, it is important to note that a smaller cut depth may result in uneven amorphous layer formation, adversely affecting the workpiece's quality. On the other hand, larger cut depths lead to thicker amorphous layers, which could potentially compromise the cutting quality of Ge. Consequently, it is crucial to select an appropriate machining depth to ensure superior cutting quality of Ge, with preference given to the crystal orientation in the (111) plane of Ge whenever possible.

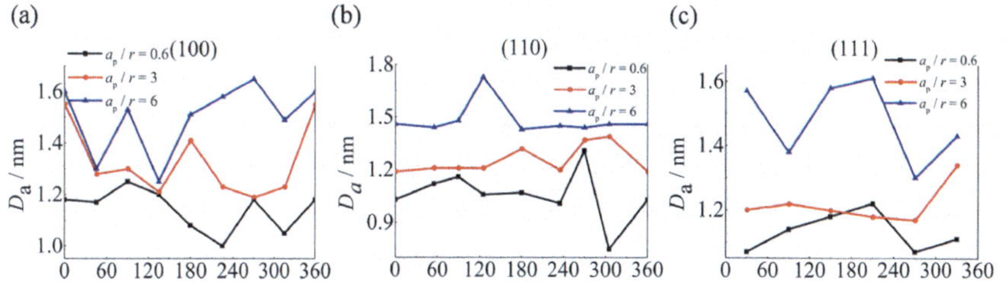


**Fig. 1.21** Variation in amorphous layer thickness ($D_a$) with respect to different crystal orientations and $a_p/r$ on the **a** (100), **b** (110) and **c** (111) crystal planes. Different colored lines represent different value of $a_p$/r, among which black, red and blue lines correspond to 0.6, 3 and 6, respectively

### 1.2.3 Visualization of Brittle-Ductile Transition

This section employed MD to simulate the BDT process of SiC due to the material easy accessibility for experimental purposes. To keep things from getting more complicated, cutting models that don't have any flaws are made so that future studies can build on them. Figure 1.22 and Table 1.14 show the MD model of cutting SiC with a diamond when the chip is 40 nm thick at the start. It is worth noting that diamond cutting of SiC is not currently useful for industrial production. Instead, ductile regime grinding is used most of the time for ultra-precision machining of this material. To keep things simple, a diamond cutting model is used in this part to look into the processes. This is because the way different grinding grits behave should be similar to how cutting edges of negatively raked diamond tools behave, especially when small-scale machining is being done. The models are made so that when the tool first touches the subject, it looks like a steady cutting process. This is done to keep the model sizes small and cut down on the number of modeling steps needed to set up a steady cutting process. During the simulation, the tool moves forward at a steady rate until either ductile deformation or brittle breakage happens. This approach allows for focusing computation power on critical aspects, such as determining the undeformed chip thicknesses at which ductile deformation occurs and when brittle fracture comes into play.

During the simulation, the boundary conditions consist of a periodic boundary in the direction of the *Z*-axis, as well as fixed borders at the bottom and both ends of the workpiece. These boundaries are used to replicate the limitations that are imposed by the materials that are next to the cutting region. A thermostat layer is implemented in close proximity to the fixed borders in order to keep the temperature at a constant 300 K. The thickness of both the fixed boundary areas and the thermostat regions is

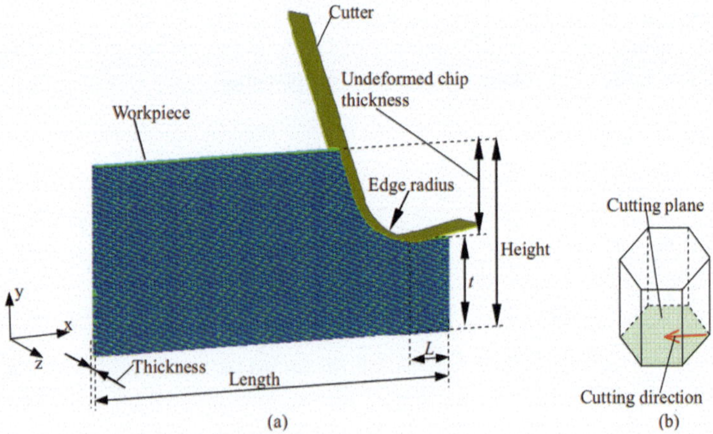


**Fig. 1.22** Diagram of **a** the MD model and **b** cutting orientation

**Table 1.14** Detailed parameters for MD modelling

| Parameter | Value |
|---|---|
| Length | 180 nm |
| Thickness | 2.5 nm |
| *t* | 40 nm |
| *L* | 20 nm |
| Underformed chip thickness | 40, 35, 30, 25, 20, 10 nm |
| Atoms in workpiece | 2.9, 2.8, 2.6, 2.5, 2.3, 2.0 million |
| Tool rake angle | – 20° |
| Tool clearance angle | 10° |
| Cutting edge radius | 30 nm |
| Total atoms in tool | 422,159 |
| Cutting speed | 100 m/s |

0.87 nm. Therefore, the temperature is kept constant. Table 1.14 provides a comprehensive breakdown of the particular simulation parameters. All of the models have the same length ($L$) and thickness ($t$) in order to guarantee that there is a constant distance between the cutting zone and the fixed boundaries. This causes the total model heights to vary depending on the thickness of the chip that was initially used. Using a cutting edge radius of 30 nm, the tool is deemed to have a rigid body since it has small cutting distances and is predicted to have little tool wear. The parameters, such as total length in the $X$-direction and thickness in the $Z$-direction, remain the same for all models.

It has been observed that the cutting edge radius of commercial diamond cutters is normally somewhere around 50 nm [46]. Therefore, the chosen cutting edge radius for the simulations is within a reasonable range. A high cutting speed of 100 m/s is employed in this section to alleviate computational demands, a common practice in previous MD studies [28]. It is important to recognize that cutting speed has a notable impact on the temperature within the cutting zone, which in turn may affect CUCT due to thermal effects [47]. Nevertheless, given the short cutting distance in this section, heat generation is not anticipated to significantly affect CUCT.

In accordance with the illustration in Fig. 1.22b, (0001) <1–210> is the cutting orientation for simulation models. It is known that CUCT of 6H SiC varies with crystal orientations [11]. The selection of the (0001) <1–210> orientation in this section is based on the fact that CUCT in this direction is the smallest, thereby reducing computational and memory requirements.

The potential function, stress calculation, and derivation process used in the MD simulation can be referenced in literature [48]. Upon deriving the equations for calculating interatomic forces from the Vashishta potential, they were integrated into the MD codes. Custom MD codes were developed for this section for two primary reasons. Firstly, the Vashishta potential is not currently supported in widely-used MD packages like LAMMPS. Secondly, the computational and memory efficiencies of these popular MD packages are insufficient for modeling the scales necessary for this section on conventional desktop computers. As these packages are general-purpose codes and not specifically optimized for machining simulations.

For instance, the NVIDIA GTX580 GPU utilized in this section with 1.5 GB of graphics memory could only handle approximately 0.5 million atoms using LAMMPS, whereas the modeling scale required for this section exceeded 2 million atoms.

The GPU-accelerated MD codes utilized in this section were developed utilizing the CUDA API from NVIDIA Corporation. The code architecture employed the link-cell type [49]. In comparison to the neighbor-list method utilized in widely-used MD packages like LAMMPS, the link-cell method offers advantages such as the avoidance of memory bottlenecks through coalesced memory access and more efficient utilization of shared memory among the GPU stream processors.

A key optimization of the link-cell method implemented for machining simulations in this section involved dynamic allocation of memory space for atom cells, with allocation only for cells containing atoms. This approach ensured efficient usage of memory, particularly in scenarios where large amounts of space within the machining models remained unoccupied by atoms. Consequently, the custom MD codes developed in this section enabled MD modeling of systems containing over 4 million atoms on a NVIDIA GTX580 GPU with 1.5 GB of graphics memory, a feat unattainable with conventional MD packages, and each model used approximately 48 h to complete.

A time step of one femtosecond was used for the MD computations, and each simulation was run for roughly one hundred fifty picoseconds until either a brittle fracture or a considerable ductile chip development was seen. The lattice constants for SiC were determined to be 3.073 and 15.08 Å, respectively, according to the research conducted by [50]. Diamond's lattice constant was found to be 3.567 Å for this particular portion of the diamond structure. In order to integrate the equations of motion, the velocity-Verlet method was utilized, and the MD computations were carried out inside a microcanonical ensemble.

In terms of analyzing stress distribution within the cutting zone, mechanical stress was computed in this section following the approach of Tanaka et al. [51]. While previous MD studies on machining typically focused on virial stress analysis [52] due to the complexities of calculating mechanical stress in atomic models and the lack of support for this feature in open-source MD packages. Virial stress may yield divergent results under certain conditions [53]. To address this issue, the necessary codes for calculating mechanical stress is needed, aiming to employ them to investigate stress distribution accurately [48].

A selection of snapshots from the MD simulations are presented in Fig. 1.23, which shows the results of the cutting operations with varying thicknesses of chips that have not been deformed ($t_0$). In the case of a chip thickness of 40 nm that has not been distorted, brittle fracture takes place before any evident plastic deformation takes place. Both ductile mode chip formation and brittle fracture are detected at undeformed chip thicknesses of 35 and 30 nm. Both forms of chip creation are observed. On the other hand, material removal for undeformed chip thicknesses of 25, 20, and 10 nm is performed primarily by the production of ductile mode chips.

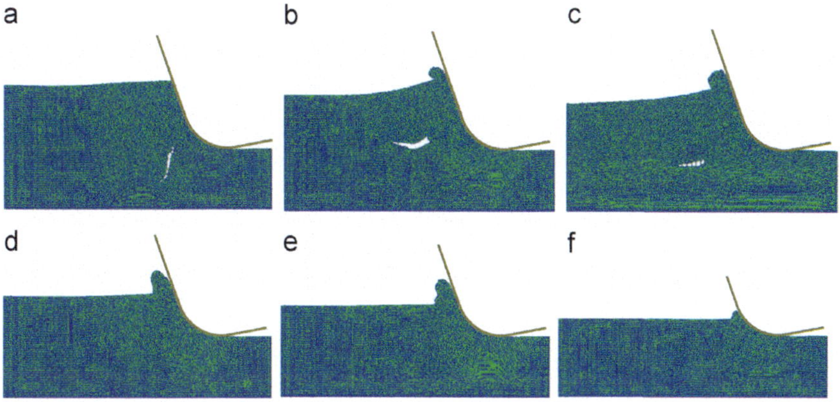


**Fig. 1.23** MD snapshots at various $t_0$

Notably, when the undeformed chip thickness is 10 nm, the chip formation resembles the extrusion process described by Fang et al. in nano-cutting of silicon, where extrusion rather than shearing occurs [54, 55]. The modeling outcomes across varying undeformed chip thicknesses suggest a transition in the cutting process: with an increase in undeformed chip thickness, the cutting response shifts from a ductile mode to a hybrid mode encompassing ductile and brittle attributes, and ultimately to a purely brittle mode.

It can be seen in Fig. 1.23 that the position of the fractures' commencement and the direction in which they propagate change depending on the thickness of the chip that has not been damaged. Cracks start in front of the cutting direction and expand in a direction that is generally parallel to it when the chip thickness is between 30 and 35 nm without any deformation. In the event that the fracture continues to advance and remove material from above, this might result in the formation of a small crater on the surface that has been machined. Conversely, when the undeformed chip thickness is 40 nm, the crack forms beneath the cutting zone and propagates downward. If the crack were to continue propagating to remove material ahead, a steeper crater would likely form, although this could not be fully realized in this section due to limitations imposed by the fixed boundary. Future MD simulations with larger model sizes are warranted to allow for unrestricted crack propagation.

When looking at the results of the modeling, it is anticipated that different morphologies would be observed for different undeformed chip thicknesses. Craters on the machined surface should be shallow if the thickness of the undeformed chip is somewhat over the critical value. However, if the thickness of the undeformed chip is much bigger, the craters may be deeper and steeper.

## References

1. Abdulkadir LN, Abou-El-Hossein K, Jumare AI, Odedeyi PB, Liman MM, Olaniyan TA (2018) Ultra-precision diamond turning of optical silicon—a review. Int J Adv Manuf Technol 96:173–208
2. Zhao CY, Cheung CF, Liu MY (2019) Nanoscale measurement with pattern recognition of an ultra-precision diamond machined polar microstructure. Precis Eng 56:156–163
3. Ding Y, Xu G, Guo F, Kou S, Lan Y, Ding Z et al (2003) Properties of single crystal copper produced by heated mould continious casting. Chin J Nonfer Met 13:1071–1076
4. Fan X, Wen Y (2000) Development and prospect of single crystal metal wire with high fidelity. Ordnance Mater Sci Eng 23:62–65
5. Niewczas M, Bian Q (2016) Influence of dislocation microstructure on low temperature magnetoresistance in copper single- and polycrystals. Mater Des 96:203–216
6. Guo TB, Li Q, Wang C, Zhang F, Jia Z (2017) Deformation characteristics and mechanical properties of single crystal copper during equal channel angular pressing by route A. Acta Metall Sin 53:991–1000
7. Liang Y, Chen G, Sun Y, Jiaxuan C, Chen W, Yu N (2014) Research status and outlook of ultra-precision machine tool. J Harbin Inst Technol 46(05):28–39 (in Chinese)
8. Dai Y, Zhang G, Luo T, Luo Q (2020) Centre cone generation and its force performance in single-point diamond turning. Int J Mech Sci 184:105780
9. Zhang G, To S, Zhang S (2016) Relationships of tool wear characteristics to cutting mechanics, chip formation, and surface quality in ultra-precision fly cutting. Int J Adv Manuf Tech 83(1–4):133–144
10. Zhang G, Dai Y, To S, Wu X, Lou Y (2019) Tool interference at workpiece centre in single-point diamond turning. Int J Mech Sci 151:1–12
11. Patten J, Gao W, Yasuto K (2005) Ductile regime nanomachining of single-crystal silicon carbide. J Manuf Sci E-T Asme 127:522–532
12. Lai M, Zhang X, Fang F (2017) Crystal orientation effect on the subsurface deformation of monocrystalline germanium in nanometric cutting. Nanoscale Res Lett 12:1–10
13. Yan J, Asami T, Harada H, Kuriyagawa T (2012) Crystallographic effect on subsurface damage formation in silicon microcutting. Cirp Ann 61:131–134
14. Zhang G, Han J, Chen Y, Wang J, Wang H (2022) Brittle-ductile transition and nano-surface generation in diamond turning of single-crystal germanium. J Manuf Process 82:628–645
15. Zhang G, Han J, Chen Y, Xiong J, Wang J, Ran J (2022) Generation mechanism and dual-dynamics simulation of surface patterns in single-point diamond turning of single-crystal copper. J Manuf Process 75:1023–1038
16. Chen H (1996) Elastic anisotropy of metals. Metallurgical Industry Press, Beijing
17. Furukawa Y, Moronuki N (1988) Effect of material properties on ultra-precise cutting processes. Cirp Ann 37:113–116
18. Xia X (2014) Researching effect of germanium single-crystal anisotropy in ultra-precision machining. Kunming University of Science and Technology
19. Mao J (2013) The research to germanium single-crystal cutting process based on the molecular dynamic. Master dissertation, Changchun University of Science and Technology, Jilin Changchun (in Chinese)
20. van Bouwelen FM, van Enckevort WJP (1999) A simple model to describe the anisotropy of diamond polishing. Diam Relat Mater 8(2):840–844
21. Son SM, Lim HS, Ahn JH (2005) Effects of the friction coefficient on the minimum cutting thickness in micro-cutting. Int J Mach Tool Manu 45:529–535
22. Zhou Y, Gu Y, Lin J, Lu M, Zheng Y, Zhou W, Zhao H, Jia R, Wu S (2022) Feasibility study of single-crystal silicon ductile-regime turning via fast tool servo. J Mater Res Technol 16:1478–1493
23. Lawn BR, Hockey BJ, Wiederhorn SM (1980) Atomically sharp cracks in brittle solids: an electron microscopy study. J Mater Sci 15:1207–1223

24. Lemaitre P (1988) Fracture toughness of germanium determined with the Vickers indentation technique. J Mater Sci Lett 7:895–896
25. Plimpton S (1995) Fast parallel algorithms for short-range molecular dynamics. J Comput Phys 117(1):1–19
26. Zhu PZ, Hu YZ, Ma TB, Wang H (2010) Study of AFM-based nanometric cutting process using molecular dynamics. Appl Surf Sci 256:7160–7165
27. Chavoshi SZ, Goel S, Luo X (2016) Influence of temperature on the anisotropic cutting behaviour of single crystal silicon: a molecular dynamics simulation investigation. J Manufac Proc 23:201–210
28. Goel S, Luo X, Reuben RL (2012) Shear instability of nanocrystalline silicon carbide during nanometric cutting. Appl Phys Letters 100:231902–231907
29. Ueda K, Sugita T, Hiraga H, Iwata K (1991) A J-integral approach to material removal mechanisms in microcutting of ceramics. CIRP Ann 40:61–64
30. Tersoff J (1989) Modeling solid-state chemistry: interatomic potentials for multicomponent systems. Phys Rev B Condens Matter 39:5566–5568
31. Inamura T, Takezawa N, Taniguchi N (1992) Atomic-scale cutting in a computer using crystal models of copper and diamond. CIRP Ann 41:121–124
32. Foiles SM, Baskes MI, Daw MS (1986) Embedded-atom-method functions for the fcc metals Cu, Ag, Au, Ni, Pd, Pt, and their alloys. Phys Rev B Condens Matter 33:7983–7991
33. Xiao G, He Y, Geng Y, Yan Y, Ren M (2019) Molecular dynamics and experimental study on comparison between static and dynamic ploughing lithography of single crystal copper. Appl Surf Sci 463:96–104
34. Kittel C (2004) Introduction to solid state physics, 8th edn. Wiley, New York, pp 15–38
35. Goel S, Luo X, Agrawal A, Reuben RL (2015) Diamond machining of silicon: a review of advances in molecular dynamics simulation. Int J Mach Tool Manu 88:131–164
36. Metropolis N (1953) Equation of state calculations by fast computing machines. J Chem Phys 21(6):1087
37. Tersoff J (1988) Empirical interatomic potential for silicon with improved elastic properties. Phys Rev B 38(14):9902–9905
38. Wang Q, Bai Q, Chen J, Guo Y, Xie W (2015) Stress-induced formation mechanism of stacking fault tetrahedra in nano-cutting of single crystal copper. Appl Surf Sci 355:1153–1160
39. Li J, Fang Q, Zhang L, Liu Y (2015) The effect of rough surface on nanoscale high speed grinding by a molecular dynamics simulation. Comput Mater Sci 98:252–262
40. Lai, M. (2016) Study on Nano-cutting Mechanism of Monocrystalline Germanium. PhD dissertation, Tianjin University (in Chinese)
41. Lai M, Zhang X, Fang F, Bi M (2017) Fundamental investigation on partially overlapped nano-cutting of monocrystalline germanium. Precis Eng 49:160–168
42. Demir E, Mercan C (2018) A physics-based single crystal plasticity model for crystal orientation and length scale dependence of machining response. Int J Mach Tool Manu 134:25–41
43. Mukaida M, Yan J (2017) Ductile machining of single-crystal silicon for microlens arrays by ultraprecision diamond turning using a slow tool servo. Int J Mach Tool Manu 115:2–14
44. Zhu PZ, Fang FZ (2012) Molecular dynamics simulations of nanoindentation of monocrystalline germanium. Appl Phys A 108(2):415–421
45. Stukowski A (2009) Visualization and analysis of atomistic simulation data with OVITO—the open visualization tool. Modell Simul Mater Sci Eng 18(1):015012
46. Gao W (2010) Precision nanometrology—sensors and measuring systems for nanomanufacturing. Springer, London. ISBN: 978-1-84996-253-7
47. Schinker MG (1991) Subsurface damage mechanisms at high-speed ductile machining of optical glasses. Precis Eng 13:208–218
48. Xiao G, To S, Zhang G (2015) Molecular dynamics modelling of brittle–ductile cutting mode transition: case study on silicon carbide. Int J Mach Tools Manuf 88:214–222
49. Brown WM, Wang P, Plimpton SJ (2011) Tharrington, Implementing molecular dynamics on hybrid high performance computers—short range forces. Comput Phys Commun 182:898–911

50. Coati A, Sauvage-Simkin M, Garreau Y, Pinchaux R, Argunova T, K, (1999) Aid, 3×3 R 30° reconstruction of the 6H-SiC (0001) surface: a simple T4 Si adatom structure solved by grazing-incidence x-ray diffraction. Phys Rev B 59:12224
51. Tanaka H, Shimada S, Anthony L (2007) Requirements for ductile-mode machining based on deformation analysis of mono-crystalline silicon by molecular dynamics simulation. CIRP Ann Manuf Technol 56:53–56
52. Cai M, Li X, Rahman M (2007) Study of the mechanism of nanoscale ductile mode cutting of silicon using molecular dynamics simulation. Int J Mach Tools Manuf 47:75–80
53. Cheung KS, Yip S (1991) Atomic-level stress in an inhomogeneous system. J Appl Phys 70:5688–5690
54. Fang F, Wu H, Liu Y (2005) Modelling and experimental investigation on nanometric cutting of monocrystalline silicon. Int J Mach Tools Manuf 45:1681–1686
55. Fang F, Wu H, Zhou W, Hu X (2007) A study on mechanism of nano-cutting single crystal silicon. J Mater Process Technol 184:407–410

# Chapter 2
# Surface Generation in Machining Single Crystal Materials

**Abstract** The precise control of surface generation in machining single crystal materials is fundamental for achieving desired surface qualities at the nanoscale. This chapter provides a detailed exploration of the factors influencing surface generation in machining single crystal materials, focusing on the mechanisms of nano-surface generation and methods for controlling surface morphology. By understanding and optimizing these factors, it is possible to achieve exceptional surface quality, essential for high-precision engineering applications.

**Keywords** Single-point diamond turning · Surface generation · Surface morphology simulation · Single crystal materials

## 2.1 Nano-surface Generation

SPDT is an advanced machining process that uses a single-crystal diamond (SCD) tool to produce extremely smooth and accurate surfaces on workpieces, typically at the nanometer level. When applied to single-crystal materials, SPDT is capable of generating nano-surfaces with unparalleled precision and efficiency, which is highly desirable in various high-technology applications such as optics, semiconductors, and nanotechnology devices.

The focus on nano-surface generation through SPDT for single-crystal materials stems from the unique properties of these materials. Single-crystal materials, such as Cu, Si, Ge etc., have a uniform lattice structure without grain boundaries, resulting in exceptional material properties like high optical clarity, high thermal conductivity, and consistent electrical characteristics.

These properties are critical for applications in the field of electronics, photonics, and precision engineering. Generating nano-surfaces on such materials can enhance performance by minimizing surface defects, reducing light scattering, improving thermal management, and refining the interface with other nanostructures. Therefore, achieving high-quality nano-surfaces is essential for the advancement of these technologies.



G. Zhang and J. Wang, *Material-Oriented Cutting Processes in Precision Machining*, Springer Tracts in Mechanical Engineering,
https://doi.org/10.1007/978-981-96-2504-8_2

The nano-surface generation of single-crystal materials through SPDT is characterized by surfaces with ultra-low roughness (often at the nano-level), long-term environmental stability, and precise geometric forms. The surface characteristics are largely influenced by the crystallographic orientation of the material being machined, since the removal mechanism during SPDT is anisotropic.

This section will provide an in-depth perspective on the nano-surface generation of single-crystal materials through the analysis of experimental results based on well-designed experiments.

### *2.1.1 Nano-surface Characteristics in SPDT*

In this section of the research, turning tests were carried out with the assistance of the ultra-precision lathe mentioned in Chap. 1. The experimental setup is depicted in Fig. 2.1a. Figure 2.1b illustrates the results of machining tests that were performed on three single-crystal copper (SCC) workpieces, each of which had different crystal planes (100), (110), and (111). In the course of the turning process, the workpieces were affixed to a cylindrical rod, which was thereafter fastened to the spindle by means of a fixture and a vacuum chuck. Diamond tools with a rake angle of 0 degrees, a clearance angle of 15 degrees, and a nose radius of 0.502 mm were utilized in the process of turning experiments.

The cutting tool and experimental parameters used in the section are detailed in Table 2.1. The Cu (100), (110), and (111) surfaces were first subjected to rough turning followed by finishing operations. For the purpose of recording the cutting

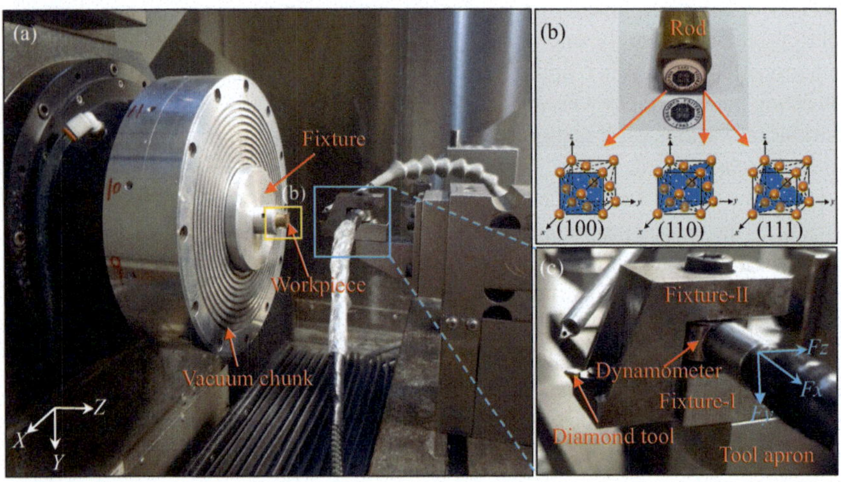


**Fig. 2.1** **a** Configuration of experiment, **b** orientation of workpiece crystal plane, and **c** installation of dynamometer

**Table 2.1** Tool and workpiece parameters in this experiment

| Tool and workpiece parameters | Values |
|---|---|
| Tool rake angle | 0° |
| Tool clearance angle | 15° |
| Tool nose radius | 0.502 mm |
| Workpiece material | Single-crystal copper |
| Workpiece dimension | $10 \times 10 \times 2\ mm^3$ |
| Spindle speed | 2000 rpm |
| Feed rate | 5 mm/min |
| Depth of cut | 4 μm |

force signals that occurred throughout the turning process, a dynamometer was employed, as shown in Fig. 2.1c. The amplified force signals were then transmitted to the DAQ-System via a charge amplifier. Subsequently, the force data was analyzed in both the time and frequency domains using the dynamometer software (Dynoware), as shown in Fig. 2.2. The chips produced during the turning process were gathered and examined using the SEM. Furthermore, the surface morphology and roughness of the machined parts were assessed with the white light interferometer mentioned in Chap. 1.

Figure 2.3 provides a visualization of the machined surface morphology of SCC, revealing the formation of fan-shaped patterns. In Fig. 2.3a, the surface morphology of Cu (100) after machining is displayed, showing four cloudy regions (2, 4, 6, and 8) and four clean regions (1, 3, 5, and 7). Each region occupies nearly the same area, and the cloudy and clean regions are distributed alternately on the surface, resulting in the formation of four quadruple-symmetric fan-shaped patterns. Figure 2.3b illustrates the surface morphology of Cu (110) after machining, presenting four cloudy regions (1, 3, 5, and 7) and four clean regions (2, 4, 6, and 8). Here, the cloudy regions vary in size: there are two large and two small cloudy regions. Specifically, regions 1 and 5 are equal in area, as are regions 3 and 7. The larger cloudy regions (1 and

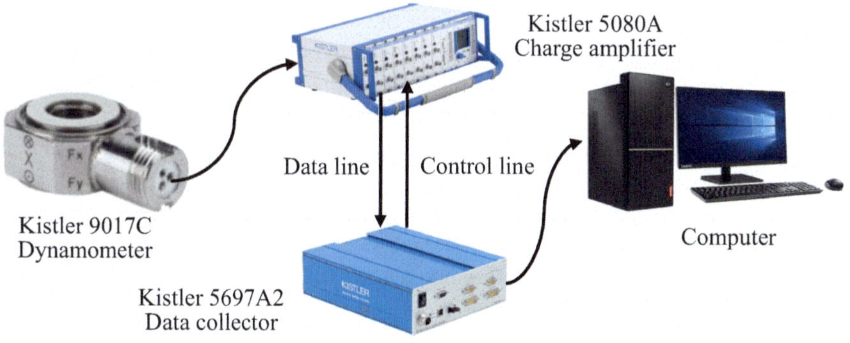


**Fig. 2.2** Cutting force capturing system

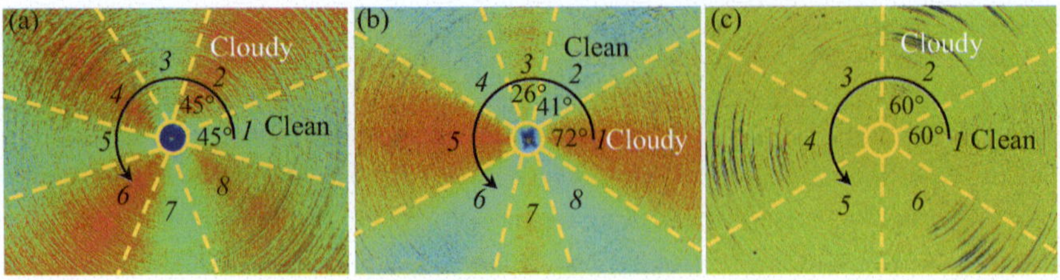


**Fig. 2.3** The surface morphology of SCC characterized by: **a** (100), **b** (110), and **c** (111) crystal planes

5) are twice the size of the smaller ones (3 and 7), while the clean regions remain uniformly sized, resulting in four double-symmetric fan-shaped patterns. Figure 2.3c demonstrates the surface morphology of Cu (111) after machining, revealing three cloudy regions (2, 4, and 6) and three clean regions (1, 3, and 5). Each region is approximately equal in size, with alternating cloudy and clean regions forming three triple-symmetric fan-shaped patterns.

Figure 2.3 provides a visualization of the surface morphology of SCC after the application of SPDT is presented. This visualisation reveals the creation of fan-shaped patterns. A presentation of the surface morphology of Cu (100) after machining is shown in Fig. 2.3a. This display includes four cloudy regions (2, 4, 6, and 8) as well as four clear regions (1, 3, 5, and 7). The cloudy and clean regions are dispersed alternatively over the surface, which results in the construction of four quadruple-symmetric fan-shaped patterns. Each region occupies roughly the same area, and the cloudy regions and clean regions are in close proximity to one another. After the machining process, the surface morphology of Cu (110) is depicted in Fig. 2.3b, which shows four cloudy regions (1, 3, 5, and 7) and four clean regions (2, 4, 6, and 8). The overcast regions in this area range in size from large to tiny, including two huge cloudy regions and two small cloudy regions. Particularly, regions 1 and 5 are identical in terms of their total area, as are regions 3 and 7. The bigger cloudy areas (1 and 5) are twice as big as the smaller ones (3 and 7), but the clean areas stay the same size. This makes four fan-shaped patterns that are twice symmetrical. On the other hand, the clear regions retain the same size throughout. The surface morphology of Cu (111) after machining is depicted in Fig. 2.3c, which shows that there are three hazy regions (2, 4, and 6) and three clear regions (1, 3, and 5). Approximately the same amount of space is occupied by each zone, and the cloudy and clear sections alternate with one another to create three fan-shaped patterns that are triple-symmetric.

Clearly, fan-shaped patterns have an impact on the surface morphology of SCC. Figure 2.4 illustrates the surface morphology of Cu along the (100), (110), and (111) crystal orientations. The surface roughness (Sa) for each crystal orientation was determined using the white light interferometer, serving as an indicator for overall Sa. Significantly, Fig. 2.4a illustrates that the residual height of machined surface (Cu (100)) in the cloudy area is greater than that in the clean area, showing a Sa of about 18 nm in the former and around 11 nm in the latter. Furthermore, Fig. 2.4b depicts the surface morphology of machined surface (Cu (110)), whereby higher surface heights are observed in cloudy regions with larger fan areas, while surface heights in regions with smaller fan areas follow suit as the clean regions show the lowest surface height. The corresponding Sa values are illustrated in Fig. 2.4e, with the two largest cloudy fan regions exhibiting the highest Sa (about 40 nm), the two smaller cloudy fan regions having a roughness value of about 30 nm, and the four clean regions demonstrating the smallest roughness at approximately 20 nm. Turning to Fig. 2.4c, the surface morphology of machined surface (Cu (111)) reveals relatively high surface heights in the three cloudy fan regions, whereas the three clean fan regions display lower surface heights. Likewise, the Sa distribution is presented in Fig. 2.4f, showing approximately 35 nm roughness in the three cloudy fan regions and roughly 11 nm roughness in the three clean fan regions. As a result, the comparison enables us to draw the conclusion that the surface quality of the machined surface made of Cu (100) is the greatest, with a Sa of 15.433 nm. Following this, the surface that has been machined with Cu (111) has the second greatest surface quality, with a Sa of 17.016 nm. In conclusion, the machined surface (Cu (110)) is found to have the poorest surface quality, indicated by a Sa of 20.102 nm.

To provide further insights, Fig. 2.5a–c indicate three-dimension (3D) topographic views of A, B and C zones in the Cu (110) sample, as depicted in Fig. 2.4b. Figure 2.5a indicates that the zone B exhibits the highest surface fluctuation, with a peak-to-valley (PV) height of around 30 nm due to its position as the boundary between two distinct regions. Figure 2.5b illustrates the zone B follows this pattern with a PV height of 26

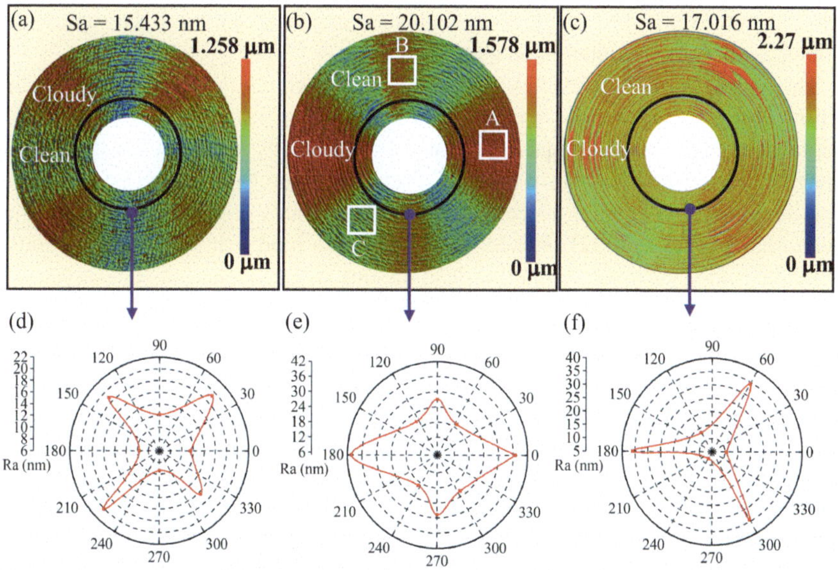


**Fig. 2.4** The surface structure of SCC is characterized by: **a** the (100), **b** the (110), and **c** the (111) crystal planes, along with the Sa analysis on single crystal copper with: **d** the (100), **e** the (110), and **f** the (111) crystal planes

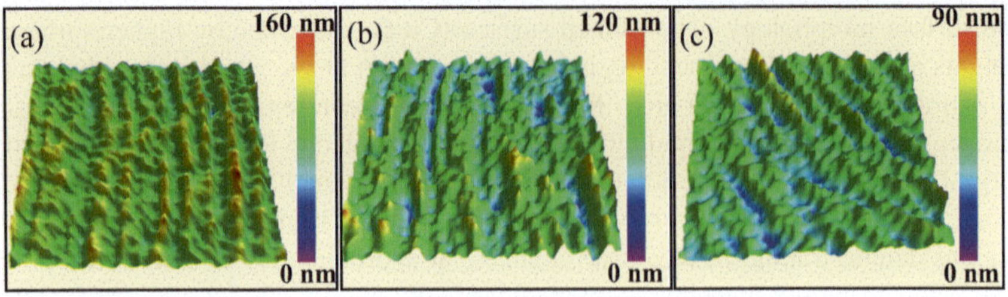


**Fig. 2.5** The 3D structure of a SCC with the (110) crystal plane in: **a** zone A, **b** zone B, and **c** zone C (as illustrated in Fig. 2.4b)

nm. On the other hand, zone C demonstrates the minimum surface fluctuation, with a PV height of 15 nm, as illustrated in Fig. 2.5c. Upon comparing the surface quality of the three zones, it is evident that the clean fan region (zone C) has the greatest level of surface quality.

Chip morphology serves as an indicator of material properties. Upon examining the chips produced from various crystal planes as depicted in Fig. 2.6, it becomes evident that during the turning process, continuous chips are formed, albeit with incomplete morphology. Notably, a small crack structure with consistent intervals appears, indicating an uneven distribution of strain within the chips. The unequal distribution can be traced to the anisotropy of the SCC. Therefore, while cutting the workpiece along different crystal orientations, there are notable differences in cutting performance, resulting in changes in the shape of the chips produced. As a result, the chips manifest as either continuous or cracked, signifying the significant influence of crystal orientation on chip formation.

The anisotropy of SCC can also be assessed through cutting force analysis. The cutting force signal and the power spectrum diagram that corresponds to it are displayed in Fig. 2.7. This signal is generated during turning procedures on the crystal planes (100), (110), and (111) of the SCC. Previous research [1] and MD simulations have shown that the cutting force is greater in the cloudy regions compared to the clean regions during SPDT. Hence, the force peaks represent the cloudy regions, while the force valleys represent the clean regions.

The force curve in SPDT of Cu (100) is shown in Fig. 2.7a. The cutting force changes several times during a turning cycle. The force is strongest at places 2, 4, 6,

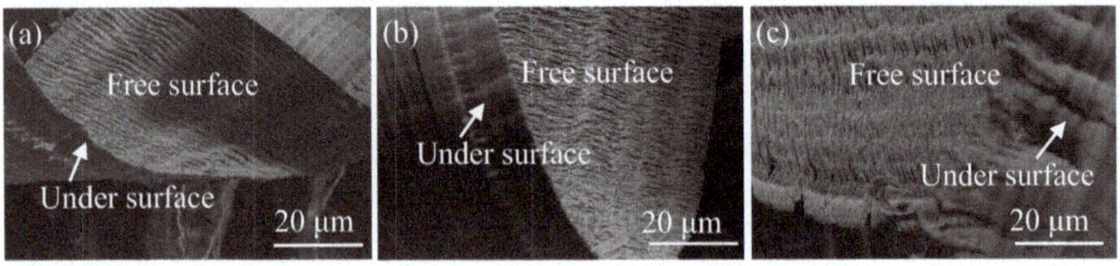


**Fig. 2.6** SCC chip shape with **a** (100), **b** (110), and **c** (111) crystal planes

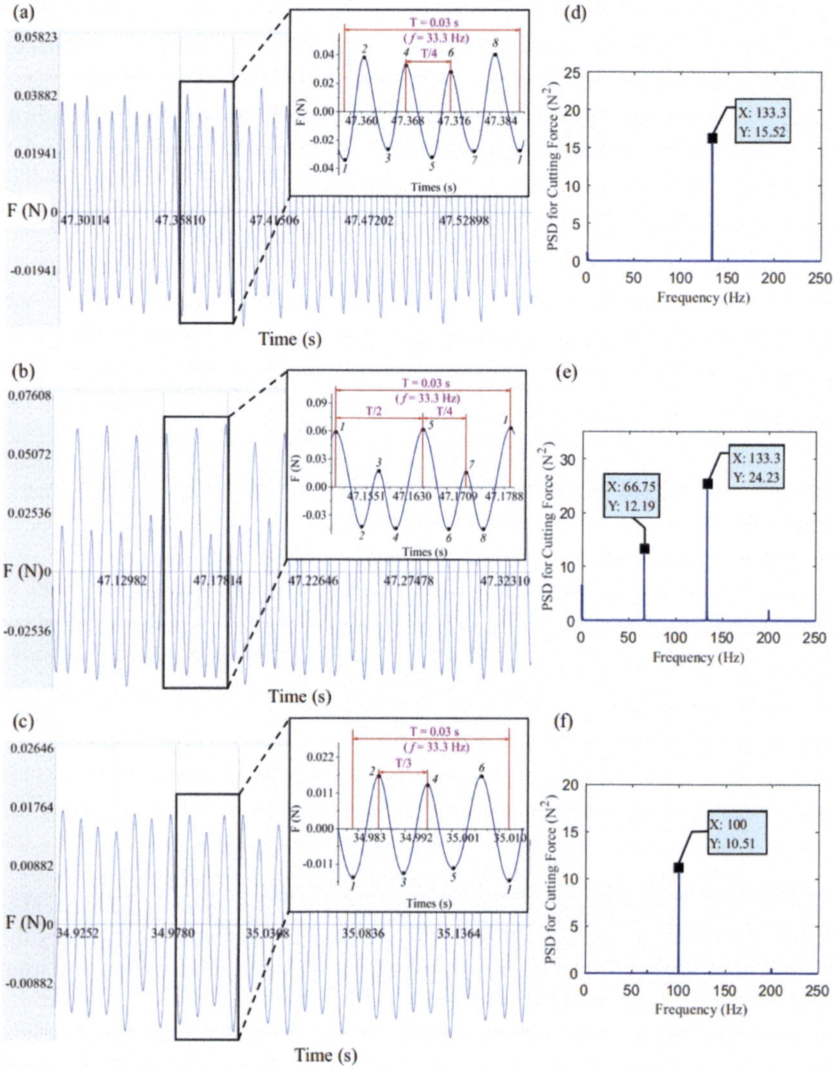


**Fig. 2.7** Diagram of cutting force and its spectrum during SPDT: crystal planes, **a**, **d** (100), **b**, **e** (110), and **c**, **f** (111)

and 8, which are the four dark areas shown in Fig. 2.3a. The forces are 0.0389786, 0.032796, 0.0282988, and 0.040967 N, in that order. Force troughs 1, 3, 5, and 7 on the other hand line up with the four clean areas in Fig. 2.3a, showing force values of − 0.0347191, − 0.0272797, − 0.0328584, and − 0.0285532 N, in that order.

In comparison to cutting in cloudy areas, cutting in clean regions produces reduced and more consistent cutting forces. The cutting force exhibits periodic fluctuations during each revolution, with a frequency of 33.3 Hz corresponding to a spindle speed of 2000 rev/min. With the four-fold symmetry of Cu (100), the fluctuation frequency of the force should be 133.2 Hz.

Analysis of the spectrum of the cutting force signal in Fig. 2.7d reveals a peak at 133.3 Hz, aligning closely with the expected value of 133.2 Hz. The cutting force signal in Fig. 2.7d has a peak at 133.3 Hz, which is very close to the number that would be expected, which is 133.2 Hz. Hence, the spectrum study of the cutting force accurately shows the four-fold symmetric distribution of Cu (100).

The force signal on the Cu (110) crystal plane exhibits periodic fluctuations during turning. These fluctuations consist of alternating big peaks, valleys, and smaller peaks that correspond to surface fan-shaped patterns, as seen in Fig. 2.7b. The corresponding force values for these patterns are given in Fig. 2.3b, and they are as follows: 0.0592252, – 0.0428885, 0.0175995, – 0.0445542, 0.0614862, – 0.0453146, 0.016166, and – 0.0454979 N. These patterns are labelled 1 through 8, and their respective force values are presented in the figure. There is a dispersion of fan-shaped patterns on the Cu (110) plane in a four-fold and double-symmetric configuration as a result of the turning frequency being 33.3 Hz and the spindle speed being 2000 rev/min.

As a result, the spectrum peaks at 66.6 and 133.2 Hz, which correspond to double-symmetric and four-fold distributions, respectively. In Fig. 2.7e, the spectrum diagram of the cutting force signals shows two spectrum peaks at frequencies of 66.75 and 133.3 Hz, which are in line with the theoretical values of 66.6 and 133.2 Hz. This shows that the fan-shaped patterns on the surface are what cause the cutting force to change regularly.

Referring to Fig. 2.7c, the graph depicts the force signal seen while the Cu (111) crystal plane is being turned on. The force curve exhibits three distinct peaks and troughs, which correspond to the fan-shaped patterns labelled 1, 2, 3, 4, 5, and 6 in Fig. 2.3c. The force values of these exhibits are – 0.0151415, 0.0163841, – 0.013944, 0.0136994, – 0.0122866, and 0.0163514 N, respectively. The Cu (111) surface rotates at a frequency of 99.9 Hz for a complete rotation, as specified by the spindle speed. The force spectrum analysis in Fig. 2.7f shows a peak in the spectrum at 100 Hz, which aligns with the estimated frequency. This showcases the capabilities of the cutting force signal and its spectrum analysis to accurately capture the surface fan-shaped patterns of SCC.

Similarly, if the soft SCC is replaced with a BSCM, under the experimental parameters described in Tables 1.1 and 1.2 in Chap. 1, the machined surface still exhibits a fan-shaped pattern, as shown in Fig. 2.8. As a result of the anisotropy of the material, the (100) crystal plane has eight fan-shaped sections that are almost the same size. These regions are similar in size. Anisotropy leads to the BDT during the turning process. There is a symmetry rule present in the surface morphology, and there are four symmetrical axes. Regions 1, 3, 5, and 7 possess brittleness (with worse surface quality) and regions 2, 4, 6, and 8 are ductile regions (with better surface quality).

As a further feature, the crystal plane (111) exhibits six fan-shaped sections of comparable dimensions, with a distribution of brittle and ductile regions that alternates around the centre point. After a rotation, the surface morphology follows a

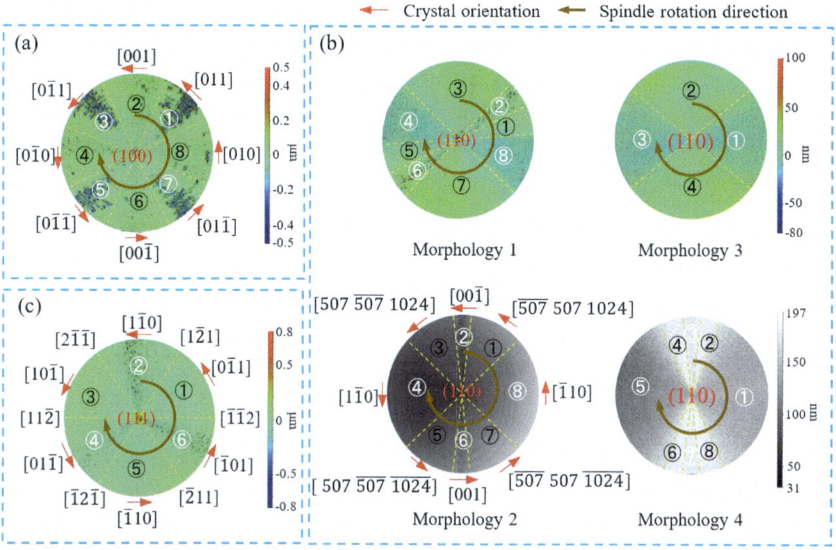


**Fig. 2.8** Anisotropies with fan-shaped morphologies of machined Ge crystal planes. Crystal planes **a** (100), **b** (110), and **c** (111). Four different surface morphologies are observed under different cutting parameters: morphology 1, morphology 2, morphology 3, and morphology 4. (Cutting parameters for morphologies 1 and 2: $n = 2000$ r/min, $f = 5$ mm/min, $a_p = 2$ μm; Cutting parameters for morphologies 3 and 4: $n = 2000$ r/min, $f = 10$ mm/min, $a_p = 2$ μm. The red arrows in the figures indicate crystal orientations, the brown arrows show spindle rotation direction, and the numbers indicate the sequence of alternating brittle and ductile regions on the surfaces.)

symmetry rule with three symmetrical axes. Areas 2, 4, and 6 indicate brittle areas, whereas regions 1, 3, and 5 represent ductile regions. This is because the symmetry rule is applicable to the surface morphology.

It is important to note that the surface morphology of the (110) crystal plane also exhibits anisotropy, but it is unstable. As shown in Fig. 2.8b, previous studies have reported two different surface morphologies, labeled as morphology 1 [2] and morphology 2 [3], but no unified statement has been made.

Face-turning tests were done on the (110) crystal plane with two different feed rates to look into this effect. The surface structure changes between pairs of brittle and ductile regions at a rate of 5 mm/min. There are ductile regions 1, 3, 5, and 7, and brittle regions 2, 4, 6, and 8. There are also two pairs of brittle regions with different sizes (2, 4, 6, and 8), and two pairs of ductile regions with the same size (1, 3, 5, and 7). On the other hand, when the feed rate is raised to 10 mm/min, the surface pattern shows pairs of brittle and ductile regions with different sizes (brittle regions 2 and 4, ductile regions 1 and 3), or pairs of brittle regions with different sizes (1, 3, 5, and 7) and ductile regions of the same size (2, 4, 6, and 8).

Based on the experimental results, the Ge's machined surface always exhibits a fan-shaped pattern known as the BDT phenomenon, which negatively affects the machined surface's quality. Understanding the mechanism behind anisotropy and the BDT is crucial for achieving desired machined surfaces.

The experimental results demonstrate noticeable cyclical fluctuations in the machined surface of SCC using various characteristic method. Such fluctuations predominantly arise from the material's anisotropy caused by variations in crystallographic orientation [4].

## 2.1.2 *Nano-surface Generation Mechanism*

During the process of SCC turning, the material anisotropy induced fan-shanpe nano-surface generation leading to variations in cutting force, shear angle, and Sa [5]. This alteration in cutting force, known as material-induced vibration, cannot be fully eliminated through lathe design or process control, which significantly hinders the performance of ultra-precision diamond lathes [6]. Therefore, it is essential to simulate the machined surface morphology of SCC according to cutting forces to understand the mechanism behind the generation of fan-shaped nano-surfaces.

### 2.1.2.1 Morphology of Ideal Turning Surface

Figure 2.9a illustrates the surface morphology that would be created on the workpiece if it were to be turned using an arc diamond tool. The equation for this surface morphology is specified as [7]:

$$R_{th} = \frac{f^2}{8R} \tag{2.1}$$

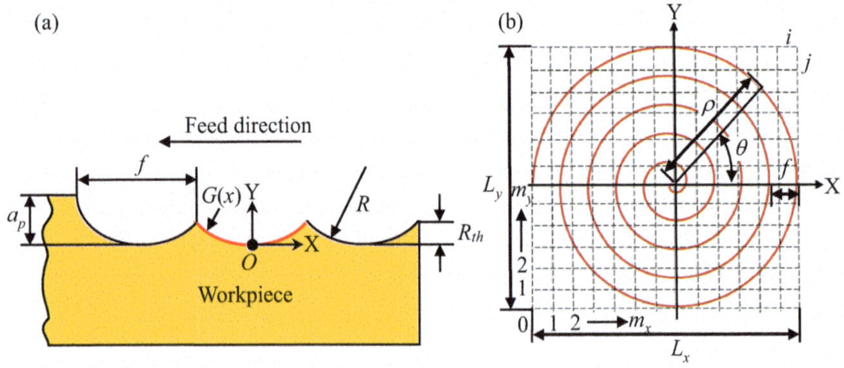


**Fig. 2.9** **a** Diagram illustrating the surface features of a machined surface using a diamond tool with round nose and **b** the theoretical tool path

where $R$ represents the tool nose radius, and $f$ denotes the feed rate.

The function $G(x)$ represents the surface contour after one complete rotation [8]:

$$G(x) = Ax^2 + C \tag{2.2}$$

Setting the $O$ (lowest point) as the origin on the two-dimension (2D) surface contour allows for the omission of the constant $C$ in $G(x)$. The simplified surface contour function must still adhere to specific boundary conditions, which can be determined by:

$$G\left(\pm\frac{f}{2}\right) = R_{th} \tag{2.3}$$

Therefore, the $A$ in Eq. (2.2) is given by:

$$A = \frac{4R_{th}}{f^2} \tag{2.4}$$

In Fig. 2.9b, the coordinates of the $(i, j)$ point inside the simulation area may be obtained by applying the following equations. This is the case if the centre of the simulation area is set to be the same as the centre of the machined surface on the workpiece.

$$\begin{cases} X_{i,j} = i \cdot m_x - \frac{L_x}{2} \\ Y_{i,j} = j \cdot m_y - \frac{L_y}{2} \end{cases} \tag{2.5}$$

where $L_x$ and $L_y$ are the length and width of the simulation area, respectively, and $m_x$ and $m_y$ represent the resolution of the simulation area in $X$- and $Y$-directions, respectively.

The Cartesian coordinate system is converted into a cylindrical coordinate system to make the process of model generation more straightforward. When the coordinates of a point within the 3D form are given as (ρ, θ, Z), the equations that are used to calculate $\rho$ and $\theta$ in the cylindrical coordinate system are as follows:

$$\begin{cases} \rho = \sqrt{X_{i,j}^2 + Y_{i,j}^2} \\ \theta = \tan^{-1}\left(\frac{X_{i,j}}{Y_{i,j}}\right) \end{cases} \tag{2.6}$$

This point's equivalent angle $\theta_v$ can be determined by:

$$\theta_v = \frac{\theta}{2\pi} \tag{2.7}$$

The tool's total rotation number $N_T$ can be determined using the following formula:

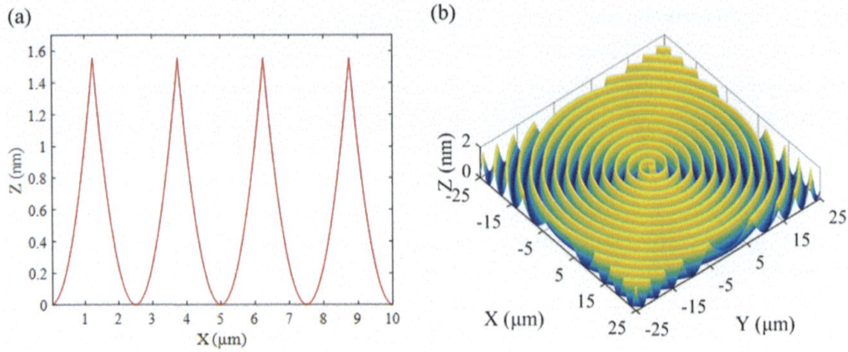


**Fig. 2.10** Ideal surface contour of **a** 2D and **b** 3D

$$N_T = \left|\frac{\rho}{f} - \theta_v\right| \quad (2.8)$$

The residual surface contour segment determines the final shape of the machined surface in turning operations. Therefore, it is crucial to determine the polar diameter that is comparable to the corresponding point in the previous, current, and subsequent cycles. The subsequent equations are generated for this specific objective:

$$\begin{cases} \rho_{v-1} = \rho - (N_T - \theta_v - 1)f \\ \rho_v = \rho - (N_T - \theta_v)f \\ \rho_{v+1} = \rho - (N_T - \theta_v + 1)f \end{cases} \quad (2.9)$$

Hence, the ultimate coordinate $Z_{th}$ of the point on the surface contour during optimal turning circumstances can be acquired in the following equation:

$$Z_{th} = \min\{G(\rho_{v-1}), G(\rho_v), G(\rho_{v+1})\} \quad (2.10)$$

Figure 2.10 displays the simulation for surface morphology in both 2D and 3D, achieved under ideal turning operations.

#### 2.1.2.2 Material Induced Vibration

A simplified one-dimensional turning system with a single degree of freedom has been designed for the purpose of investigating the influence that material-induced vibration has on the surface morphology during the turning process. The reaction of the tool to the cutting force is the primary concentrate of this system, which is founded on the 3D surface morphology that is observed under ideal turning operations.

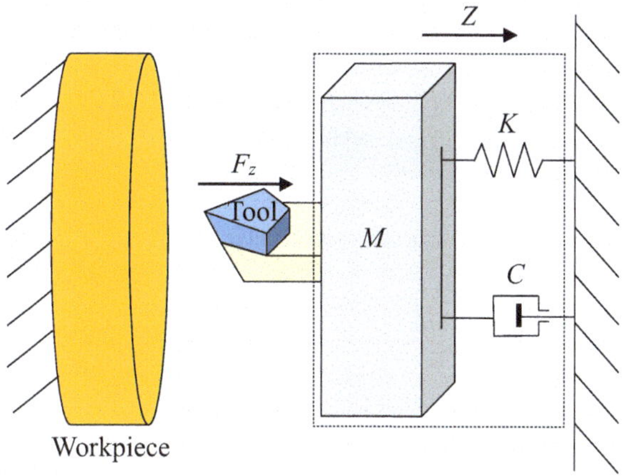


**Fig. 2.11** A simplified model of the dynamic cutting system using a set of parameters

As depicted in Fig. 2.11, the system comprises a diamond tool and a SCC workpiece. To facilitate the study of the effect of material vibration on the relative displacement of the tool, this system is modeled as a spring-mass-damping system. This section assumes the workpiece to be rigid, while representing the tool with parallel connections of a spring, mass, and damping elements.

The derived equations aim to describe this cutting system and are presented as follows:

$$\begin{cases} \frac{d^2Z(t)}{dt^2} + 2\zeta\omega_n\frac{dZ(t)}{dt} + \omega_n^2Z(t) = \frac{F_z(t)}{M} \\ \omega_n = \sqrt{\frac{K}{M}} \\ \zeta = \frac{C}{2\sqrt{KM}} \end{cases} \tag{2.11}$$

where, $Z$ represents the displacement of the cutting system, $F_z(t)$ denotes the thrust force applied during machining, $M$ corresponds to the equivalent mass of the cutting system, $\zeta$ stands for the damping ratio of the cutting system, $\omega_n$ indicates the natural frequency of the cutting system, $K$ is the spring constant of the cutting system, $C$ refers to the cutting constant specific to the turning system. Finally, $t$ represents the machining time.

Fan-shaped patterns that are noticed on the surface of the machined object are impacted by the relative displacement that occurs between the tool and the workpiece. This displacement is brought about by vibration that is created by the material. This relative displacement can be represented as $Z_{miv}(t)$. Since this displacement follows the trigonometric function rule, it can be expressed accordingly. Noted that the parameters in the relative displacement equations $Z_{miv}(t)$ differ for various crystal planes, as the induced displacements by these planes vary. The equations for the relative displacement $Z_{miv}$ of Cu (100), (110), and (111) crystal planes are as follows:

$$Z_{miv}(t) = H + A_1\cos(\omega_1 t + \varphi_1) + A_2\sin(\omega_2 t + \varphi_2) \tag{2.12}$$

Table 2.2 The parameters of the relative displacement function for Cu (100) crystal plane

| | $H_0$ | $A_1$ | $\omega_1$ | $\varphi_1$ | $A_2$ | $\omega_2$ | $\varphi_2$ |
|---|---|---|---|---|---|---|---|
| $Z_{miv}$ | $1.257 \times 10^{-3}$ | $-1.146 \times 10^{-2}$ | $8.4 \times 10^2$ | 0 | $5.254 \times 10^{-3}$ | $8.4 \times 10^2$ | 0 |
| $F_z$ | $9.18 \times 10^{-4}$ | $-3.035 \times 10^{-2}$ | $8.4 \times 10^2$ | 0 | $9.36 \times 10^{-3}$ | $8.4 \times 10^2$ | 0 |

Table 2.3 The parameters of the relative displacement function for Cu (110) crystal plane

| | $H_0$ | $A_1$ | $\omega_1$ | $\varphi_1$ | $A_2$ | $\omega_2$ | $\varphi_2$ |
|---|---|---|---|---|---|---|---|
| $Z_{miv}$ | 0 | $1.876 \times 10^{-2}$ | $4.1 \times 10^2$ | $-0.05$ | $1.852 \times 10^{-2}$ | $8.35 \times 10^2$ | 1.5 |
| $F_z$ | 0 | $2.531 \times 10^{-3}$ | $4.1 \times 10^2$ | $-0.05$ | $3.432 \times 10^{-2}$ | $8.35 \times 10^2$ | 1.5 |

Table 2.4 The parameters of the relative displacement function for Cu (111) crystal plane

| | $H_0$ | $A_1$ | $\omega_1$ | $\varphi_1$ | $A_2$ | $\omega_2$ | $\varphi_2$ |
|---|---|---|---|---|---|---|---|
| $Z_{miv}$ | $-3.765 \times 10^{-3}$ | $-1.518 \times 10^{-2}$ | $4.5 \times 10^2$ | 0 | $-1.949 \times 10^{-3}$ | $4.5 \times 10^2$ | 0 |
| $F_z$ | $2.989 \times 10^{-3}$ | $-7.435 \times 10^{-2}$ | $4.5 \times 10^2$ | 0 | $1.431 \times 10^{-2}$ | $4.5 \times 10^2$ | 0 |

where $H$ represents the height at balance position, $A_1$ and $A_2$ denote the amplitude, $\omega_1$ and $\omega_2$ stand for the angular frequency, and $\varphi_1$ and $\varphi_2$ are the initial phase difference.

The parameters $H$, $A$, $\omega$, and $\varphi$ are inherent material properties. Their respective values for Cu (100), (110), and (111) crystal planes can be found in Tables 2.2, 2.3, and 2.4.

As a consequence of this, the real contour height $Z$ that is influenced by material-induced vibration may be described using the equation that is presented below:

$$Z = Z_{th} + Z_{miv} \tag{2.13}$$

where $Z_{th}$ represents the surface contour height under ideal turning conditions, and $Z_{miv}$ denotes relative displacement under material induced vibration.

#### 2.1.2.3 Surface Morphology Simulation

When the data points of the surface morphology of Cu (100) are substituted into Eq. (2.12), it is possible to estimate the associated parameters, which are then provided in Table 2.2.

Similarly, Eq. (2.12) may be used to compute the respective parameters, which are presented in Table 2.3. This can be done by putting the data points of the surface morphology of Cu (110) into the equation.

Finally, Eq. (2.12) also can be employed to compute the respective parameters presented in Table 2.4. This can be done by putting the data points of the surface morphology of Cu (111) into the equation.

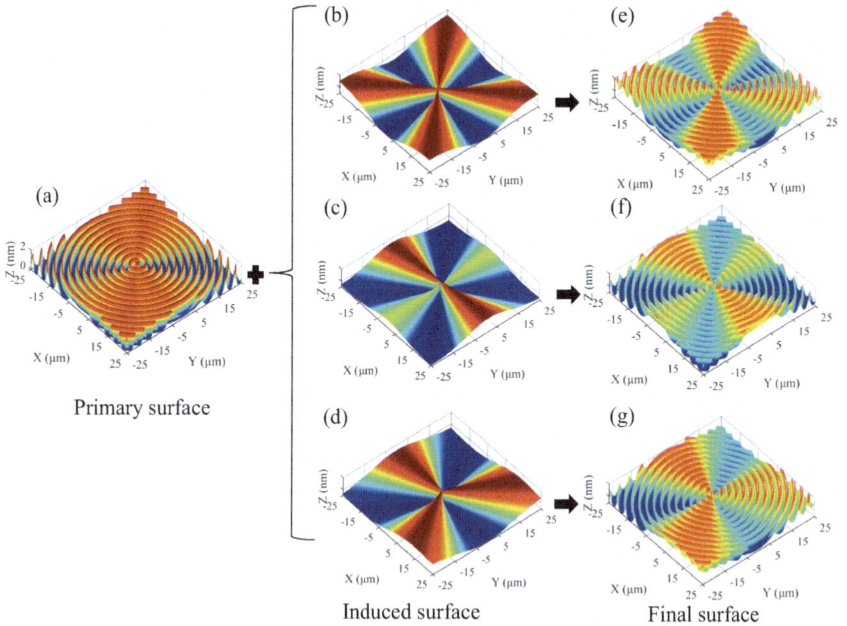


**Fig. 2.12** **a** Primary surface, vibration-induced surface morphology on: crystal planes **b** (100), **c** (110), and **d** (111), with final overlapped surface morphology on: crystal planes **e** (100), **f** (110), and **g** (111)

According to the cutting dynamic model, the final surface morphology is the consequence of overlaying the primary surface morphology from perfect turning conditions with the surface morphology generated by vibration, as depicted in Fig. 2.12. This is the case because the two surface morphologies are overlapping. The surface morphology of these crystal planes is depicted in detail by the simulation results, which may be found in Fig. 2.12e–g. It should be brought to your attention that the theoretical results are in good agreement with the findings of the experiments. To be more specific, Cu (100) exhibits four fan-shaped patterns that are quadruple-symmetrical, Cu (110) displays two fan-shaped patterns that are double-symmetrical, and Cu (111) displays three fan-shaped patterns that are triple-symmetrical.

## 2.2 Surface Morphology Controlling

This section will cast a light on the strategies that govern the manipulation of surface features and the maintenance of structural integrity during the machining process. Controlling the surface morphology is a delicate balancing act that requires a deep understanding of material properties, tooling dynamics, and the environmental conditions that influence the final outcome. This section will discuss the role of crystallographic orientation and machining parameters controls in achieving the desired nano-surface.

## 2.2.1 *Cutting Depth Controlling*

This section, taking BSCM Ge as an example, observes the phenomenon of BDT through experiments and propose reasonable surface control strategies based on the experimental results.

Plunge-cutting experiments (see Fig. 1.1c in Chap. 1) are an effective method to observe the BDT phenomenon. This section conducted twice plunge-cutting experiments. Firstly, an 8-grooves experiment was performed, aiming to test the cutting performance in different crystal planes. Then, a 36-grooves plunge-cutting experiments were operated further with the purpose of getting more accuracy experimental data about BDT in different crystal plans.

### 2.2.1.1 The 8-Grooves Experiment

Using different feed rates, plunge-cutting tests were done on three different crystal surfaces of single crystal Ge (SCG). Table 1.2 in Chap. 1 lists the characteristics of the experiment, and Fig. 2.13 shows the findings. It was found that the average BDT levels in the (100) crystal plane were 318.64, 232.19, and 313.48 nm when the feed rates were 0.3, 3, and 30 mm/min. The average depths of the BDT in the ductile regions were 206.80, 217.51, and 313.48 nm. The depths of the BDT in the (100) crystal plane tend to get deeper as the increasing of plunge-cutting rate.

Similarly, on the (110) crystal plane, the average depths of BDT in brittle areas were 199.45, 325.39, and 411.04 nm at the same feed rates, respectively. In the ductile regions, the average transition depths were 342.21, 359.63, and 362.03 nm, respectively. The trend observed indicates that the BDT depths on the (110) crystal plane generally increase with higher feed rates during plunge-cutting.

In addition, on the (111) crystal plane, the average BDT depths in the brittle regions were 383.41, 345.04, and 299.82 nm at the same feed rates, respectively. In the ductile regions, the average BDT depths were 310.33, 314.63, and 278.74 nm, respectively. Here, the critical depth of the BDT in the (111) crystal plane exhibited an inverse relationship with the feed rate, and the critical depth remained uniform during cutting.

The anisotropy of the BDT critical depth that was found in each of the three crystal planes is consistent with the predictions that were contained within the theoretical model.

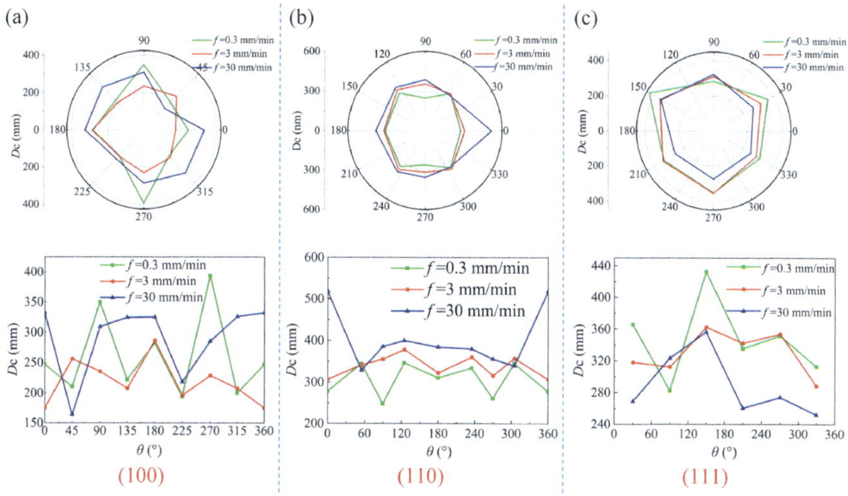

Fig. 2.13 The relationship between the critical depth ($D_c$) of BDT for the crystal planes **a** (100), **b** (110), and **c** (111) under various crystal orientations and feed rates

If the surface morphologies after plunge-cutting are regular, it is expected that the changes in the BDT will also exhibit regular patterns. The results of multiple plunge-cutting experiments using worn tool edges are presented in Fig. 2.14.

During the cutting process, the critical lengths and depths of the BDT for the three crystal planes of SCG display unique anisotropy. Among them, the (110) crystal plane demonstrates behaviour that is both irregular and unstable. The predictions of the theoretical model are consistent with the fact that the critical depths and lengths of the brittle zones are less in comparison to those of the ductile regions shown by each crystal orientation.

The influence of the tool edge on the BDT can be observed from the aforementioned analysis. Figure 2.15 illustrates that crystal planes with well-maintained tool edges exhibit less noticeable anisotropy compared to those with worn tool edges. Notably, the (100) and (111) crystal planes display a stable yet inconspicuous anisotropy in terms of the critical depth of the BDT. On the other hand, the (110) crystal plane showcases deeper and longer critical depths in the ductile zones, as well as a more pronounced anisotropy during cutting.

The anisotropy of the critical depth and length of the BDT sees insignificant variations as a result of the low tool edge radius that occurs when there is no wear. On the other hand, it is essential to take into consideration the fact that the critical depth and length of the BDT are, in general, shallower and shorter in the brittle areas in comparison to the ductile sections. When compared to cutting with worn tool edges, the cutting trials illustrate that cutting with fresh tool edges results in critical depths and lengths that are shallower and shorter. This is the case when compared to cutting with worn tool edges. It is important to note that the tool edge radius has an

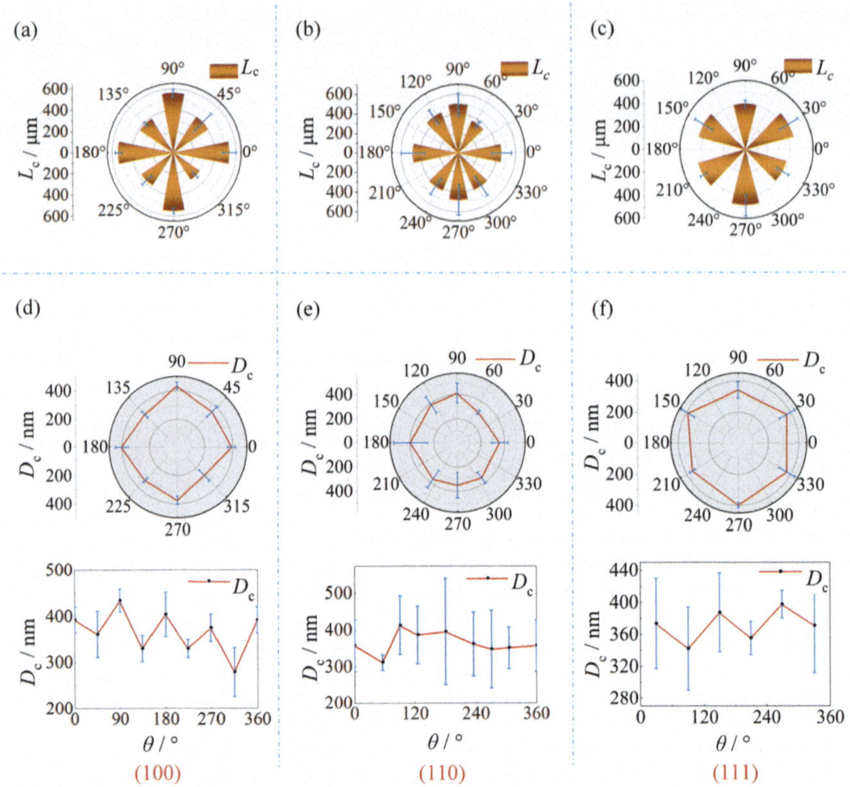


**Fig. 2.14** Changes in the critical length ($L_C$) and depth ($D_C$) of the BDT on the crystal planes, **a**, **d** (100), **b**, **e** (110) and **c**, **f** (111) (worn tool edge)

effect on the disparities that are found in the critical depths and lengths of the BDT. This is true regardless of the conditions in which the plunge-cutting is performed.

#### 2.2.1.2 The 36-Grooves Experiment

In order to obtain detailed experimental results, a 36-groove experiment was conducted to further investigate. The 3D morphology of the machined 36-grooves through plunge-cutting is depicted in Fig. 2.16a. It can be observed that the depths of the grooves increase gradually from 0 to 3 μm.

Following the plunge-cutting procedure, the overall morphology of the Ge (111) crystal plane is depicted in Fig. 2.16a, which includes all of the grooves of the crystal. Upon zooming into a single groove, as shown in Fig. 2.16b, it becomes evident that the coordinates ranging from 0.17 to 0.30 mm do not exhibit any signs of brittle deformation, as the cutting depth at this particular location does not exceed $D_C$. However, as the cutting depth increases and surpasses $D_C$, brittle fractures start to

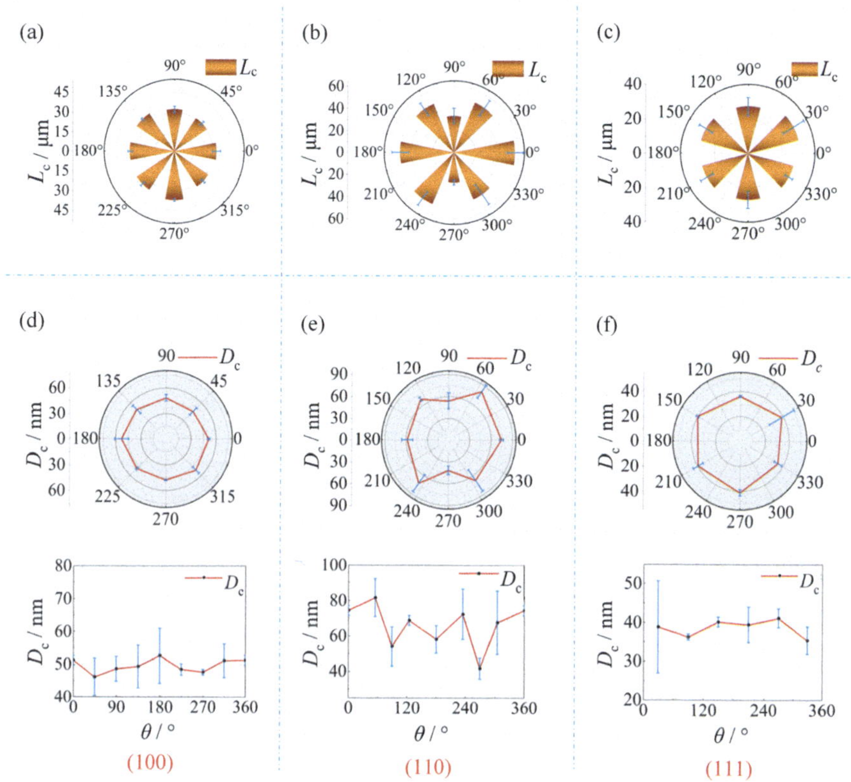


**Fig. 2.15** Changes in the critical length ($L_c$) and depth ($D_c$) of the BDT on the **a**, **d** (100), **b**, **e** (110) and **c**, **f** (111) crystal planes (fresh tool edge)

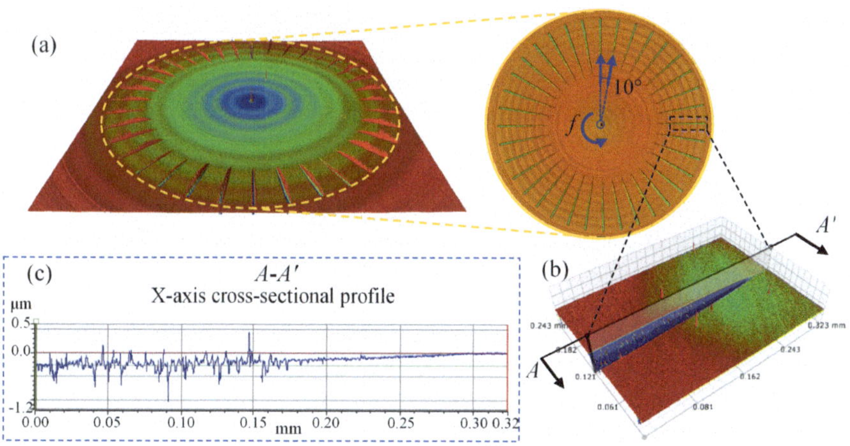


**Fig. 2.16** Captured morphology of **a** machined surface, **b** single groove; **c** cross-section

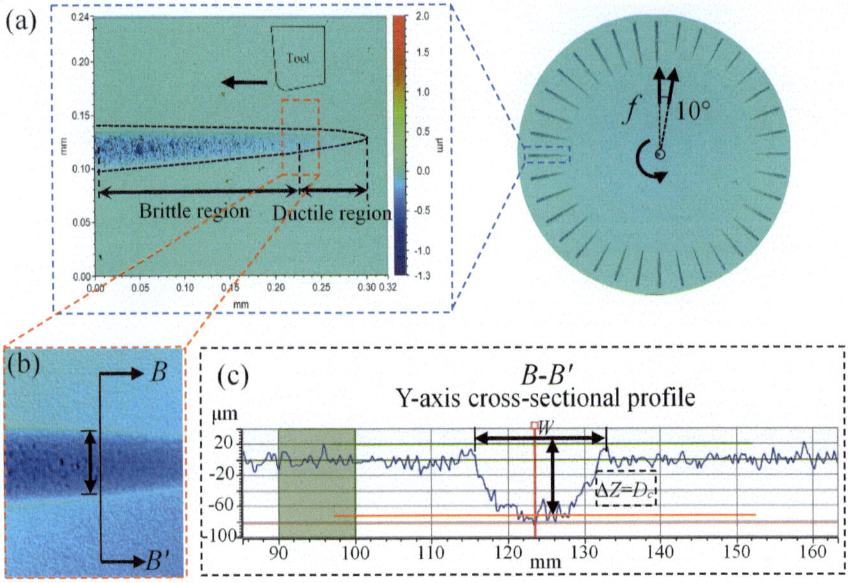


**Fig. 2.17** Captured morphology of **a** groove surface, **b** single groove, and **c** cross-section curve

appear on the groove's left side, as denoted by the black dotted brittle region in Fig. 2.17a. Detailed examination using a 3D optical microscope reveals that these black spots correspond to brittle fracture pits.

As shown in Fig. 2.16c, the surface that is located within the 0.17–0.30 mm coordinates of the groove is referred to as the ductile zone. This region is generally flat and is present throughout the groove. On the other hand, the coordinates of the groove surface, which range from 0.00–0.17 mm, appear to be uneven, which is representative of the brittle zone. The 2D cross-section curves of groove profile A-A' and boundary profile B-B', which are depicted in Figs. 2.16c and 2.17c, respectively, provide a further illustration of this boundary that separates the two areas. Because of this, the boundary $\Delta Z$ is considered to be the $D_c$ in a straightforward manner.

Furthermore, the critical depth ($D_c$), is influenced by the crystal planes and crystal orientations. Figure 2.18a–c demonstrates the variation of $D_c$ with different cutting directions. As the $D_c$ increases, the size of the brittle fracture pits also increases. Moreover, when the cutting edge comes into contact with the edge of the groove, the small cutting depth leads to the ductile removal of the groove edge.

The ductile transition at the edge of the brittle region is effectively elucidated by this phenomenon. To systematically explore the $D_c$ of BSCM at various cutting locations, experimental $D_c$ curves denoted by $\Delta Z$ were captured for each groove on the (111), (110), and (100) crystal planes, as depicted in Fig. 2.18d–f.

Table 2.5 demonstrates the average Dc values for the (111), (110), and (100) crystal planes, which are found to be 98.82, 113.38, and 113 nm, respectively. These values align closely with the theoretical $D_c$ range of Ge, which is 56.99–145.80 nm.

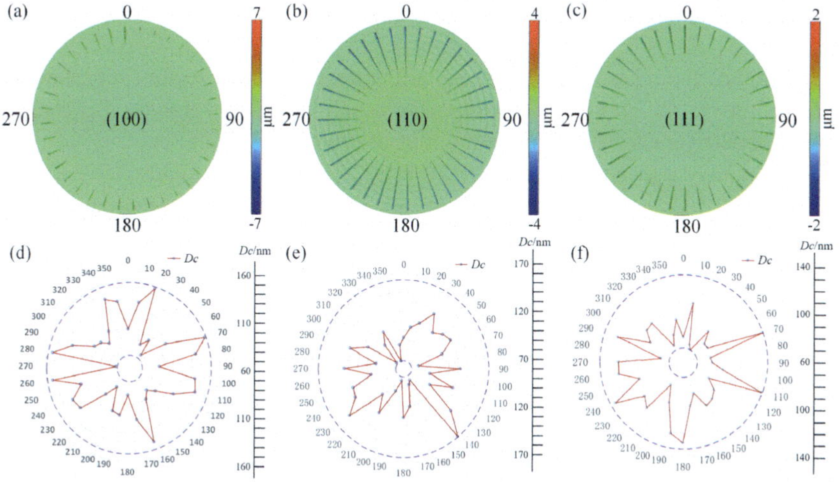


**Fig. 2.18** Surface morphology after plunge-cutting experiments of crystal planes, **a** (100), **b** (110), and **c** (111), and their $D_c$ of crystal planes **d** (100), **e** (110), **f** (111)

Notably, both the numerical values and the trend of fluctuation remain consistent between theory and experiment. Furthermore, there is an anisotropy in the $D_c$ values for different positions. The measured $D_c$ curve for the (100) crystal plane exhibits four evenly distributed valley regions, while the (110) crystal plane showcases greater $D_c$ values and a more uneven valley distribution compared to (111) and (100) crystal planes. The measured $D_c$ of the (111) crystal plane is relatively small and exhibits minimal fluctuation across all cutting directions. Overall, these experimental results align well with the calculated model predictions.

The anisotropic nature of the $D_c$ values in the (110) is the most unstable and prominent, whereas in the (111), it is relatively inconspicuous. A noteworthy observation is that the cutting performance of the (111) surpasses that of the other crystal plane. From this, it can be concluded that the cutting performance improves as the anisotropy of the crystal plane in BSCM decreases. It can be hypothesized that by controlling and reducing the anisotropy of the machined surface in BSCM, their machinability can be enhanced.

**Table 2.5** $D_c$ values of Ge in various crystal planes

| Crystal plane (nm) | (100) | (110) | (111) |
|---|---|---|---|
| Maximum $D_c$ | 148.80 | 162.70 | 133.5797 |
| Minimum $D_c$ | 73.03 | 78.56 | 72.31 |
| Average $D_c$ | 113.00 | 113.38 | 98.82 |
| Theoretical $D_c$ | 81.10–145.80 | 56.99–130.423 | 82.75 |

#### 2.2.1.3 Optimization of Cutting Depth-Oriented Surface Morphology

To investigate the relationship between cutting depth ($a_p$) and surface morphology in SPDT of Ge, different $a_{pi}$ values were selected based on different $D_{ci}$ (critical cutting depth) values. To minimize the impact of surface elastic recovery, $a_{pi}$ was chosen slightly larger than $D_{ci}$.

Referring to the experimental $D_c$ curve, $a_{p1}$ was chosen to be greater than $D_{cmax}$, $a_{p3}$ was chosen to be smaller than $D_{cmin}$, and $a_{p2}$ was selected to lie between $D_{cmax}$ and $D_{cmax}$. When the cutting depth is $a_{p3}$ ($a_{p3} < D_{cmin}$), the material removal mode is entirely ductile, as illustrated in Fig. 2.19a3, b3, c3. With the $a_p$ increases, cleavage failure occurs, cracks initiate and propagate beneath the surface, and the material removal transitions from ductile to brittle. In the case of $a_{p1}$ ($a_{p1} > D_{cmax}$), the material removal mode becomes entirely brittle, as indicated in Fig. 2.19a1, b1, c1.

Based on these findings, different surface properties emerge in Ge after facing turning with varying cutting depths: a total brittle surface, a ductile–brittle alternating surface, and a total ductile surface. The total brittle surface exhibits defects such as

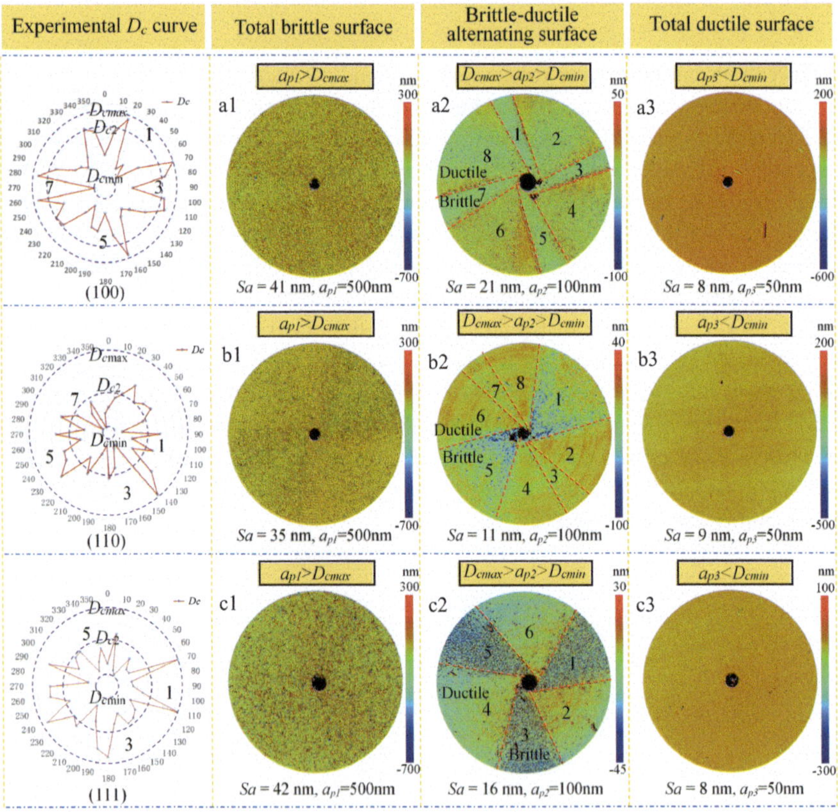


**Fig. 2.19** Machined surface controlling of various crystal planes of Ge in SPDT

step-like cracks, micro-craters, and micro-pits. Conversely, the total ductile surface does not display any defects, even at the nanometer scale, and possesses a Sa of less than 10 nm. The DBT surface demonstrates anisotropic characteristics and features regions of both brittle and ductile behavior, showcasing an alternating distribution. The entire brittle surface displays the greatest Sa, the total ductile surface displays the lowest Sa, and the ductile–brittle alternating surface lies somewhere in the middle. This is the case independent of the crystal plane, as seen in the images. It is important to note that all three of these surfaces exhibit nanoscale mirror properties.

As depicted in Fig. 2.19a2, b2, c2, a lower value of $D_c$ for a certain crystal orientation of BSCM suggests a higher possibility of brittle fracture happening during the machining process. This is because the brittle fracture is more likely to occur after the machining process. The DBT surface, machined at $a_{p2}$ ($D_{cmax} > a_{p2} > D_{cmin}$), exhibits alternating regions of ductile and brittle behavior. Within the experimental $D_c$ curve, the areas that correlate to the brittle regions are the valley regions, whereas the regions that correspond to the peak regions are the ductile regions. Both surface A2 and surface B2 showcase eight DBAP regions. In each DBAP, there are four ductile regions and four brittle regions. When this is compared to the theoretical $D_c$ curve (which can be seen in Chap. 1's Fig. 1.4), it is interesting to note that the brittle portions align with the highest values on the theoretical curve, and the ductile parts align with the minimum values.

It is important to note that although while the theoretical $D_c$ curve for the (111) crystal plane does not change, the occurrence of DBAP is still visible on the machined surface in the experiment. In addition to this, the ductile and brittle zones display patterns that are practically identical to one another. Generally speaking, the DBAP in BSCM may be efficiently regulated and anticipated by setting the cutting depth based on the theoretical $D_c$ curve that is generated. It is at the point where the selected cutting depth intersects with the hypothesized $D_c$ curve that the occurrence of DBAP should be considered to have occurred. This intersection point also acts as the border between ductile and brittle behavior, which is another important function. A surface that is brittle will be produced if the theoretical $D_c$ curve is greater than the specified cutting depth. On the other hand, a surface that is ductile will be reached if the theoretical $D_c$ curve is lower than the set cutting depth.

The experimental findings highlight the significance of adjusting the cutting depth to achieve surfaces with varied properties in BSCM, based on the theoretical and experimental $D_c$ curves. By selecting a cutting depth that is lower than the $D_{cmin}$ or higher than the $D_{cmax}$, the occurrence of ductile–brittle alternating surfaces in BSCM during SPDT can be avoided. Conversely, by choosing a cutting depth that lies between the $D_{cmax}$ and $D_{cmin}$, the DBT surfaces can be obtained on the various crystal planes.

Furthermore, by utilizing the theoretical $D_c$ curves, the alternating frequency of DBT surfaces can be predicted. This provides a valuable tool for anticipating the frequency at which the transition between ductile and brittle behavior occurs during the machining process in BSCM.

### 2.2.2 *Cutting Edge Controlling*

In addition to cutting depth, tool radius is another factor that affects the BDT. Figure 2.20a depicts the surface morphology of the workpiece after plunge-cutting with a tool edge that has been worn. In contrast, Fig. 2.20b illustrates the longitudinal section diagram of the workpiece. Upon comparison with Fig. 1.5 in Chap. 1, it is evident that the length of the ductile region is longer when using a worn tool edge, which may be attributed to the altered tool parameters depicted in Fig. 2.21.

Nevertheless, when examining the ductile region in Fig. 2.20a–b, it is clearly observable that the dense shutter-type surface morphology appears more irregular compared to the ductile region shown in Fig. 1.5a, b in Chap. 1, which exhibits a more regular morphology. The removal mode of Ge transitions from plowing in the ductile region to shearing in the brittle region. It is important to mention that in Fig. 2.20a, there is a little raised area at the conclusion of the ductile zone, right before entering the brittle region.

According to Fig. 2.20, there is a certain extent of cracking in the brittle region, resulting in a more irregular surface morphology compared to Fig. 1.5 in Chap. 1. It is apparent that the machined surface obtained after plunge-cutting using a new tool closely matches the desired groove surface morphology, especially with regards to the BDT area illustrated in Fig. 2.21b. Based on the findings of the experiment, it appears that the crucial transition depth and length are affected by the sharpness of the tool edge.

As an additional point of interest, the findings of the studies indicate that there is a connection between the cutting qualities and the tool edge radius. Cutting models with varying tool edge radii have been built in order to further examine the influence of the tool edge radius on the critical depth and length of the BDT. These models are depicted in Fig. 2.21a–b, which can be found here. As an additional step toward

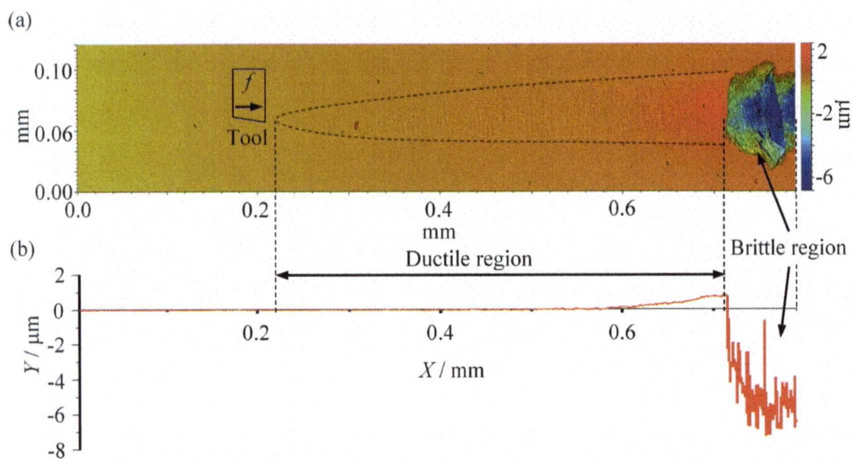


**Fig. 2.20** **a** Plunge-cutting surface using a worn tool and **b** longitudinal section diagrams

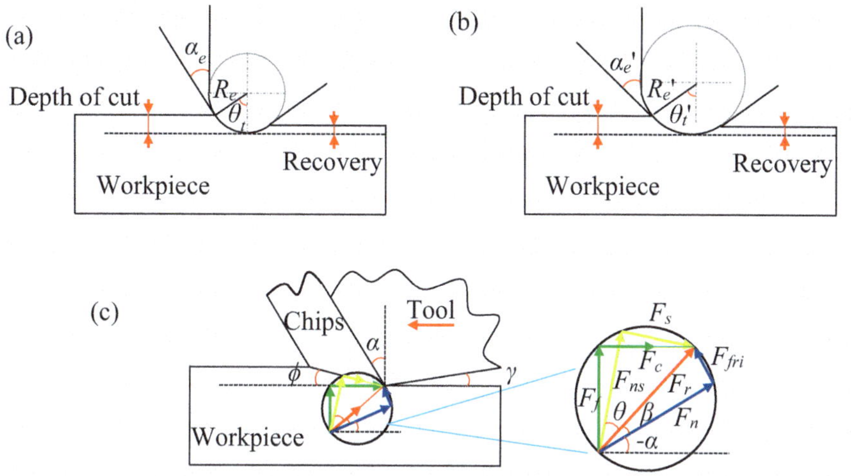


**Fig. 2.21** Schematic diagram of **a** fresh tool cutting model, **b** worn tool cutting model, and **c** cutting force components in orthogonal cutting

conducting a more comprehensive investigation, the cutting force is calculated by employing the orthogonal decomposition model shown in Fig. 2.21c.

It is possible to ascertain the friction angle ($\beta$) between the rake face and the chip by employing the model given before, as described in the following manner:

$$\beta = \arctan(\mu) = \arctan(F_{fri}/F_n) = \alpha + \tan^{-1}\frac{F_f}{F_c} \tag{2.14}$$

The variables in the equation are defined as follows: $\mu$ represents the friction coefficient, $F_{fri}$ represents the friction force between the tool rake face and the chip, $F_n$ represents the positive pressure on the chip, $\alpha$ represents the tool rake angle, $F_f$ represents the cutting force in the axial direction, and $F_c$ represents the cutting force in the tangential direction.

This internal frictional angle can be determined by using the following equation [9]:

$$\theta = \arctan(F_s/F_{ns}) = \frac{\pi}{2} - (\phi + \beta - \alpha) \tag{2.15}$$

where $\phi$ is the included angle between the feed rate direction and the shear plane.

According to Merchant's shear angle prediction equation [2, 10], if $\phi > 0$,

$$\begin{aligned} \phi + \beta - \alpha &< \frac{\pi}{2}, \\ \phi &= \frac{\pi}{4} + \frac{\alpha}{2} - \frac{\beta}{2} \end{aligned} \tag{2.16}$$

As the cutting depth lowers and the negative rake angle of the tool steadily drops to a particular value, the shear angle of the tool will eventually become zero. At this moment, the rake angle is defined as the critical negative rake angle, which may be represented in the following manner:

$$\alpha c = \beta - \frac{\pi}{2} \tag{2.17}$$

In the plunge-cutting experiment, the cut depth starts from zero, and it is important to consider the tool edge radius. Therefore, the effective negative rake angle can be defined as follows:

$$\alpha_e = \theta_\tau - \frac{\pi}{2} = \cos^{-1}\left(\frac{R_e - D}{R_e}\right) - \frac{\pi}{2} \tag{2.18}$$

where $\theta_\tau$ corresponds to the angle at a specific point on the tool edge, $R_e$ represents the radius of the tool edge, and $D$ is the cut depth.

The critical negative rake angle associated with the tool edge gets less than the effective negative rake angle when the cutting depth drops to a certain value. The Ge surface's removal technique now switches to plowing. Given that the minimal cut depth is represented by $D_{min}$ and the critical angle on the tool edge is specified by $\theta_c$, we can conclude that when the cut depth is greater than $D_{min}$, Ge is eliminated by shear and the material flows onto the rake face to create chips. On the other hand, Ge is eliminated by plowing when the cut depth is smaller than $D_{min}$, and the material flows onto the tool clearance face to produce the machined surface. Material accumulates as a result of the plowing motion, and the ductile region is extended and the plowing removal process is prolonged as the tool edge wears down. As a result, it leaves a raised surface when it suddenly switches to shear removal, failing to remove the material entirely.

The material's separation or stagnation point is shown when the critical angle at a particular spot on the tool edge equals the friction angle $\beta$. The critical effective negative rake angle in this instance is defined as follows:

$$\alpha_{ec} = \beta - \frac{\pi}{2} \tag{2.19}$$

$$D_{min} = R_e(1 - \cos\theta_c) = R_e(1 - \cos\beta) \tag{2.20}$$

The minimum cutting depth is exactly proportional to the tool edge radius while the friction angle between the tool and the workpiece remains constant. During plunge-cutting, the friction angle for plowing is positive, whereas for shear removal, it is negative. However, both angles have almost the same absolute values. An augmentation in the transition depth between plowing and shear removal will result in a proportional increase in the BDT depth of Ge.

Figure 2.22a–b depict microschematic schematics of the cutting edges, where the small yellow shapes denote varying tool edge radii. These diagrams reveal that the

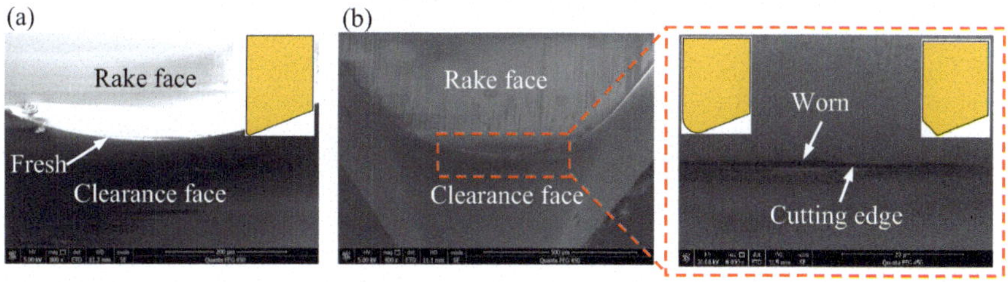


**Fig. 2.22** Morphologies of **a** fresh tool edge and **b** worn tool edge

radius of the worn tool edge is greater than that of the new tool edge. Therefore, under identical conditions, the critical BDT depth and length in Ge are directly related to the tool edge radius. As a result, in practical cutting scenarios, employing a tool with a larger edge radius will result in a greater critical BDT depth and length.

## References


1. Yuan ZJ, Lee WB, Yao YX, Zhou M (1994) Effect of crystallographic orientation on cutting forces and surface quality in diamond cutting of single crystal. CIRP Ann 43(1):39–42
2. Wang Z, Zhang J, Xu Z, Zhang J, Li G, Zhang H, Li Z, Hu, Hassan, Fang F, Hartmaier A, Yan Y, Sun T (2020) Crystal anisotropy-dependent shear angle variation in orthogonal cutting of single crystalline copper. Precis Eng 63:41–48
3. Tersoff J (1988) Empirical interatomic potential for silicon with improved elastic properties. Phys Rev B 38(14):9902–9905
4. Lee WB, Cheung CF (2001) A dynamic surface topography model for the prediction of nano-surface generation in ultra-precision machining. Int J Mech Sci 43:961–991
5. Zhang SJ, To S, Zhang GQ, Zhu ZW (2015) A review of machine-tool vibration and its influence upon surface generation in ultra-precision machining. Int J Mach Tool Manu 91:34–42
6. Lee WB, Cheung CF, To S (1999) Materials induced vibration in ultra-precision machining. J Mater Process Technol 90:318–325
7. Chen X, Xiao JF, Zhu Y, Tian RJ, Shu XW, Xu JF (2017) Micro-machinability of bulk metallic glass in ultra-precision cutting. Mater Des 136:1–12
8. Lin J (2018) Research on surface topography modeling and scattering properties of ultra-precision machining aluminum mirrors. Harbin Inst Technol
9. Demir E (2008) Taylor-based model for micro-machining of single crystal fcc materials including frictional effects—application to micro-milling process. Int J Mach Tool Manu 48(14):1592–1598
10. Merchant ME (1944) Basic mechanics of the metal-cutting process. J Appl Mech 11:A168–A175

# Chapter 3
# Cutting Process in Machining Polycrystalline Materials

**Abstract** Polycrystalline materials plays a crucial role in the production of optical components due to its wide range of applications. However, the presence of irregular micro-structures (IMS) on the machined surface poses challenges to achieving high-quality results in SPDT. Understanding the underlying mechanism responsible for the formation of these IMS is therefore essential for obtaining superior machined surfaces in polycrystalline copper. In this chapter, a comprehensive theoretical and experimental investigation is conducted to unravel the generation mechanism of IMS.

**Keywords** Cutting process · Ultra-precision machining · Surface generation · Polycrystalline materials

## 3.1 Property of Irregular Microstructures

Polycrystalline materials, composed of numerous small crystalline grains with unique properties, are widely used in various applications [1]. One such application is in ultra-precision machining, a rapidly advancing technology that utilizes SCD as the primary tool material [2–5]. Among the workpiece materials commonly chosen for ultra-precision machining are non-ferrous metals like copper and aluminum [6, 7]. In particular, polycrystalline copper is known for its excellent machinability with diamond tools and desirable optical properties, making it a popular choice for the fabrication of optical components through processes like SPDT [8]. Therefore, the unique physical, chemical, and mechanical properties exhibited by polycrystalline materials play a crucial role in the success of ultra-precision machining technologies.

The exceptional optical properties of polycrystalline copper are highly dependent on its cutting performance, as it can be easily machined to achieve a mirror-like surface with roughness values below 10 nm (Ra) [8]. However, the surface quality during SPDT can still be influenced by various factors, including machining strategies, material recovery etc. [5]. Extensive research has been conducted by scholars to enhance the surface quality [9–12]. Despite the abundance of research



G. Zhang and J. Wang, *Material-Oriented Cutting Processes in Precision Machining*, Springer Tracts in Mechanical Engineering,
https://doi.org/10.1007/978-981-96-2504-8_3

on diamond turning of copper, none has adequately elucidated the underlying mechanism responsible for the formation of IMS during SPDT.

The presence of IMS on the surface of polycrystalline copper poses challenges to achieving high-quality machined surfaces. However, the underlying generation mechanism of these IMS remains unclear. In this section, the properties of irregular microstructures will be explored through appropriately designed polycrystalline copper SPDT experiments.

### *3.1.1 Irregular Microstructures Generation*

In this section, SPDT experiments were conducted using the same ultra-precision lathe mentioned in Chap. 1, as also depicted in Fig. 3.1. The cutting parameters utilized in this experiment can be found in Table 3.1. Key information regarding the chemical compositions and specifications of the copper material investigated is presented in Table 3.2. During turning, the workpiece was securely held in place by a vacuum chuck on the lathe's spindle, ensuring rotational movement to facilitate the primary cutting action. To accurately measure the cutting force, a dynamometer (Kistler, 9017C) was used, as illustrated in Fig. 3.1.

A corrosion experiment was carried out on the polycrystalline copper samples in order to evaluate the influence that various grain orientations have on the surface roughness of the material. For the purpose of generating the corrosion solution, ferric nitrate crystals were combined with anhydrous ethanol at a proportion of 2 g: 5 ml. After the completion of SPDT, the workpiece surfaces were immersed in the corrosion solution, ensuring that the machined surfaces did not come into contact with the container wall to minimize any potential confounding factors.

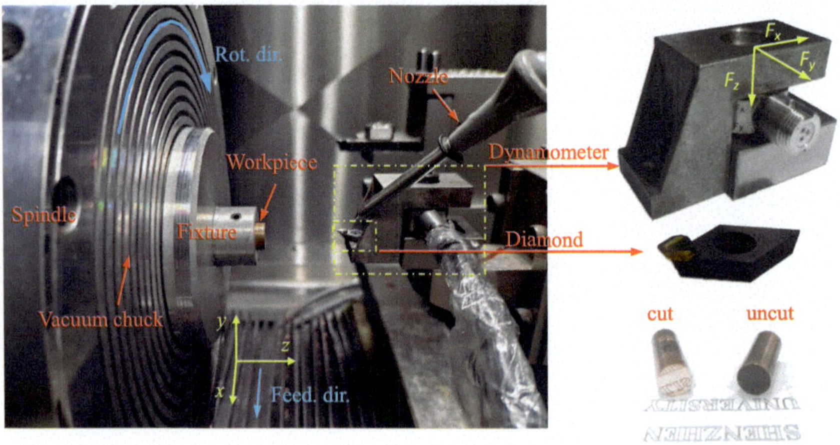


**Fig. 3.1** Experimental setup of SPDT

**Table 3.1** Turning parameters

| Parameter | Value |
|---|---|
| Spindle speed ($S_p$) (r/min) | 500, 1000, 2000, 3000, 3500 |
| Feed rate ($F_r$) (mm/min) | 1, 15 |
| Cutting depth ($a_p$) (mm) | 0.0005, 0.005 |
| Tool material | Single crystal diamond |
| Tool rake angle ($\alpha$) (°) | 0 |
| Tool clearance angle ($\delta$) (°) | 10 |
| Tool radius ($R$) (mm) | 0.502 |
| Workpiece material | Polycrystalline copper (pure copper) |

**Table 3.2** Parameters of the workpiece

| Brand | Chemical composition | Diameter | Tensile strength | Vickers hardness | Elastic modulus |
|---|---|---|---|---|---|
| TU2 | Cu + Ag: $\geq$ 99.95%<br>Sn: $\leq$ 0.002%<br>Zn: $\leq$ 0.003%<br>Pb: $\leq$ 0.004%<br>P: $\leq$ 0.002%<br>Ni: $\leq$ 0.002%<br>Fe: $\leq$ 0.004%<br>Sb: $\leq$ 0.002%<br>S: $\leq$ 0.004%<br>As: $\leq$ 0.002%<br>Bi: $\leq$ 0.001%<br>O: $\leq$ 0.003% | 10 mm | > 195 MPa | $\leq$ 70 HV | 107.9 GPa |

Following the turning process, the machined surfaces' roughness and topography were quantitatively evaluated using the white light interferometer mentioned in Chap. 1. Additionally, the optical microscope (Leica, DM2700M) was employed for visual inspection of the machined surfaces. Furthermore, energy spectrum analysis of the machined surfaces was performed using an energy dispersive spectrometer (EDS) (JEOL, JSM-5910LV). Moreover, the microscopic morphology of the resulting chips was examined using the SEM.

Figure 3.2 depicts the pervasive presence of IMS on the machined surface of the polycrystalline copper during the SPDT experiments. These IMS have a detrimental impact on the overall quality of the machined surface and compromise its intended optical functionality. The sizes of these IMS predominantly range from 20 to 45 μm, closely resembling the grain sizes of less than 50 μm [13]. It is plausible to attribute the emergence of these IMS to the diamond cutting process, which is known to interact differently with varying grain orientations.

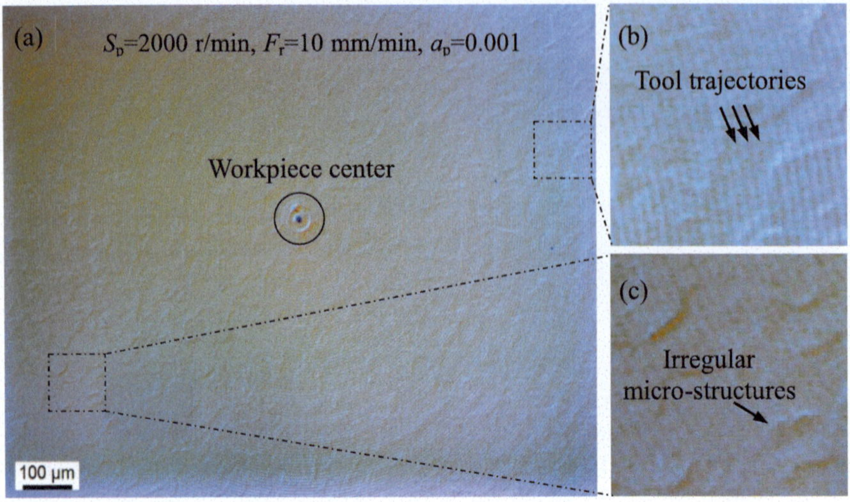


**Fig. 3.2** IMS on the workpiece's machined surface

Figure 3.3 demonstrates the fluctuation of the surface contour within the range of − 13.5 to 10.6 nm using a faster feed rate. The tool trajectory effectively encompasses the IMS present on the surface. Conversely, with a slower feed rate, resulting in more pronounced IMS. Typically, a larger feed rate is used during SPDT to achieve a superior machined surface. However, this decision comes at the cost of the IMS becoming the primary factor influencing the surface quality.

Oxidation on the surface of polycrystalline copper can result in material recovery. To mitigate the influence of oxidation, an element analysis experiment was conducted. It is worth noting that oxidation may occur post-SPDT, leading to a deterioration of surface quality [14]. In this section, EDS was employed to analyze the element composition of the IMS in order to identify their sources. Figure 3.4 displays the findings of the elemental scanning that was performed. A significant contribution to the elimination of the uneven distribution impact of oxygen on surface topography is made by the distribution of Cu and O elements, as seen in Fig. 3.4a–b. Additionally, the hypothesis regarding the grain orientation is further supported, given the single-phase metallographic structure of polycrystalline copper.

To gain a deeper understanding of the underlying mechanisms responsible for the formation of IMS, a corrosion experiment was conducted on the machined surface of the polycrystalline copper samples subsequent to the EDS analysis. The comparison diagrams showcasing the surface contour before and after corrosion are presented in Fig. 3.5. Upon comparing the surface topography before and after corrosion, it becomes evident that the IMS closely resemble the grain shapes, exemplified in parts A-E in Fig. 3.5a–b. Therefore, it can be concluded that the material recovery induced by the various grain orientations is indeed the primary factor contributing to the generation of IMS.

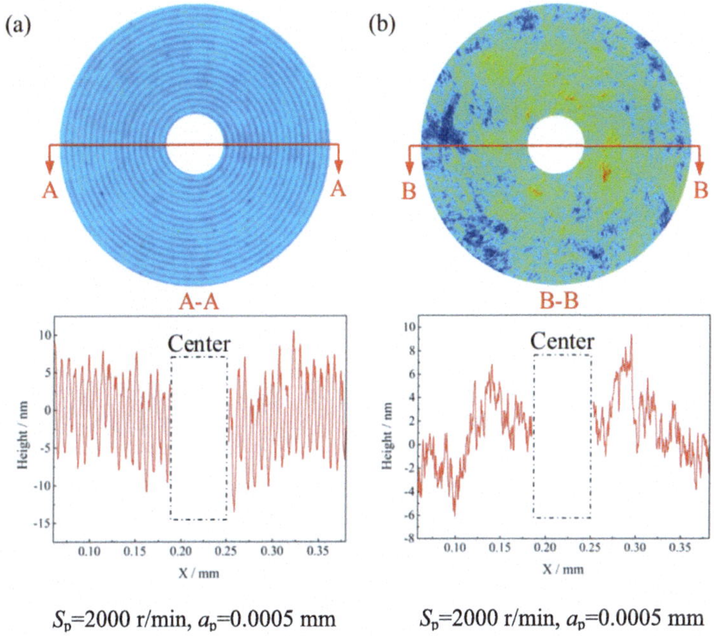


**Fig. 3.3** Machined surface contour using different feed rate: **a** $F_r = 15$ mm/min, **b** $F_r = 1$ mm/min

## *3.1.2 Cutting Force Characterization of IMS*

The arrangement of grains on the surface can be detected by monitoring the variation in cutting force while machining. Table 3.3 presents the cutting force parameters post signal filtering. Furthermore, Fig. 3.6 illustrates the cutting force signal component $F_x$ throughout the cutting operation. At spindle speeds of 1000, 2000, and 3000 revolutions per minute, respectively, the spindle completes one full rotation in 0.06, 0.03, and 0.02 s, respectively. There is a complete rotation period of the spindle represented by the symbol 1r in Fig. 3.6.

As the cutting tool moves across various grain orientations on a polycrystalline copper surface, it leads to fluctuations in the force signal. The majority of grains typically measure between 0.02 and 0.045 mm. A circular route with a radius of 0.065 mm is followed by the tool as it moves closer to the center of the workpiece. This path extends across around 9–20 grains and reaches a peak frequency of approximately 1000 Hz.

The number of changes in the force output in Fig. 3.6 ranges from 12 to 16 as the spindle speed changes, which corresponds to the number of grains covered by a circular trajectory of the cutting tool. Within each frequency range, the power spectrum appears suitable and aligns well with the frequency of the grain count.

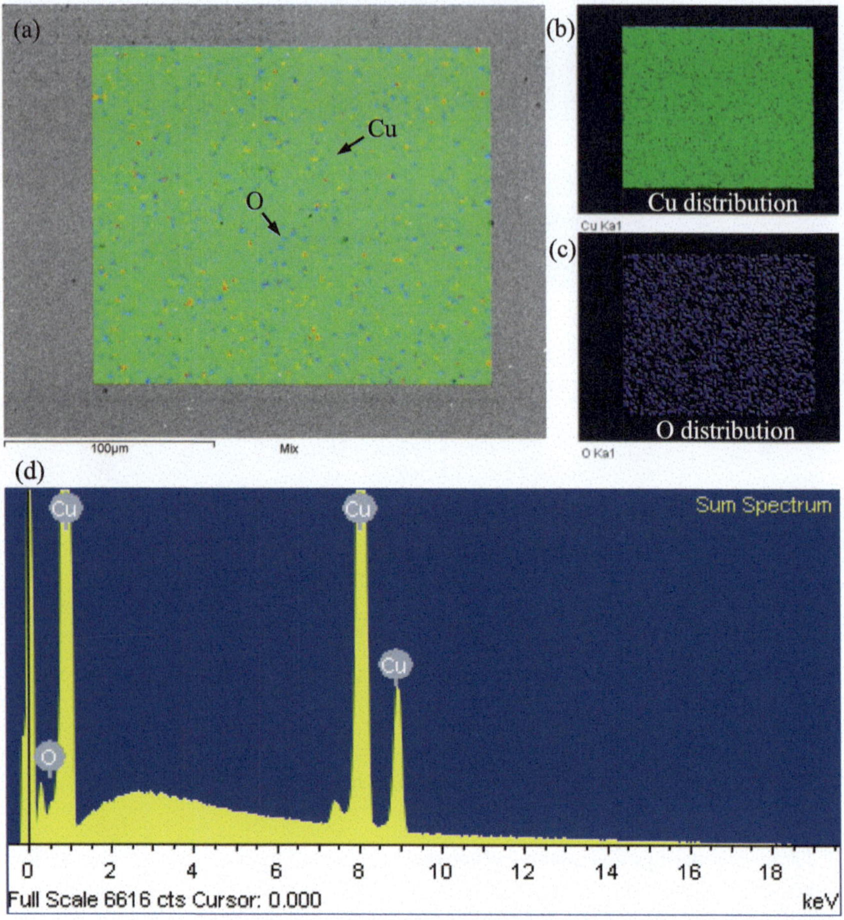


**Fig. 3.4** EDS mapping of the workpiece machined surface. **a** elements distribution, **b–c** Cu and O distribution, **d** the sum spectrum

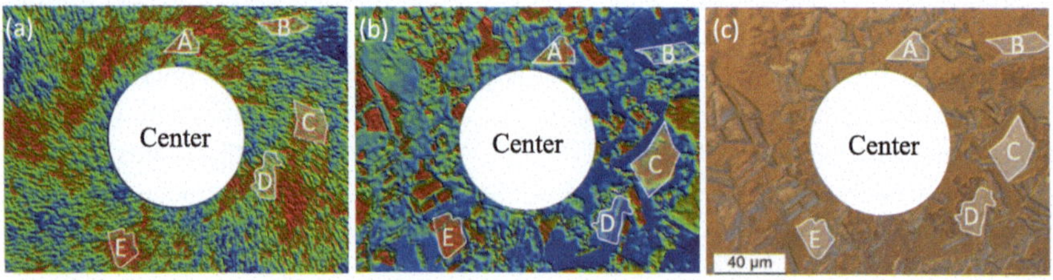


**Fig. 3.5** Grain distribution on the machined surface of the workpiece: **a** captured morphology of machined surface using the white light interferometer, **b** the morphology after corrosion, and **c** examined with an optical microscope following corrosion

**Table 3.3** Characteristic parameters of cutting force signal

| $S_p$ (r/min) | $f$ (Hz) | $T_r$ (s) | Grain numbers covering |
|---|---|---|---|
| 1000 | 16.7 | 0.06 | 14–16 |
| 2000 | 33.3 | 0.03 | 12–16 |
| 3000 | 50 | 0.02 | 14–16 |

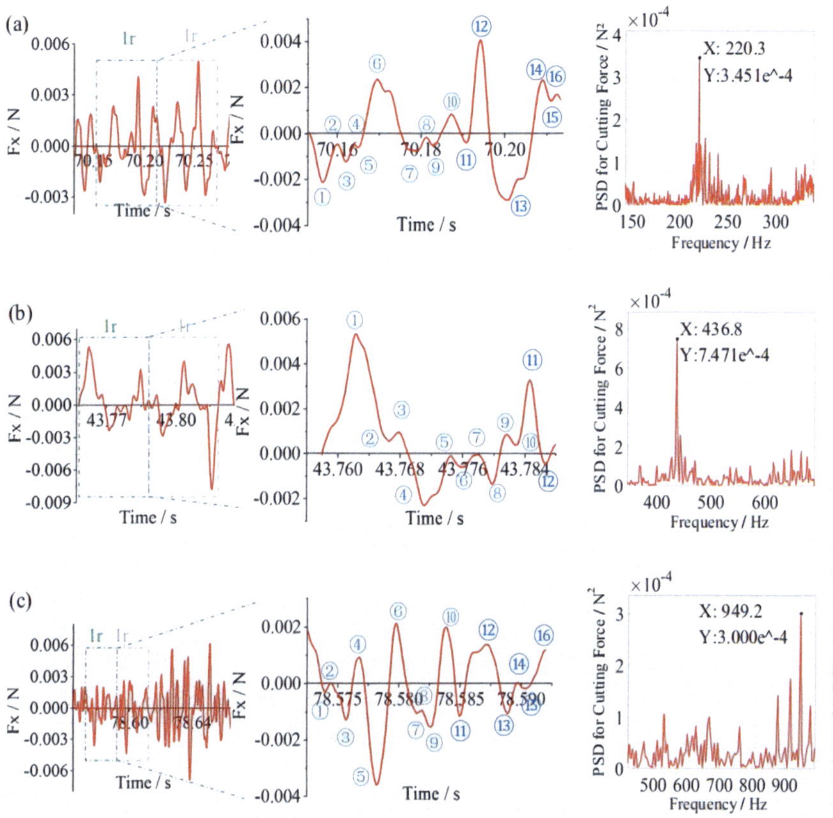


**Fig. 3.6** Variations in cutting force signals at spindle speeds of **a** 1000 revolutions per minute, **b** 2000 revolutions per minute, and **c** 3000 revolutions per minute

The force measurement system has a natural frequency of 50,329.212 Hz, which is considerably higher than the filtering frequency of the force signals. Consequently, the fluctuations observed in the cutting force demonstrate the indirect relationship between grain orientations and surface accuracy.

### 3.1.3 Chip Morphology Characterization of IMS

The chip morphology can provide insights into the material properties of different grain orientations. Upon closer examination of the Fig. 3.7, it is evident that the circled areas exhibit a distinct laminar structure, which differs from the surrounding regions. These circled areas appear darker in color and display variances in laminar thickness and direction.

By taking into account the shapes and sizes of these IMS, as well as the grains, it becomes apparent that the marked area on the chip reflects different cutting performances based on grain orientations. This marked area represents the removal of the IMS present on the machined surface.

In contrast, the chip morphology in single-crystal materials is influenced by crystallographic orientations and planes, exhibiting a sawtooth-like surface feature and not generating large irregular microstructures. Conducting in-situ cutting experiments on single-crystal copper using a micro-cutting apparatus integrated with SEM allows for the effective and visually intuitive observation of the micro-mechanism of chip formation. The detail of this part please refer to [15].

The in-situ cutting experiments are shown in Fig. 3.8a. The size, stiffness, and stability of the apparatus were enhanced to achieve a closed-loop resolution of 0.6 nm in both cutting and DOC directions. The tool-workpiece configuration depicted in Fig. 3.8b involves a process where the copper specimen is initially attached to a

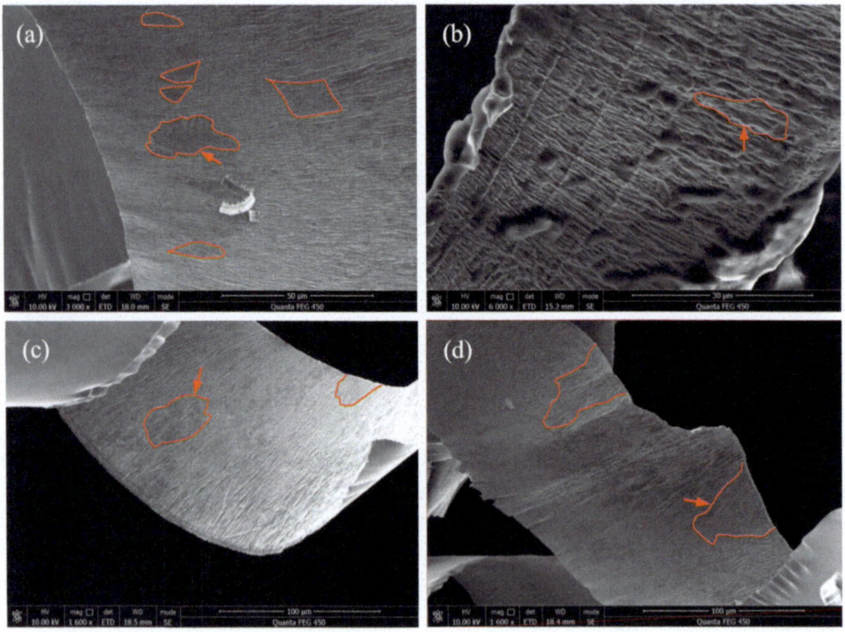


**Fig. 3.7** Captured chip morphologies using SEM

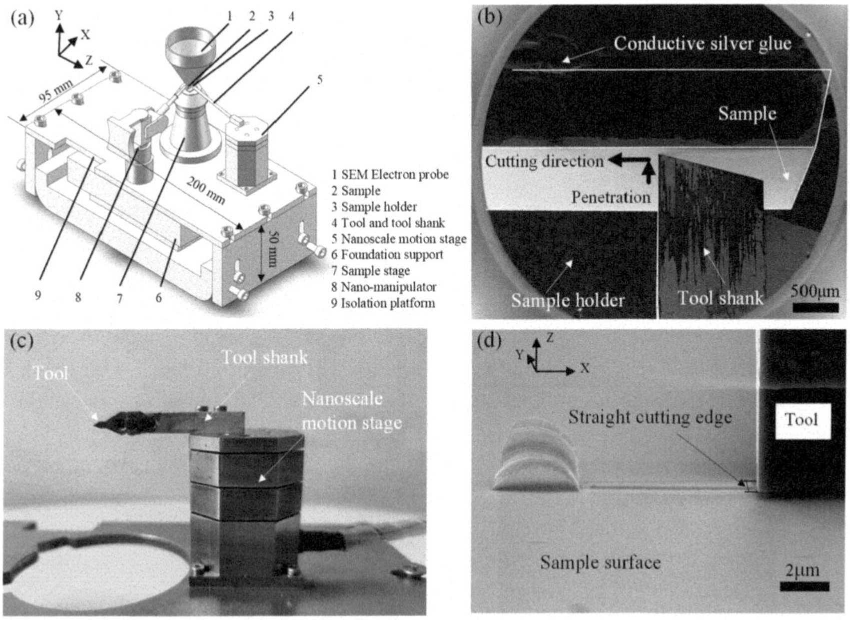


**Fig. 3.8** Experimental setup for in-situ micro-cutting: **a** configuration of diamond micro-cutting, **b** SEM examination of sample and tool handle, **c** tool device and **d** captured morphology of the diamond micro-cutting

specimen holder using conductive silver glue. A diamond cutting tool is fabricated through focused ion beam technology following a controllable sharpening method.

The diamond tool in Fig. 3.8c is attached to one end of the tool shank using conductive silver glue, while a nanometer-precise motion stage driven by a piezoelectric ceramic actuator is fixed to the opposite end of the tool shank. This setup enables tool movement in both cutting and depth of cut directions with a closed-loop resolution of 0.6 nm.

In Fig. 3.8d, the micro-cutting device demonstrates the orthogonal cutting setup for single crystalline Cu. The straight cutting edge is positioned perpendicular to the longitudinal *Z*-direction of the specimen before cutting. By adjusting the SEM micropositioner axes, the specimen surface is aligned parallel to the cutting direction with an accuracy of 0.1°. The cutting procedure begins by penetrating the specimen to a depth of 500 nm, followed by a 7 μm cut along the longitudinal *X*-direction at a constant speed of 0.07 μm/s. The slow cutting speed enables thorough observation of chip formation over the short cutting distance. Precise control of the cutting tool is maintained using a nanometer-precision motion stage and visual feedback from SEM imaging. Prior to the actual cutting at the predetermined depth, several ultra-shallow cutting trials with a 2 nm feed are performed to accurately document the tool position at zero depth of cut.

The process of chip creation that was seen by in-situ SEM during the micro-cutting of Cu(110) is depicted in Fig. 3.9. With extrusion largely impacting the top surface material, Fig. 3.9a first illustrates the parallel compression-induced swelling of displaced material ahead of the tool edge during the early cutting stage. This swelling occurs during the early cutting stage. Shearing takes place in the material that is displaced as a result of extrusion on active slip systems as the tool moves forward. This results in the formation of the first chip lamella, which is seen in Fig. 3.9b. At the beginning of the process of chip production, the size of this first chip lamella is quite modest. Following this, the first chip lamella is moved down the rake face of the cutting tool, which ultimately results in the formation of a second chip lamella, as shown in Fig. 3.9c. This process proceeds in a steady manner, which ultimately leads to the development of subsequent chip lamella. Figure 3.9d is the last illustration that demonstrates the existence of a saw-tooth chip that was generated in front of the tool edge.

The profile of the chip that was formed during the diamond cutting experiment on Cu(110) is depicted in Fig. 3.10a. This chip was cut at a distance of 7 μm, displaying periodic chip lamellas. The shearing action is responsible for arranging the material that has been distorted in front of the cutting tool into periodic lamellas that have a

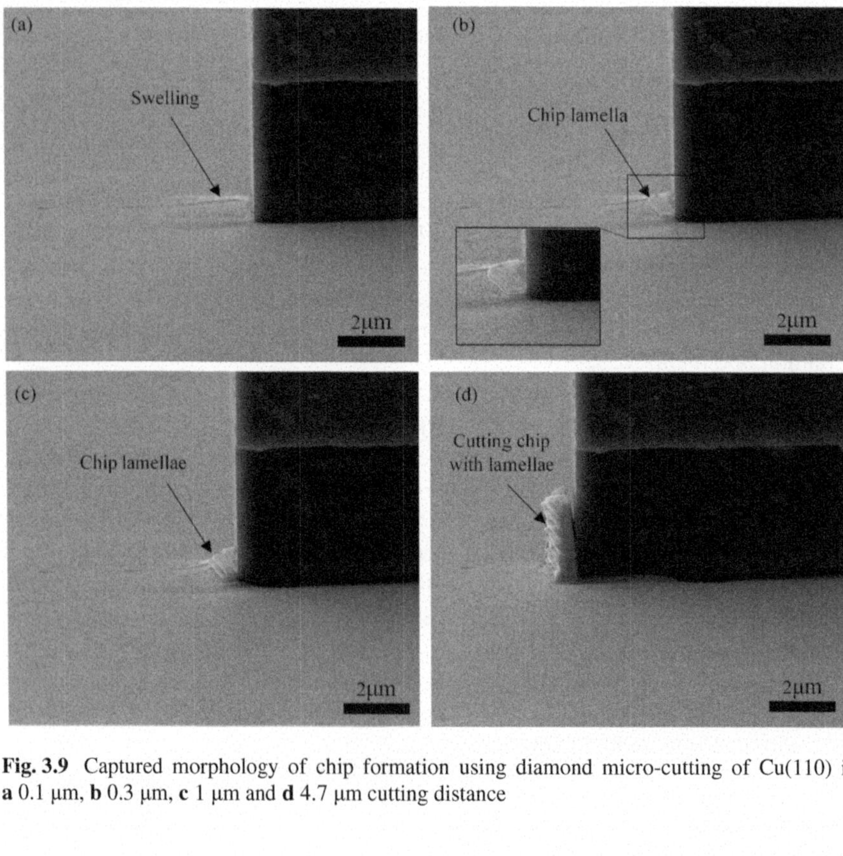


**Fig. 3.9** Captured morphology of chip formation using diamond micro-cutting of Cu(110) in **a** 0.1 μm, **b** 0.3 μm, **c** 1 μm and **d** 4.7 μm cutting distance

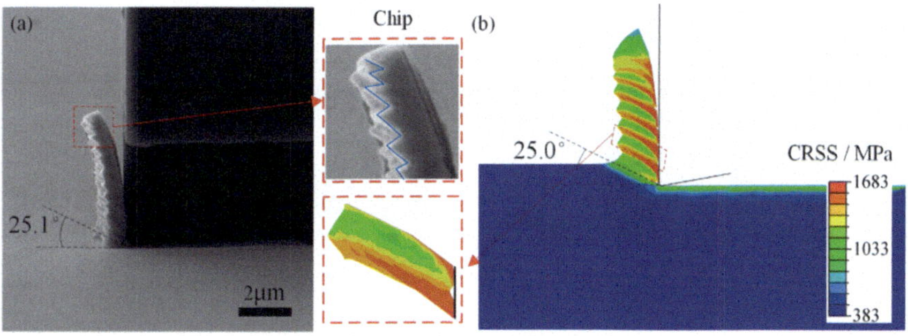


**Fig. 3.10** Chip contour comparison on Cu(110): **a** In-situ SEM and **b** FE simulation

spacing of roughly 0.5 μm. A clear demonstration of the fact that the real primary shear zone (PSZ) is an exceptionally narrow and well defined slip line between neighboring chip lamellas can be seen in the morphology of the shear planes in the chip profile. It is the geometric orientation of this slip line in the PSZ that determines the shear angle. This slip line extends from the tool-contact surface to the free surface in relation to the cutting direction. The standard saw-tooth chip has a shear angle of 25.1°, according to the measurements.

The findings of the finite element (FE) simulation of diamond cutting under identical conditions to the experiment are shown in Fig. 3.10b, where the specimen is color-coded according to the critical resolved shear stress that is dependent on work hardening. Detailed descriptions of the CPFE simulation for the cutting process are available in recent studies. The simulated chip profile characteristics, including the shape and curvature of the chip lamellae, closely resemble the experimental results, as shown in the two enlarged views of chips within the red dashed box in Fig. 3.10. The FE simulation makes a prediction of a shear angle of 25.0°, which is in good agreement with the actual results. This demonstrates that the established FE model for orthogonal cutting of single-crystalline copper is among the most accurate models available.

The chip characterization at a cutting distance of 7 μm reveals a significant influence from crystallographic orientation, as shown in Fig. 3.11. Cu(111) displays a more complex chip profile compared to Cu(010) and Cu(100). The shear angle for Cu(010) is 25.2° experimentally and 29.5° by CPFE simulation, while for Cu(111), it is 18.9° in experiment and 16.6° in simulation. Cu(010) has the maximum shear angle, whereas Cu(111) has the lowest shear angle. The shear angle changes depending on the crystallographic orientation. Method used for shear angle determination may have notable error range due to variations in chip profile images at different cutting distances. Uncertainties in shear angle calculation may arise from potential lamellae curvature evolution in the cutting zone. Nevertheless, the crystallographic orientation dependency of shear angle and chip profile is corroborated by both tests and FE simulations. This is especially true for the chip profile. It is a reflection of the geometry of activated slip systems on the surface of the cutting element. It's possible

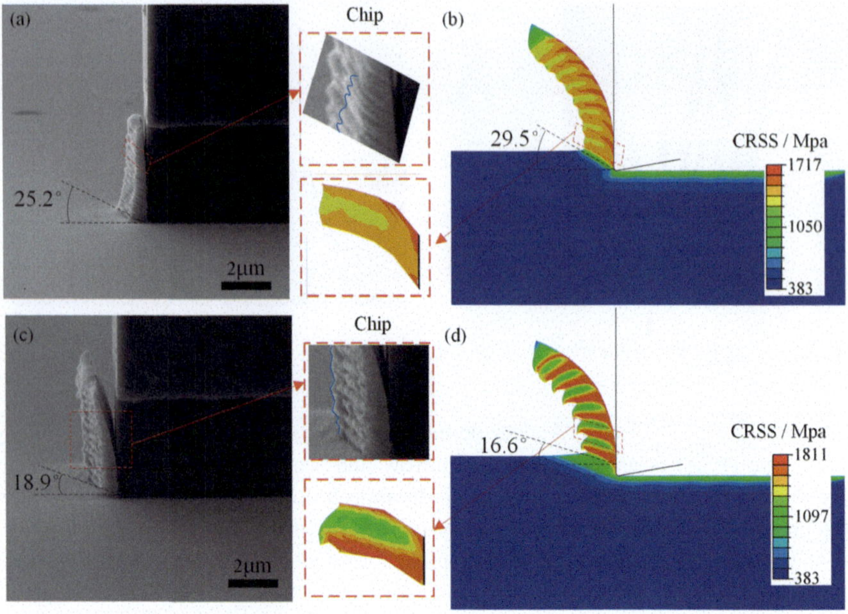


**Fig. 3.11** Chip contour comparison on Cu(010): **a** In-situ SEM and **b** FE simulation, and on Cu(111): **c** In-situ SEM and **d** FE simulation

that different friction conditions and workpiece size are to blame for the differences in chip curvature that are seen between models and tests.

The contrast in chip morphology described above highlights the unique chip characteristics of polycrystalline materials, and the generation of irregular microstructures will have a significant impact on nanoscale machining precision.

## 3.2 Surface Generation Mechanism

MD simulation has emerged as a vital tool in the study of SPDT since its introduction to micromachining in the previous century [16–19]. However, previous research on SPDT of polycrystalline copper has predominantly based on the machining and associated deformation mechanisms occurring at grain boundaries, while relatively neglecting the impact of crystal orientations and the optimization of machining parameters to enhance surface quality. This gap in knowledge has resulted in inadequate insights into the underlying mechanisms and the improvement of surface quality, which holds implications for the optical properties of polycrystalline copper. Therefore, further investigations into crystal orientations and the optimization of machining parameters will help fill these research gaps.

To further understand the variation in machined surfaces, MD simulations were employed to simulate the nanocutting process of single-crystal copper particles with different orientations, utilizing a double-tool model.

By combining experimental observations and MD simulations, the section aims to elucidate the differences in machined surface characteristics during SPDT of polycrystalline copper. This section also aims to explore the underlying machining mechanisms responsible for the surface morphology of polycrystalline copper and identify potential approaches to suppress the formation of IMS. Through this comprehensive analysis, a deeper understanding of the generation mechanism of IMS will be achieved, ultimately contributing to the enhancement of surface quality in polycrystalline copper machining.

### *3.2.1 MD Modeling of Nano-cutting*

In SPDT experiments, it is possible to discretize polycrystalline copper into single-crystal copper particles where the orientations of the particles might change [20]. Hence, the workpiece in this section consists of single-crystal copper with different crystal planes. The dimensions of the workpiece are $100a \times 50a \times 10a$, where $a$ represents the lattice constant of the copper material. For the sake of simplicity, the interaction of the coolant with other atoms was disregarded in the weeny models.

To effectively model the workpiece, three representative crystal planes were selected. The workpiece models were configured with shrink-wrapped boundary conditions in the $X$- and $Y$-axes, while periodic boundary conditions were applied in the $Z$-axis. This approach is employed to minimize the influence of size effects on the simulation.

It was decided to separate the workpiece model into three distinct layers in order to reduce the amount of difference that existed between the simulated results and the real outcomes. These levels were the boundary layer, the thermostat layer, and the Newtonian layer. It was decided to designate the boundary layers as fixed layers in order to reduce the impact of the boundary effects. On the other hand, the Thermostat layers and the Newtonian layers were identified as the moving layers. In these layers, atomic motion was carried out in accordance with the rules of physics that were established by classical Newton. At a temperature of 298K, the temperature of the thermostat layers was maintained at room temperature.

Additionally, appropriate initial conditions were established, encompassing suitable initial coordinates and velocities for the particles. Given the substantial disparity in hardness between diamond and copper, the simulation overlooked tool wear. Instead, the tool was treated as a rigid body, assuming the same angle as the experimental tool depicted in Fig. 3.12. A description of the pertinent simulation parameters can be found in Table 3.4.

To establish an appropriate nanocutting model, it is essential to consider the lattice constant. In this regard, simulations and calculations were executed to determine the relationship between the cohesive energy and lattice constant of carbon (C) with

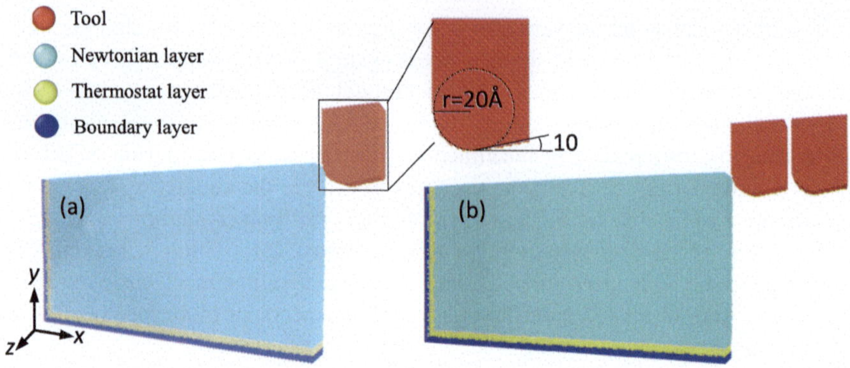


**Fig. 3.12** MD model of single-crystal copper nano-cutting

**Table 3.4** Detailed parameters employed in MD simulation

| Item | Parameter |
|---|---|
| Ensemble | NVT (Relaxation)<br>NVT (Thermostat layers)<br>NVE (Newtonian layers) |
| Simulation software | LAMMPS [21, 22] |
| Workpiece material | Single-crystal copper, (010)(110)(111) |
| Tool material | Single-point diamond |
| Tool rake angle ($\alpha$) (°) | 0 |
| Tool clearance angle ($\delta$) (°) | 10 |
| Tool cutting edge radius (Å) | 20 |
| Cutting depth (Å) | 20 |
| Cutting velocity (Å/ps) | 0.1, 1.0, 5.0 |
| Equilibration temperature (K) | 298 |
| Time step (ps) | 0.001 |
| Cutting direction | $[\bar{1}00]$ on the (010) plane<br>$[\bar{1}10]$ on the (110) plane<br>$[\bar{1}\bar{1}2]$ on the (111) plane |

a diamond structure and copper (Cu) with a FCC structure. This analysis yielded a curve, as depicted in Fig. 3.13, through nonlinear fitting. The obtained lattice constant and cohesive energy values closely align with the values referenced in [23]. Furthermore, reference [24] confers that the cohesion energy of C and Cu is 7.37 and 3.49 eV, respectively, further affirming the reliability of our results.

The lattice constant of the model was derived from the lattice constant at the system's lowest cohesive energy and is presented in Table 3.5. Notably, the visualization software OVITO [25] confirmed the absence of dislocation defects in the

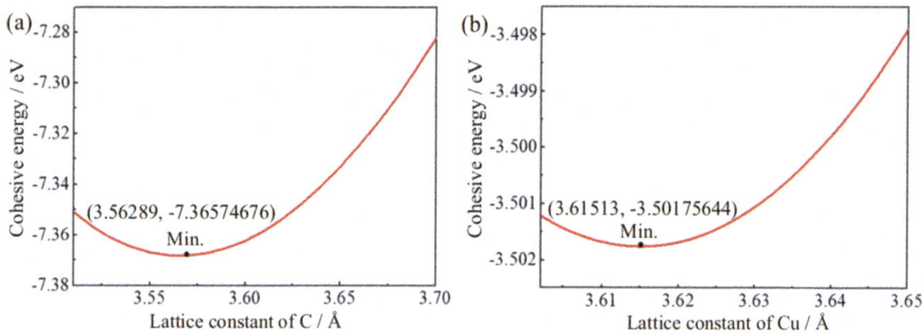


**Fig. 3.13** Minimum lattice constant of **a** C and **b** Cu

**Table 3.5** Experiment and equilibrium lattice constant

| Element | Experiment (eV) | Simulation (eV) |
|---|---|---|
| C | 3.6149 | 3.56289 |
| Cu | 3.56683 | 3.61513053 |

relaxed structure, indicating an absence of internal defects in the ideal FCC crystal structure. In addition, the time step was set to be 0.00 ps in the simulation.

In the MD model, the interaction between copper atoms was described using the Embedded Atom Method (EAM) potential function [26]. The EAM potential can be expressed as:

$$E = \sum_{i} Fi\left(\sum_{j \neq i} \rho j(rij)\right) + \frac{1}{2}\sum_{j \neq i} \Phi ij(rij) \tag{3.1}$$

where $E$ represents the total energy (unit: eV), $F_i$ denotes the embedding energy (unit: eV), $\rho j$ represents the atomic electron density at site $j$ induced by other atoms (unit: g/cm$^3$), $rij$ denotes the distance between atomic $i$ and atomic $j$ (unit: Å) and $\Phi ij$ represents the pair potential interaction (unit: eV).

Since the C–C bond is a covalent bond, the interaction between carbon atoms was described using the Tersoff potential function [27–29]:

$$E' = \frac{1}{2}\sum_{i \neq j} fC(rij)[\varphi^R(rij) + bij\varphi^A(rij)] \tag{3.2}$$

where $E'$ represents the potential energy (unit: eV), $fC(rij)$ denotes the truncation function (unit: Å), $\varphi^R(rij)$ and $\varphi^A(rij)$ are the repulsive and attractive parts of the potential energy respectively (unit: eV), $bij$ represents the three-body term, a modulation function (dimensionless), and $rij$ denotes the atomic distance between $i$ and $j$ (unit: Å).

**Table 3.6** Morse potential parameters for C and Cu

| Element | $D$ (eV) | $\alpha'$ (Å$^{-1}$) | $r_0$ (Å) | $r_c$ (Å) |
|---|---|---|---|---|
| C-Cu | 0.087 | 1.7 | 2.05 | 10 |

Taking into account references [30], the interaction between C and Cu atoms can be described using the Morse potential function:

$$E = D\{\exp[-2\alpha'(r - r0)] - 2\exp[-\alpha'(r - r0)]\} \quad r < rc \tag{3.3}$$

where $D$ represents binding energy (unit: eV), $\alpha'$ denotes a constant associated with the properties of a material (unit: Å$^{-1}$), $r0$ represents the equilibrium separation (unit: Å), $rc$ denotes the cutoff (unit: Å).

In the simulation, it is crucial to select a cutoff distance that balances calculation accuracy and efficiency. The cutoff distance should not exceed half of the minimum side length of the central primitive cell to adhere to the nearest neighbor mirror convention. By minimizing errors between the simulation calculations and the actual system, the cutoff radius is determined first. Subsequently, the model size is set to be $10a$, as indicated in Table 3.6. The table also presents the parameters used in the Morse potential function.

### 3.2.2 *Visualization of Surface Generation Mechanism*

Figure 3.14 illustrates the Cu's deformation during nano-cutting with various orientations. This experiment was performed at a cutting depth of 2 nm and a cutting distance of 16.36 nm. Specifically, Fig. 3.14a–i demonstrates the material deformation and chip generation in the single-tool diamond cutting model, while Fig. 3.14j–l presents the same phenomena in the double-tool diamond cutting model, exhibiting different cutting velocities.

When analyzing chip morphology, chip thickness that varies based on changes in crystal planes at a cutting velocity of 10 m/s was observed. Notably, the shear slip region of the material exhibits the largest chip thickness while having the shortest height in the (010) crystal plane. Conversely, the (110) crystal plane shows the smallest chip thickness but the highest height. This information is illustrated in Fig. 3.14a–c.

At a cutting velocity of 100 m/s, Fig. 3.14d–f demonstrates that chip thickness remains at its maximum while height remains the shortest due to the accumulation of material deformation in the cutting direction within the (010) crystal plane. On the other hand, the (111) crystal plane exhibits the smallest chip thickness and the highest height, albeit with diminishing differences.

Upon reaching a cutting velocity of 500 m/s, the thickness and height of chips become approximately equal in all three crystal planes. It is worth noting that due to

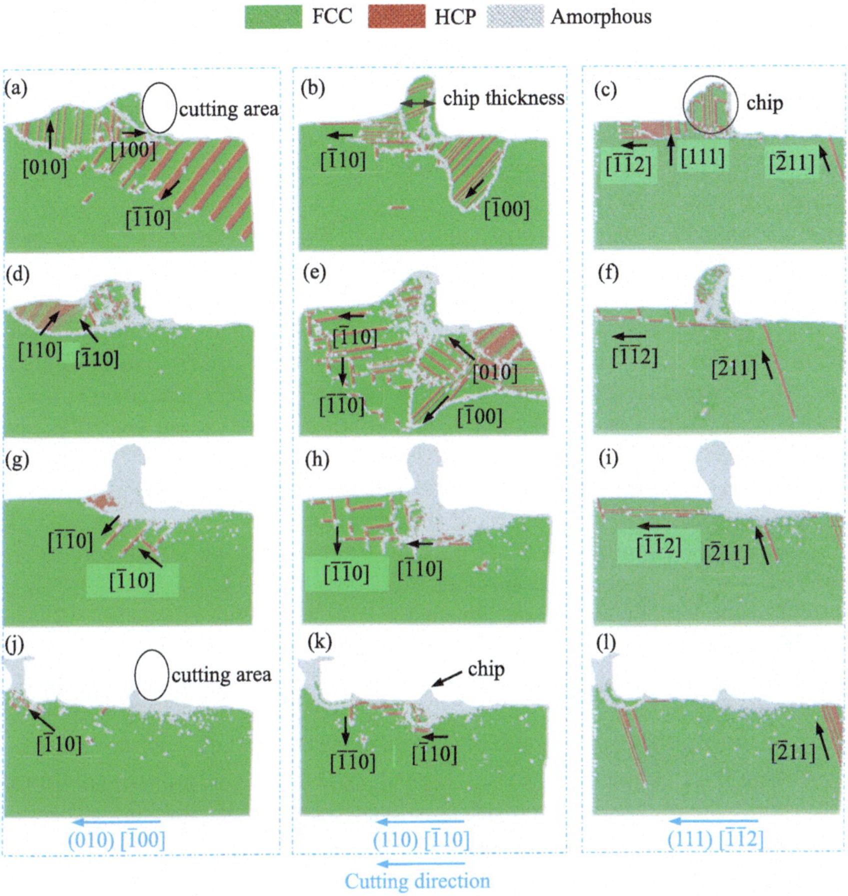


**Fig. 3.14** Material deformation in various cutting speed: **a–c** 10 m/s, **d–f** 100 m/s and **g–l** 500 m/s

the high velocity, amorphous atoms are generated in the cutting area. Figure 3.14g–i highlights that the (110) crystal plane contains the most amorphous atoms, with the deepest amorphous layer found in the working region of the tool flank face. In contrast, the (111) crystal plane exhibits the fewest amorphous atoms and the thinnest amorphous layer.

In Fig. 3.14j–l, when considering high velocity and low feed rate, the number of amorphous atoms that present in the three crystal planes was observed and the secondary chips are similar. These chips are generated as a result of the material recovering after cutting the workpiece surface. Additionally, taking into account the various grain orientations mentioned earlier, it's apparent that the chip morphology varies to some extent. Specifically, the chips' thickness and height differ at different velocities. The experimental results align with the theoretical models, which effectively explain the topographic differences on the machined surface caused by grain orientations during the machining process.

To analyze the shear and slip directions of the material, the correlative environment algorithm was employed in LAMMPS and calculated the common neighbor analysis values for each atom in the model. The atoms were then colored based on the CNA method, as depicted in Fig. 3.14. At a cutting speed of 10 m/s, the (010) crystal plane predominantly exhibits shear slip in the [010], $[\bar{1}\bar{1}0]$ and [100] directions. In the working region of the tool flank face, the material primarily slips along the $[\bar{1}\bar{1}0]$ direction. Upon increasing the cutting velocity to 100 m/s, shear slip primarily occurs in the area where the tool rake face contacts the material. The main slip directions observed are [110] and $[\bar{1}10]$. When the cutting speed escalates to 500 m/s, due to the presence of amorphous atoms on the secondary machined surface, no specific slip direction is observed. Considering different cutting velocities, the shear slip directions of the (010) crystal plane mainly lie within the <110> and <010> families of directions. Now directing our attention to the (110) crystal plane, at a cutting speed of 10 m/s, shear and slip directions manifest primarily as $[\bar{1}10]$ and $[\bar{1}00]$. As the cutting velocity increases to 100 m/s, multiple slip directions are observed, with the predominant slip direction being along $[\bar{1}10]$, $[\bar{1}\bar{1}0]$, [010] and $[\bar{1}00]$. At a cutting speed of 500 m/s, the material primarily slips along the $[\bar{1}\bar{1}0]$ and $[\bar{1}10]$ directions. However, within the tool action region under a double-tool model, only a few $[\bar{1}\ 10]$ directions exist. The shear slip directions of the (110) crystal plane, for different cutting velocities, mainly fall within the <110> and <100> families of directions. Lastly, focusing on the (111) crystal plane, at a cutting velocity of 10 m/s, the material shear slips along $[\bar{2}\ 11]$, $[\bar{1}\bar{1}2]$ and [111] directions, primarily observed in the contact area between the tool rake face and the workpiece. At a cutting velocity of 100 m/s, the dominant slip directions are $[\bar{2}11]$ and $[\bar{1}\bar{1}2]$, aligning with the cutting direction. Similarly, for a velocity of 500 m/s, $[\bar{2}11]$ and $[\bar{1}\bar{1}2]$ remain the main slip directions, with slipping predominantly along $[\bar{2}11]$ in higher velocities and lower feed rates. The shear slip directions of the (111) crystal plane, considering different cutting velocities, mostly occur within the <112> family of directions.

In crystal materials, the crystal plane (111) exhibits the greatest distance between neighboring cleavage planes, indicating weaker covalent bonding forces between atoms. Consequently, this plane is more susceptible to slip and cleavage phenomena, as confirmed by the results of MD simulation. The presence of IMS resulting from different grain orientations further supports the observed variations in slip directions and cleavage phenomena across different crystal planes.

Analyzing the subsurface defects, the material based on the centro-symmetry parameter was calculated and colored. Figure 3.15a–i showcase the subsurface defect structures of single-crystal copper particles at various velocities and orientations with a cutting distance of 30 nm. Figure 3.15j–l depict the subsurface defect structures of three crystal planes using a double-tool model with a cutting distance of 23 nm.

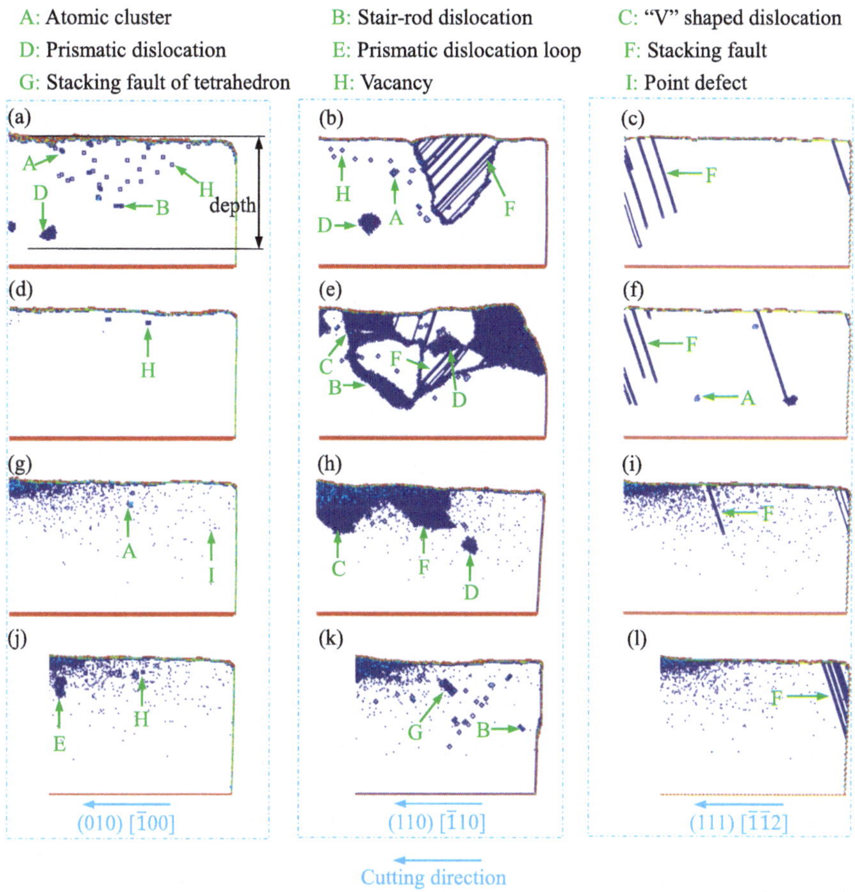


**Fig. 3.15** Dislocation in various cutting speed: **a–c** 10 m/s, **d–f** 100 m/s and **g–l** 500 m/s

When considering the (010) crystal plane, as the velocity increases, the types of subsurface defect structures decrease. At a cutting velocity of 10 m/s, the predominant defect structures include atomic clusters, stair-rod dislocations, prismatic dislocations, and vacancies. At 100 m/s, vacancies become the main defect structure. When the cutting velocity reaches 500 m/s, the prevalent defect structures are point defects, atomic clusters, and vacancies. In the double-tool model, the prominent defect structures include prismatic dislocation loops and vacancies.

When examining the crystal plane at different cutting velocities, various subsurface defect structures are observed. At a cutting velocity of 10 m/s, vacancies, atomic clusters, prismatic dislocations, and stacking faults are predominantly present. Increasing the velocity to 100 m/s leads to the dominance of stair-rod dislocations, "V" shaped dislocations, prismatic dislocations, and stacking faults, resulting in decreased cutting performance. Further increasing the velocity to 500 m/s reveals

8. Yuan Z, Wang X (eds) (2016) Precision and ultraprecision machining technology, 3th ed. China Machine Press (in Chinese)
9. Liu YX, Suslov S, Han QY, Xu C, Hua L (2012) Microstructure of the pure copper produced by upsetting with ultrasonic vibration. Mater Lett 67(1):52–55
10. Gu Z, Niu X, Liu Y, Gao B, Wang C, Deng H (2017) Removal of residual metal oxide after Cu-CMP in ultra-precision machining. Chin J Rare Metals 41(02):146–154 (in Chinese)
11. Wang ZF, Zhang JJ, Xu ZW, Zhang JG, ul Hassan H, Li G, Zhang HJ, Hartmaier A, Fang FZ, Yan YD, Sun T (2019) Crystal plasticity finite element modeling and simulation of diamond cutting of polycrystalline copper. J Manuf Process 38:187–95
12. Chaudhari A, Soh ZY, Wang H, Kumar AS (2018) Rehbinder effect in ultraprecision machining of ductile materials. Int J Mach Tools Manuf 133:47–60
13. Lee WB, To S, Cheung CF (2000) Effect of crystallographic orientation in diamond turning of copper single crystals. Scripta Mater 42(10):937–945
14. Han JD, Li LH, Lee WB (2018) Investigation on the effects of laser heating on surface quality of OFHC copper in diamond cutting process. Precis Eng 54:294–301
15. Ding X, Jarfors AEW, Lim GC, Shaw KC, Liu YC, Tang LJ (2012) A study of the cutting performance of poly-crystalline oxygen free copper with single crystalline diamond micro-tools. Precis Eng 36(1):141–152
16. Wang Z, Sun T, Zhang H, Li G, Li Z, Zhang J, Yan Y, Hartmaier A (2019) The interaction between grain boundary and tool geometry in nanocutting of a bi-crystal copper. Int J Extreme Manuf 1(4):045001–045012
17. Wang ZF, Zhang JJ, Zhang JG, Li G, Zhang HJ, ul Hassan H, Hartmaier A, Yan YD, Sun T (2020) Towards an understanding of grain boundary step in diamond cutting of polycrystalline copper. J Mater Process Technol 276:116400–116410
18. Wang ZF, Zhang JJ, Xu ZW, Zhang JG, ul Hassan H, Li G, Zhang HJ, Hartmaier A, Fang FZ, Yan YD, Sun T (2019) Crystal plasticity finite element modeling and simulation of diamond cutting of polycrystalline copper. J Manuf Process 38:187–195
19. Liu DH, Wang G, Yu JC, Rong Y (2017) Molecular dynamics simulation on formation mechanism of grain boundary steps in micro-cutting of polycrystalline copper. Comput Mater Sci 126:418–425
20. Zhang JJ, Wang ZF, Yan YD, Sun T (2016) Concise review: recent advances in molecular dynamics simulation of nanomachining of metals. Curr Nanosci 12(6):653–665
21. Plimpton S (1995) Fast parallel algorithms for short-range molecular dynamics. J Comput Phys 117(1):1–19
22. Stadler J, Mikulla R, Trebin HR (2011) IMD: a software package for molecular dynamics studies on parallel computers. Int J Mod Phys C 08(05):1131–1140
23. Sharma A, Datta D, Balasubramaniam R (2018) Molecular dynamics simulation to investigate the orientation effects on nanoscale cutting of single crystal copper. Comput Mater Sci 153:241–250
24. Kittel (ed) (2007) Introduction to solid state physics, 7th edn. Wiley, pp 37–38
25. Stukowski A (2010) Visualization and analysis of atomistic simulation data with OVITO—the open visualization tool. Model Simul Mater Sci 18(1):015012–015019
26. Foiles SM, Baskes MI, Daw MS (1986) Embedded-atom-method functions for the fcc metals Cu, Ag, Au, Ni, Pd, Pt, and their alloys. Phys Rev B Condens Matter 33(12):7983–7991
27. Tersoff J (1989) Modeling solid-state chemistry: Interatomic potentials for multicomponent systems. Phys Rev B Condens Matter 39(8):5566–5568
28. Wang QL, Bai QS, Chen JX, Guo YB, Xie WK (2015) Stress-induced formation mechanism of stacking fault tetrahedra in nano-cutting of single crystal copper. Appl Surf Sci 355:1153–1160
29. Li J, Fang QH, Zhang LC, Liu YW (2015) The effect of rough surface on nanoscale high speed grinding by a molecular dynamics simulation. Comput Mater Sci 98:252–262
30. Li J, Fang QH, Liu YW, Zhang LC (2014) A molecular dynamics investigation into the mechanisms of subsurface damage and material removal of monocrystalline copper subjected to nanoscale high speed grinding. Appl Surf Sci 303:331–343

# Chapter 4
# Cutting Process in Machining Amorphous Materials

**Abstract** Amorphous materials (AM), which lack a long-range order typical of crystalline structures, presents unique challenges and opportunities in precision manufacturing. This chapter delves into the machining performance of amorphous materials, with a particular focus on SPDT and the mechanisms of surface generation as revealed through MD simulations. By exploring the machining performance, analyzing results, and leveraging advanced simulation techniques, it aims to enhance the understanding and optimization of machining practices for these unique materials, ultimately contributing to their broader application in high-precision industries.

**Keywords** Machining performance · Single-point diamond turning · Surface generation · Amorphous materials

## 4.1 Machining Performance of Amorphous Materials

### *4.1.1 Advanced Application of AM*

AM, also known as metallic glasses, are novel multifunctional advanced materials combining the properties of conventional metals and glass. They possess excellent overall characteristics, including a wide range of applications, cost-effectiveness, and high recyclability. The introduction of amorphous alloys has had a significant impact on various industries, including industrial production [1, 2], aerospace [3], medicine [4, 5], and defense equipment [6]. They are also used in a wide range of coatings, including those for thermal, optical, electromagnetic, and corrosion-resistant devices [7–9]. Unlike crystalline materials, AM have more active sites on their uneven surfaces. This means they have better electrochemical qualities for uses like splitting water and making anodes for lithium-ion batteries [10–13].

Furthermore, because there are no grain limits in amorphous alloys (AAs) to enhance their resistance to corrosion in both acidic and alkaline environments. These alloys exhibit high wear resistance compared to crystalline materials. Thin films of AAs demonstrate exceptional flexibility [14–17], making them well-suited for

 
G. Zhang and J. Wang, *Material-Oriented Cutting Processes in Precision Machining*, Springer Tracts in Mechanical Engineering,
https://doi.org/10.1007/978-981-96-2504-8_4

use in wearable electronic devices, micro-scanners, and coatings [18–21]. Despite their different structures, AAs share characteristics with metallic crystals, including high reflectivity of visible light and surface plasmon resonance effects [22–24]. As a result, they are employed as color coatings and substrates for surface-enhanced Raman spectroscopy [25, 26]. AAs based on Zr, Ti, and Au have proven to be effective biomedical materials for nerve probes, surgical tools etc. due to their favorable biocompatibility, manufacturability, and mechanical durability [27–30]. Figure 4.1 illustrates the properties and applications of AAs [31].

The acquisition of AAs first took place in 1960 [32]. Subsequently, the research on AAs witnessed rapid advancements. In the 1980s, scholars conducted successive studies on amorphous alloys and achieved certain results. They developed various types of high glass-forming ability and high-performance AAs, such as Zr-, Ti-, Cu-, Fe-, Ni-, and Pd-based alloys, as shown in Fig. 4.2 [33].

AAs exhibits the hardness akin to ceramics, which is suitable for molding application [34]; however, when subjected to heat and transformed into its supercooled liquid state, the substance exhibits a plastic-like softening behavior and displays the characteristics of a Newtonian fluid.

In the realm of 3D printing, Shen et al. utilized 3D-printed Zr-based AAs to produce highly intricate and remarkably elastic components [35]. Furthermore, Zhang et al. successfully fabricated gears with exceptional mechanical properties using thermally-sprayed 3D-printed Fe-based AAs [36].

For mold manufacturing, Pan et al. introduced an innovative preparation method employed Mg-based AAs rapid prototyping material as a supplementary mold for

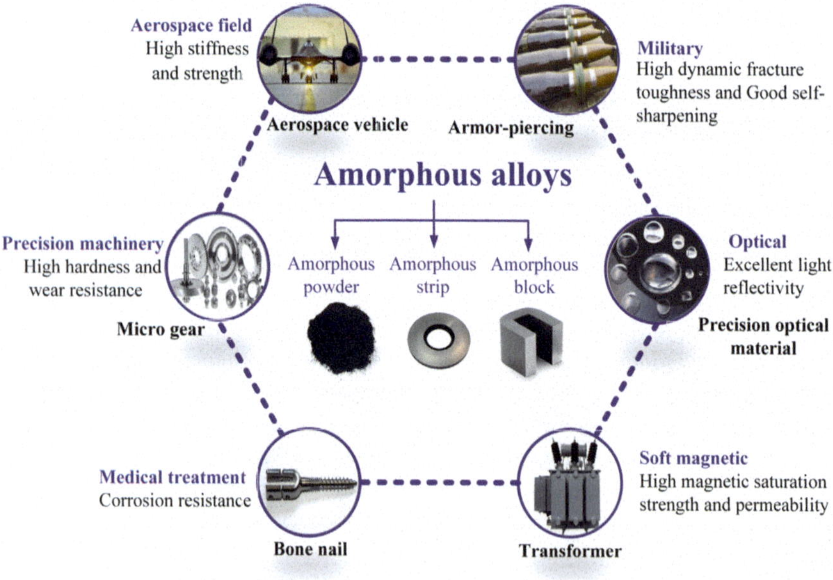


**Fig. 4.1** Properties and applications of amorphous alloys

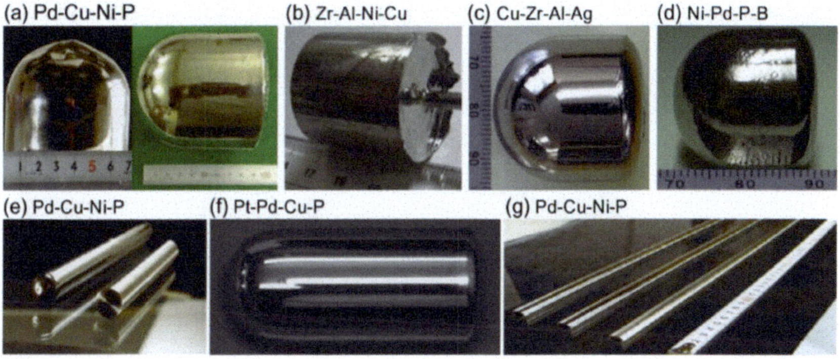


**Fig. 4.2** Different amorphous materials

aspheric lenses, boasting a AAs mold hardness of up to 3.445 GPa [37]. In the area of coatings, Lee et al. used magnetron sputtering deposition to put ZrCu AAs on a pet substrate. This created an ITO/ZrCu bilayer transparent conductive electrode metal layer with a smooth ZrCu film thickness below 6 nm, which improved certain functions [38]. Additionally, Chu et al. employed AAs as a coating material, known for its high elastic limit, resistance to corrosion, and friction, enhancing not only the mechanical properties of the workpiece but also providing superior wear and oxidation resistance compared to traditional materials [39].

Regarding ultra-precision machining, Fujita et al. studied the cutting behavior of AAs using various cutting parameters. The investigation involved Zr-based AAs, Pd-based AAs, steel, and brass as cutting materials [40]. Even though AAs had twice the tensile strength of steel, it had only half the cutting force of steel, according to the results of the tests. AAs had a much smoother surface than steel and just a little smoother than brass. Chen et al. investigated the cutting performance of Zr-based AAs using diamond and boron nitride tools [7]. Their research highlighted the machined surface of Zr-based AAs exhibited noticeable softening and nanocrystallization effects, leading to subpar cutting performance and difficulties in achieving a mirror-like surface with a depth of cut of 0.6 μm. Therefore, AAs should be classified as difficult-to-machine materials. Understanding its machining performance in depth and exploring the intrinsic reasons for its machinability challenges are very important for the ultra-precision machining of AAs.

Pd-based AAs exhibits exceptional oxidation resistance, corrosion resistance, and amorphous forming ability, making it easily accessible [42–47]. These properties, combined with its excellent cutting performance and the ability to produce optical mirror surfaces, position Pd-based AAs as a significant material in the optics field. Therefore, ultra-precision machining has become the dominant method for producing precise components and optical parts [48–50]. Nevertheless, there is limited research resulting in a lack of theoretical insights and experimental references.

This section focuses on exploring the influence of cutting parameters on the surface quality of Pd-based AAs in SPDT, specifically $Pd_{40}Ni_{10}Cu_{30}P_{20}$. It delves into the theoretical analysis of oxidation mechanisms and strategies to prevent oxidation, while also examining the impact of tool wear and chip morphology on surface oxidation. The subsequent sections provide valuable insights for SPDT research on AAs, offering essential theoretical and technical support for the turning and manufacturing processes of AAs in optical mold applications.

### 4.1.2 SPDT of Amorphous Materials

The $Pd_{40}Ni_{10}Cu_{30}P_{20}$ AAs rods were cut using an ultra-precision lathe mentioned in Chap. 1, as shown in Fig. 4.3a. The specimens had a diameter of 2 mm and a length of 16 mm, as depicted in Fig. 4.3b. Cutting was done with a natural diamond tool with specific geometric parameters (refer to Table 4.1). Three experimental groups (group 1 to 3) with different cutting parameters (see Table 4.2) were used to study the machinability of the Pd-based AAs. Oil-based lubricant was used during cutting.

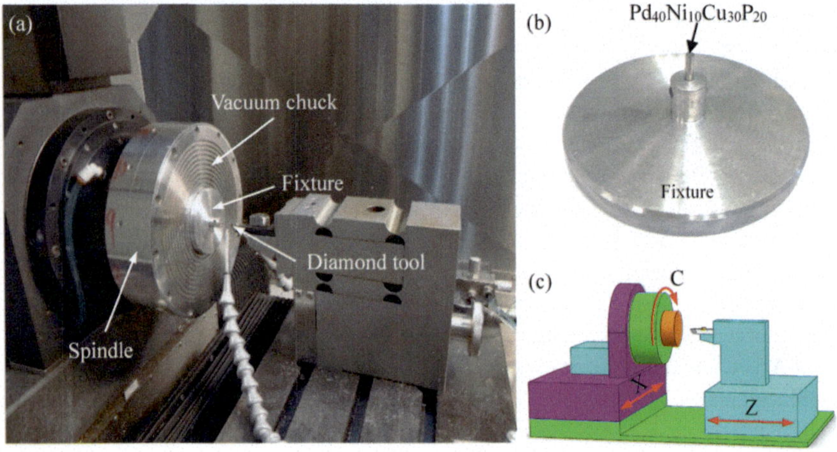


**Fig. 4.3** **a** The experimental setup for SPDT, **b** the fixture and workpiece, and **c** the configuration of 3-axis ultra-precision lathe

**Table 4.1** Details of tool parameters

| Tool parameters | Tool 1 | Tool 2 |
|---|---|---|
| Tool material | Nature diamond | |
| Tool rake angle ($\alpha$) | 0° | |
| Tool clearance angle ($\beta$) | 15° | |
| Tool nose radius ($R$) | 0.502 mm | 0.101 mm |

**Table 4.2** Details of cutting parameters

| Items | Feed rate ($f$) (mm/min) | Tool nose radius ($R$) (mm) | Spindle speed ($v$) (rpm) | Depth of cut ($ap$) (μm) |
|---|---|---|---|---|
| Group 1 | 5 | 0.502 | 1000, 1500, 2000, 3000 | 4 |
| Group 2 | 5 | 0.502 | 2000 | 2, 4, 6, 8 |
| Group 3 | 5 | 0.101 | 2000 | 2, 4, 6, 8 |

### *4.1.3 Machining Performance and Analysis*

During the turning process, the AAs had a propensity to crystallize, necessitating the examination of its crystallization status. Crystallization not only alters the structure of AAs but also significantly affects its mechanical properties, and machinability. Thus, an X-ray diffraction (XRD) analysis was conducted on the machined surface, as illustrated in Fig. 4.4. The XRD pattern clearly indicates the amorphous structure on the machined surface, displaying an amorphous diffuse diffraction peak. No crystalline phase was observed, affirming that ultra-precision machining could be performed on AAs without inducing any crystalline transformation. Although crystallization may have occurred during the cutting process, it did not impact the amorphous state of the machined surface due to the excellent glass-forming ability of AAs. Therefore, further discussion on crystallization is not explored in this section.

Within this part, a more in-depth investigation of the impact that cutting parameters have on the surface quality of AAs was investigated. The variations in Ra shown by spindle speeds of 1000, 1500, 2000, and 3000 revolutions per minute are depicted in Fig. 4.5a. The surface roughness (Ra) initially decreases and then increases with rising spindle speed, reaching its lowest value at 2000 rpm. This phenomena is associated with the AAs being in a temperature range that is known to as the supercooled

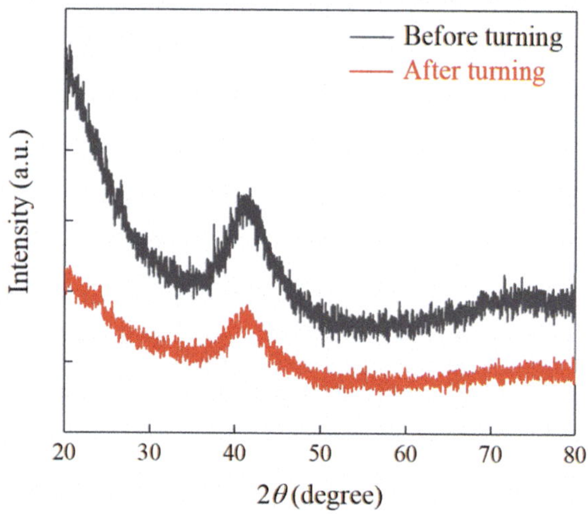


**Fig. 4.4** XRD pattern of Pd-based AAs

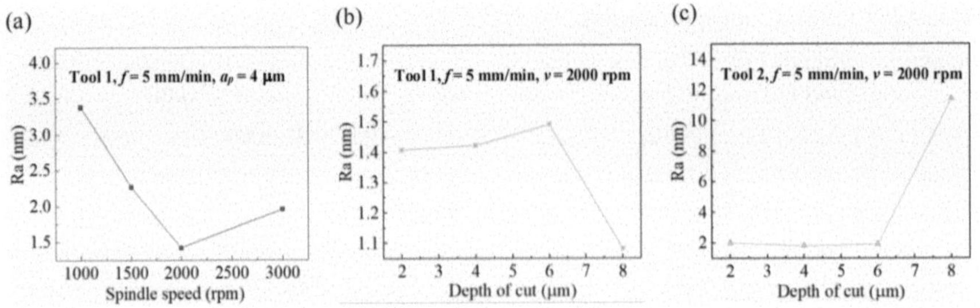


**Fig. 4.5** The impact of cutting parameters on the surface roughness of machined surfaces, including **a** Tool 1 with $a_p = 4$ μm and varying spindle speeds, **b** Tool 1 operating at 2000 rpm with different $a_p$, and **c** Tool 2 also at 2000 rpm with varying $a_p$

liquid state. This temperature range is between the glass transition temperature ($T_g$) and the first crystallization temperature ($T_x$) [61]. In this state, AAs demonstrates superplasticity and self-healing properties [52], or it transitions into a true liquid state (melting). As a result, AAs, which is typically brittle at room temperature [53], becomes highly malleable. As spindle speed increases, the workpiece's surface temperature rises due to AAs being in the supercooled liquid or true liquid state.

Figure 4.5b shows the changes in Ra for different $a_p$ of 2, 4, 6, and 8 μm indicating Ra initially increases and then decreases, with the optimum Ra achieved at a $a_p$ of 8 μm. When the $a_p$ is less than 6 μm, the cutting temperature is below $T_g$, resulting in the AAs being brittle. Therefore, the surface quality deteriorates, and the Ra increases with smaller $a_p$. However, when the $a_p$ exceeds 6 μm, the cutting temperature surpasses $T_g$, causing the AAs to exhibit superplasticity. Consequently, the Ra decreases correspondingly.

Figure 4.5c illustrates the variations in Ra in relation to $a_p$ values of 2, 4, 6, and 8 μm, with a consistent tool nose radius of 0.101 mm. Notably, a gradual decrease in Ra was observed for $a_p$ below 6 μm. However, upon increasing the $a_p$ to 8 μm, Ra abruptly rose, defying conventional expectations.

The extraordinary superplastic characteristics of AAs led to an exceptionally smooth surface, as shown in Fig. 4.6. Additionally, the cutting process produced continuous chips, as demonstrated in Fig. 4.7. With the increase of DOC, the wrinkles on the free surface of the chips became more evident, yet they retained a continuous band form. This phenomenon occurs because when the cutting temperature exceeded the $T_g$ and entered the supercooled liquid state, the initially brittle AAs displayed superior cutting performance.

Surface tension gives the AAs the power to "self-heal" when it is in the condition of a supercooled liquid or a true liquid. Surface tension is responsible for this ability. Unlike aluminum and copper, which typically exhibit burrs and grooves on the machined surface (Fig. 4.8a), the AAs responds to surface tension similar to glass. The surface tension softens and smoothens any burrs, allowing them to flow into the grooves and manifest the self-healing ability of the AAs. This process ultimately improves the surface quality. Moreover, increasing cutting temperature leads

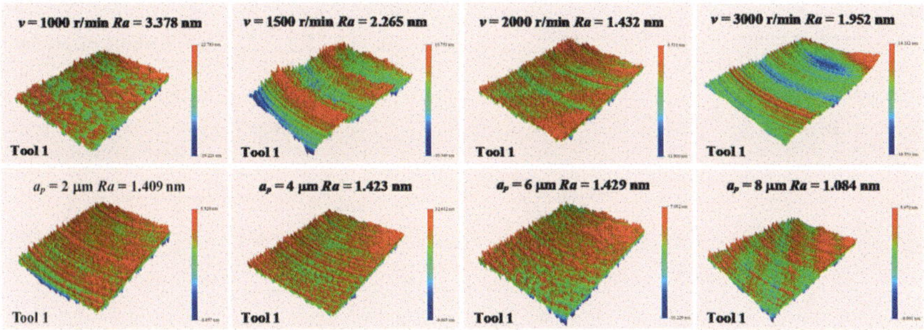


**Fig. 4.6** Surface morphology was assessed using a white light interferometer after machining

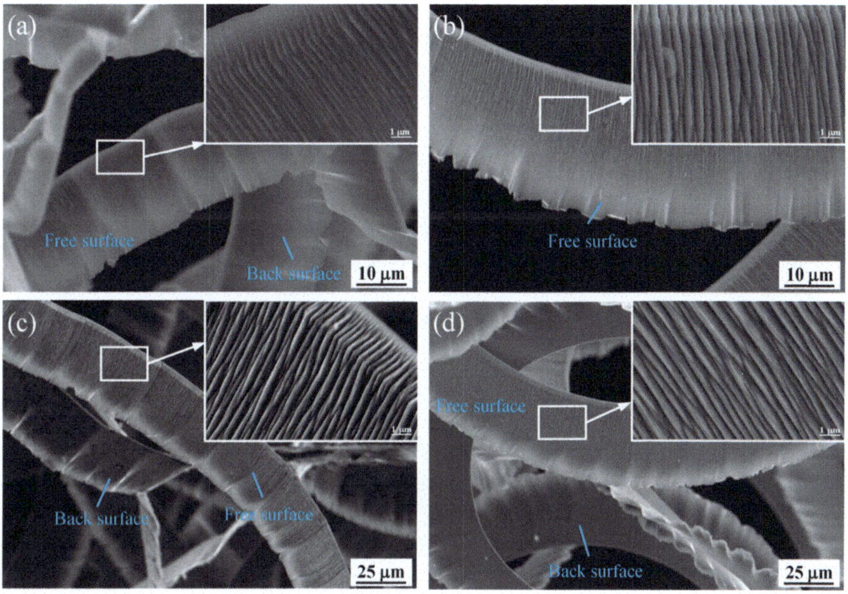


**Fig. 4.7** SEM images of chip morphology with various $a_p$ **a** 2 μm, **b** 4 μm, **c** 6 μm, and **d** 8 μm with $v = 2000$ rpm, $f = 5$ mm/min

to reduced fracture toughness and enhanced machinability [54–58]. The result is a better machined surface with improved cutting performance (Fig. 4.8b).

Based on these observations, when the $a_p$ is 4 μm and the $v$ is below 2000 rpm, an increase in $v$ raises the cutting temperature. Once this temperature surpasses $T_g$, the surface gradually softens, thereby enhancing the cutting performance and reducing Ra (Fig. 4.9). The optimal self-healing effect and cutting performance are achieved at the $v$ of 2000 rpm, resulting in the best surface quality and smallest Ra. However, as the $v$ further increases, the workpiece surface becomes sticky, leading

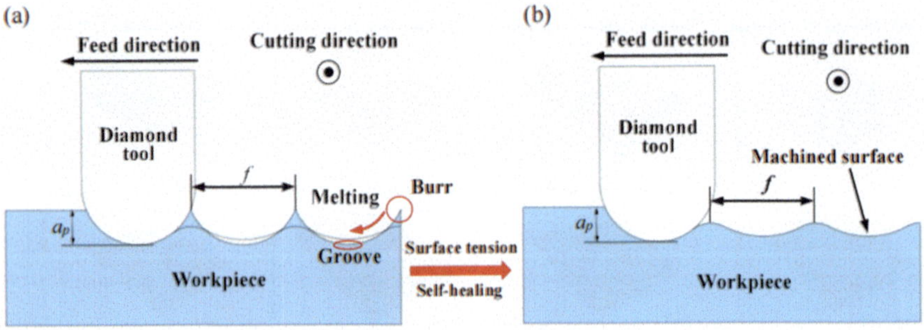


**Fig. 4.8** Schematic diagram of the machined surface morphology of AAs: **a** initial state and **b** final state

to the formation of a prominent built-up edge on the tool. This, in turn, diminishes the cutting performance (Fig. 4.9).

To further explore the unusual phenomena observed in the Ra and understand their origins, the topography of the AAs surface was characterized using AFM. When the $a_p$ was set at 2, 4, or 6 μm, the machined surface displayed only noticeable tool marks, as illustrated in Fig. 4.10a–c. However, when the $a_p$ was increased to 8 μm, irregular micro/nanostructures were observed, as depicted in Fig. 4.10d. These structures were responsible for the increase in Ra and the subsequent decline in surface quality of the AAs.

SEM was employed to investigate the irregular micro/nanostructures' distribution on the machined surface of Pd-based AAs. Figure 4.11a reveals that these irregular micro/nanostructures are primarily distributed along the outer circular region of the machined surface, following the tool mark traces. This observation is further elucidated in the magnified image depicted in Fig. 4.11c. In contrast, the center area of the machined surface appears relatively clean, devoid of any irregular micro/nanostructures, as illustrated in Fig. 4.11b.

EDS was employed to get the composition of micro/nanostructure region and the clean region, as shown in Fig. 4.12a. Notably, the presence of oxygen elements was

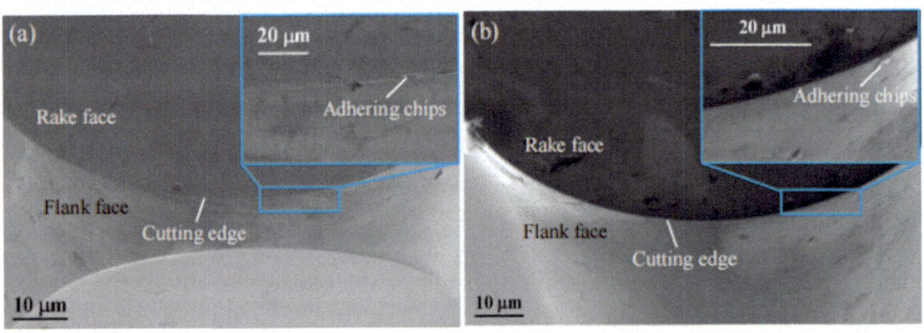


**Fig. 4.9** SEM images of tool edge: **a** Tool 1 and **b** Tool 2

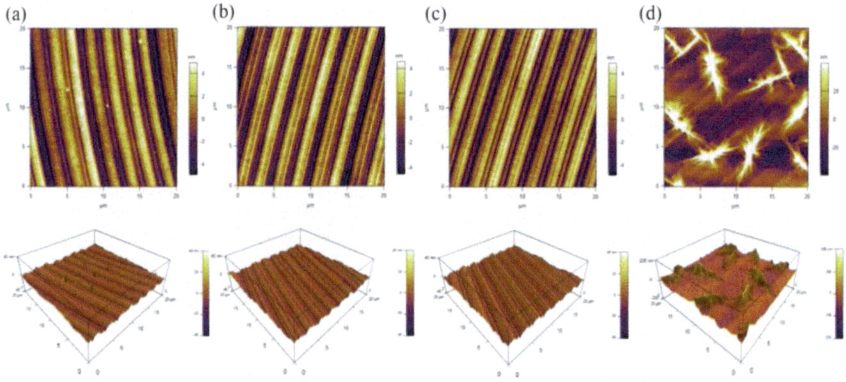


**Fig. 4.10** AFM images of the machined surface using tool 2 with various $a_p$: **a** 2 μm, **b** 4 μm, **c** 6 μm, and **d** 8 μm with $v = 2000$ rpm, $f = 5$ mm/min

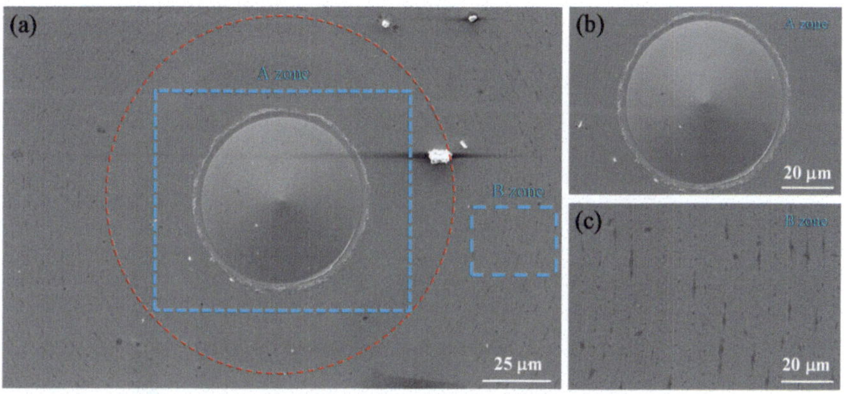


**Fig. 4.11** SEM images of machined surface: **a** the whole area, **b** the A zone, and **c** the B zone

detected in spectrum 1, as depicted in Fig. 4.12b, while spectrum 2 did not exhibit any oxygen signal, as shown in Fig. 4.12c. This indicates that the irregular micro/nanostructures were a result of oxidation, which was predominantly localized along the tool mark traces in the outer circular region.

Figure 4.13 illustrates that the Ra value for the irregular micro/nanostructures on the machined surface was measured at 26 nm, while that of the smooth surface was only 3.8 nm. This stark difference confirms that oxidation significantly impacts the surface quality, leading to an increase in Ra. Despite being a material with high amorphous forming ability and exceptional oxidation resistance, AAs can be susceptible to oxidation when subjected to high temperatures above $T_g$ for extended periods. It is important to note that the possibility of the AAs entering a true liquid state (melting) can be ruled out, as in such a case, the machined surface would be completely smooth. However, as seen in Fig. 4.13, distinct tool marks are clearly visible.

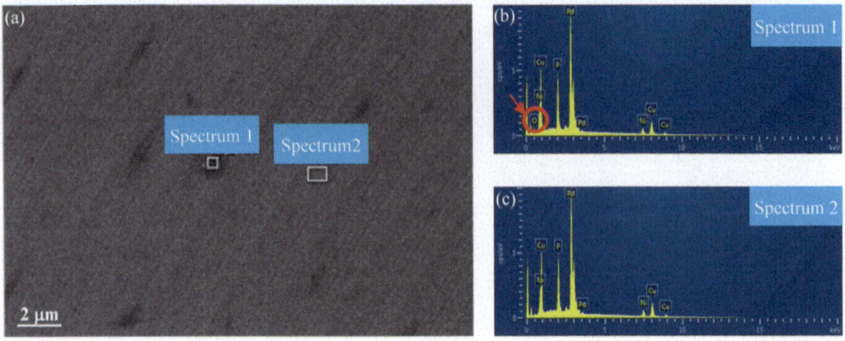


**Fig. 4.12** The EDS analysis of the surface condition of the workpiece along with **a** spectrum 1 and **b** spectrum 2

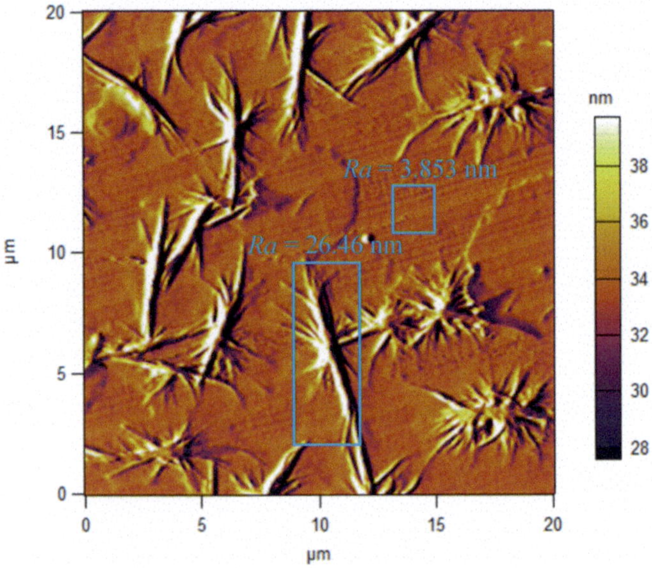


**Fig. 4.13** Irregular micro/nanostructures on the machined surface

The experimental data above demonstrates that amorphous alloys belong to difficult-to-machine materials for SPDT, meaning that ideal surfaces cannot be achieved through conventional SPDT. Therefore, non-traditional machining techniques are typically employed for current amorphous alloys. With the advancement of new technologies, scholars have introduced a broad definition of non-traditional methods. These methods involve the application of electrical, magnetic, sound, light, chemical energy, or a combination thereof to the workpiece for material removal, deformation, performance modification, or plating. Such methods are collectively referred to as non-traditional machining methods.

### 4.1.4 *Non-traditional Machining Methods for Amorphous Alloys*

Ultra-precision machining is not always suitable for all applications involving AAs machining. Consequently, recent years have seen a growing interest in exploring non-traditional machining methods for AAs. These methods are depicted in Fig. 4.14 [31]. By carefully selecting the appropriate cutting parameters, non-traditional techniques can avert the crystallization and oxidation, facilitating the machining of complex amorphous components that conventional methods cannot achieve.

The primary focus of non-traditional machining of amorphous alloys is to preserve the amorphous nature of the materials in the machined region. In addition, the cutting temperature is the key factor for amorphous materials machining, Fig. 4.15 illustrates the impact of temperature when subjected to various non-traditional machining methods [31]. In addition, Table 4.3 provides a comparative analysis of the economic efficiency and machining performance [31].

Hence, it is possible to achieve mirror surface machining of amorphous alloys by combining SPDT with non-traditional machining methods, such as utilizing ultrasonic vibration-assisted machining or laser-assisted machining on ultra-precision lathes.

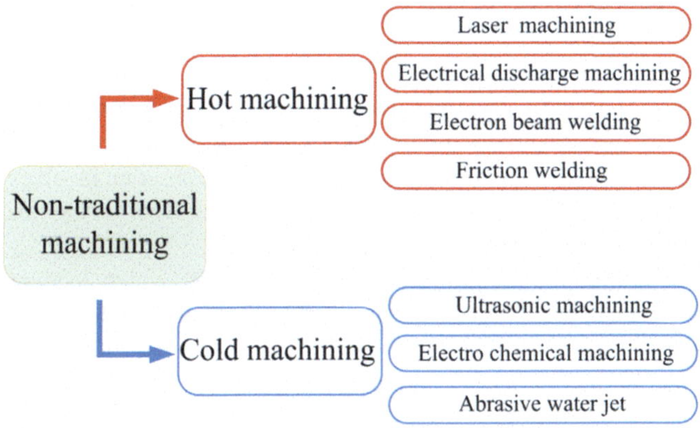


**Fig. 4.14** Non-traditional machining methods

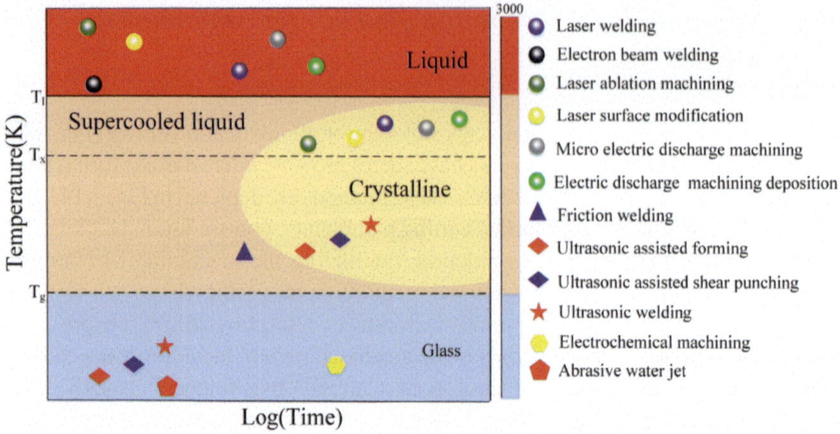


**Fig. 4.15** Temperature effect on the crystallization of amorphous alloys employed various non-traditional machining methods

## 4.2 Surface Generation Mechanism

### *4.2.1 Modelling*

The cutting heat in the machining process primarily originates from two zones, as depicted in Fig. 4.16 [59]. The first zone is where the tool shears the workpiece material, causing a sudden rise in temperature near the workpiece and the primary shear plane, known as the shear heat source. The secondary zone involves frictional contact between the chip and the tool's rake face, causing wear on the rake face due to the friction heat source [60].

To calculate the total cutting heat source intensity ($Q$), which accounts for both the shear and friction heat sources, Smith's orthogonal cutting heat source intensity definition can be employed. The formula for the cutting heat sources is as follows [61, 62]:

$$\begin{cases} Q = Q_{shear} + Q_{friction} \\ Q_{shear} = \frac{F_s v_s}{L_s w} = \frac{Pv\cos(\varphi+\beta-\gamma)\cos\gamma}{L_s w\sin(90^\circ+\gamma-\varphi)} \\ Q_{friction} = \frac{F_f v_c}{L_c w} = \frac{Pv\sin\beta\sin\varphi}{L_c w\sin(90^\circ+\gamma-\varphi)} \end{cases} \tag{4.1}$$

Equation (4.1) demonstrates that both $Q_{shear}$ and $Q_{friction}$ are contingent upon the cutting resultant force ($P$) and the cutting velocity ($v$), while keeping other parameters constant. Based on the cutting force signal diagram shown in Fig. 4.17, it is clear that the value of $P$ remains rather stable throughout the cutting operation. Thus, $P$ may be regarded as a constant, resulting in the cutting heat source being predominantly dependent on the cutting velocity ($v$). For the purpose of analyzing the relationship between cutting velocity and the cutting heat source, four spots were selected along

**Table 4.3** Comparisons of different non-traditional machining processes

| Machining process | Capital investment | Toolings/fixtures | Power requirements | Tool wear | Machining efficiency | Structure dimensions | Surface quality |
|---|---|---|---|---|---|---|---|
| Laser machining | Medium | Low | Low | Very low | High | Middle | Low |
| Ultrasonic machining | Low | Low | Low | Medium | Low | Small | High |
| Electrical discharge | Medium | High | Low | High | Medium | Middle | Medium |
| Electro-chemical machining | Very high | Medium | Medium | Very low | Medium | Large | Medium |
| Electron beam welding | Very high | High | Medium | Very low | Very low | Small | Medium |
| Abrasive water jet machining | High | Low | Medium | Very low | High | Large | Medium |
| Friction welding | Low | Low | Low | Very low | Low | Small | Low |

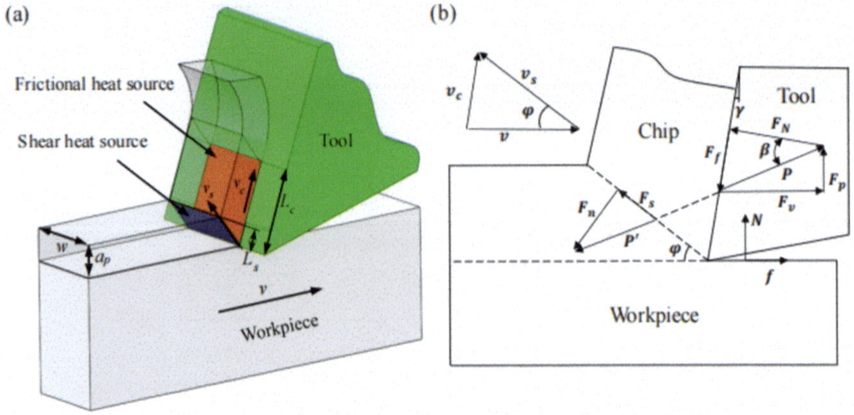


**Fig. 4.16** Schematic diagram of **a** cutting force components and **b** material flows in orthogonal cutting

the *X*-axis of the workpiece surface. These points, called A, B, C, and D, are evenly spaced and may be seen in Fig. 4.18.

For simulating the cutting process, the MD simulation software LAMMPS was used [63]. The cutting model is shown in Fig. 4.19 [64–66]. The diamond tool was configured to be rigid, with a rake angle of 0°, a clearance angle of 15°, and a tool nose radius of 2 nm. The workpiece had dimensions of 35 nm × 14 nm × 2.5 nm. To mitigate size effects due to the simulation scale, a periodic boundary condition was applied in the *Z*-direction. Although the temperature and energy of the system changed continuously during the cutting process, the volume remained relatively stable. Therefore, the NVE was chosen to simulate the real environment, with a time step of 1 fs.

**Fig. 4.17** The cutting force feature in the SPDT process

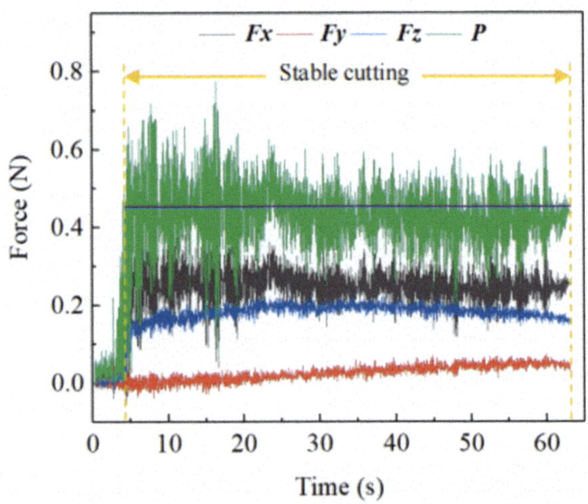

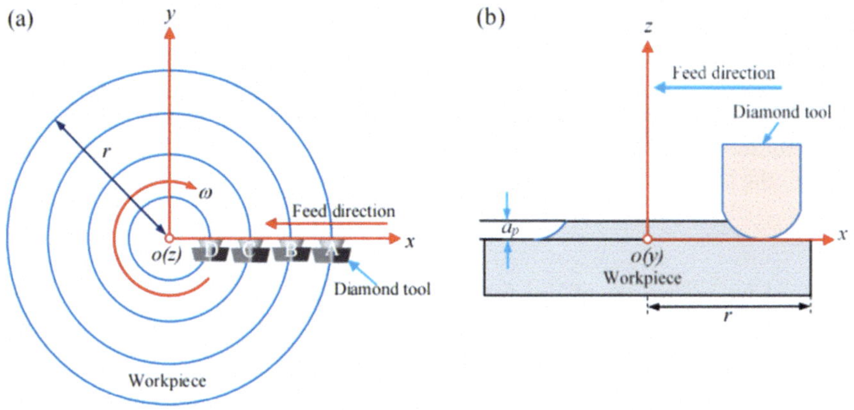


**Fig. 4.18** Schematic of SPDT: **a** top view and **b** side view

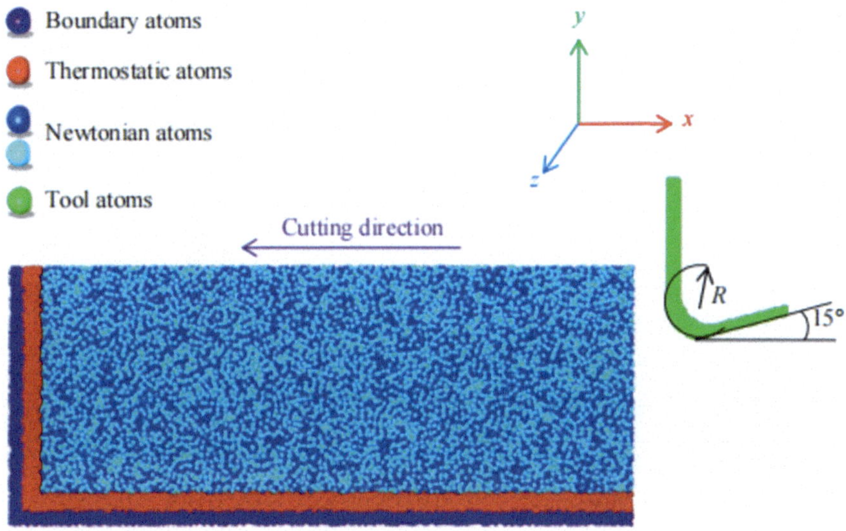


**Fig. 4.19** Schematic of MD simulation model

A series of cutting velocities were established at 20, 40, 60, and 80 m per second in the *X*-axis direction in order to investigate the impact that cutting speeds had on the temperature of the workpiece. The cutting distance was kept at a constant 30 nm throughout the process. Within the context of the cutting simulation, the atomic temperature was determined from the transformation of kinetic energy, as indicated in Eq. (4.2).

$$\frac{1}{2}\sum_{i} m_i v_i^2 = \frac{3}{2} N k_b T \tag{4.2}$$

where $N$ represents the number of atoms; $v_i$ denotes the velocity of the ithatom; $k_b$ stands for the Boltzmann constant, which is equal to $1.3806503 \times 10^{-23}$ J/K; and $T$ refers to the temperature of the atom.

### 4.2.2 Visualization of Surface Generation

Figure 4.20 displays the temperature distribution in the workpiece, with the chip action area as the focal point. It is evident that the highest temperature is observed at the chip, followed by a zone of elevated temperature in the friction region of the tool flank. Furthermore, both the workpiece's subsurface and machined surface exhibit relatively high temperatures. Conversely, the temperature near the boundary inside the workpiece appears to be lower, demonstrating a significant temperature gradient.

The remarkable rise in temperature at the chip is attributed to the substantial deformation occurring during the cutting process. This area experiences maximum deformation energy, consequently influencing the atomic energy and, in turn, the temperature. Similarly, in the friction zone of the workpiece, significant compression

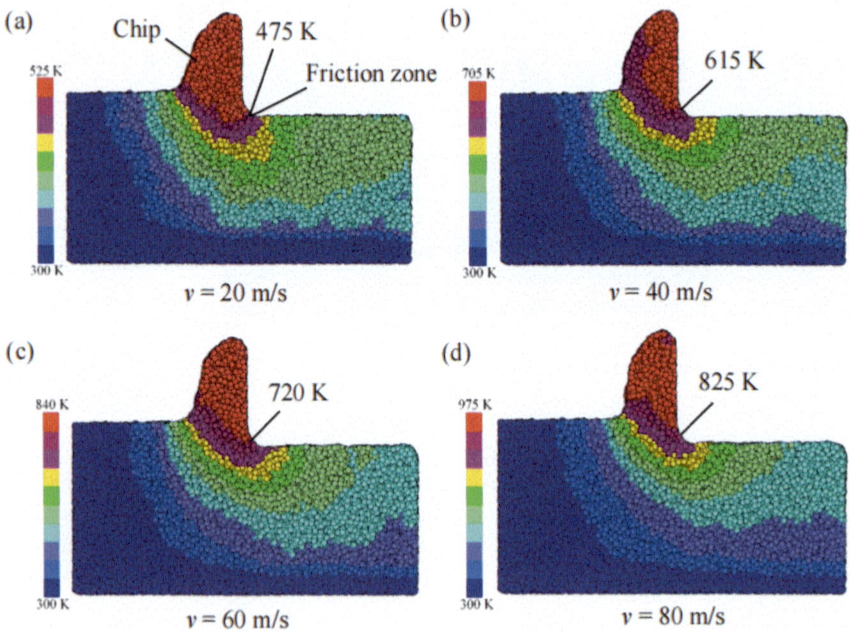


**Fig. 4.20** Temperature distribution evolution at various cutting speeds: **a** 20 m/s, **b** 40 m/s, **c** 60 m/s, and **d** 80 m/s

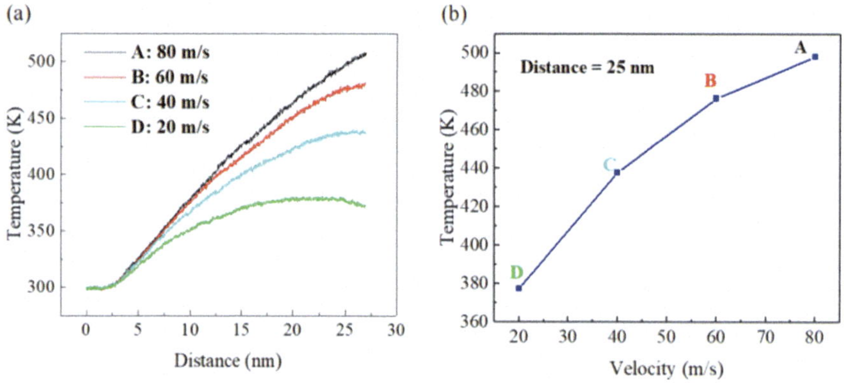


**Fig. 4.21** **a** Variation in the temperature of the AAs at different cutting speeds as the cutting distance increases and **b** the relation between AAs temperature and cutting speed at a cutting distance of 25 nm

due to the tool results in substantial atom deformation and corresponding energy build-up, leading to high temperatures in this specific region.

For a cutting velocity of 20 m/s, the temperature within the friction zone reaches 475 K, which is well below the $T_g = 670$ K of AAs. Subsequently, as the cutting velocity increases to the range of 20–80 m/s, the AAs temperature in the friction zone escalates to 720 K while remaining below the initial $T_x = 720$ K. Upon further increasing the cutting velocity to 80 m/s, the temperature in the flank's friction zone reaches 825 K. Consequently, the workpiece surpasses the $T_x$ threshold, resulting in the commencement of AAs crystallization.

Figure 4.21a depicts the temperature variation of the workpiece with cutting distance at different cutting velocities, while Fig. 4.21b illustrates the relationship between workpiece temperature and various cutting velocities at a fixed cutting distance of 25 nm. It is evident that cutting velocity plays a significant role in determining the workpiece temperature. As the cutting velocity increases, the workpiece temperature exhibits a corresponding rise, and the curve slope becomes steeper, signifying an accelerated temperature increase.

Initially, when the tool has not yet made contact with the workpiece, the temperature remains constant. However, once the tool comes into contact, a rapid temperature surge is observed. This temperature surge is attributed to the heat generated in the cutting area, which intensifies with an increase in cutting velocity. If the heat is not adequately diffused, the temperature within the cutting area exceeds the $T_g$, rendering the area viscous. Consequently, additional friction arises between the tool and the workpiece, leading to heightened temperatures on the workpiece surface. Sustained elevation of the workpiece temperature beyond $T_g$ can result in the generation of surface oxide.

## References

1. Inoue A, Takeuchi A (2011) Recent development and application products of bulk glassy alloys. Acta Mater 59(6):2243–2267
2. Li N, Chen W, Liu L (2016) Thermoplastic micro-forming of bulk metallic glasses: a review. Jom 68(4):1246–1261
3. Ismagilov FR, Papini L, Vavilov VE, Gusakov DV (2019) Design and performance of a high-speed permanent magnet generator with amorphous alloy magnetic core for aerospace applications. IEEE Trans Ind Electron 67(3):1750–1758
4. Khan MM, Deen KM, Haider W (2019) Combinatorial development and assessment of a Zr-based metallic glass for prospective biomedical applications. J Non-Cryst Solids 523:119544
5. Jabed A, Khan MM, Camiller J, Greenlee-Wacker M, Haider W, Shabib I (2019) Property optimization of Zr-Ti-X (X = Ag, Al) metallic glass via combinatorial development aimed at prospective biomedical application. Surf Coat Technol 372:278–287
6. Fujita K, Morishita Y, Nishiyama N, Kimura H, Inoue A (2005) Cutting characteristics of bulk metallic glass. Mater Trans 46(12):2856–2863
7. Chen X, Xiao J, Zhu Y, Tian R, Shu X, Xu J (2017) Micro-machinability of bulk metallic glass in ultra-precision cutting. Mater Des 136:1–12
8. Kumar P, Kumar M, Bajpai V, Singh NK (2017) Recent advances in characterization, modeling and control of burr formation in micro-milling. Manuf Lett 13:1–5
9. Trexler MM, Thadhani NN (2010) Mechanical properties of bulk metallic glasses. Prog Mater Sci 55(8):759–839
10. Wang X, Pawar G, Li Y, Ren X, Zhang M, Lu B, Banerjee A, Liu P, Dufek EJ, Zhang J-G et al (2020) Glassy Li metal anode for high-performance rechargeable Li batteries. Nat Mater 19(12):1339–1345
11. Wang Z-J, Li M-X, Yu J-H, Ge X-B, Liu Y-H, Wang W-H (2020) Low-iridium content IrNiTa metallic glass films as intrinsically active catalysts for hydrogen evolution reaction. Adv Mater 32(4):1906384
12. Jia Z, Wang Q, Sun L, Wang Q, Zhang L-C, Wu G, Luan J-H, Jiao Z-B, Wang A, Liang S-X et al (2019) Attractive in situ self-reconstructed hierarchical gradient structure of metallic glass for high efficiency and remarkable stability in catalytic performance. Adv Funct Mater 29(19):1807857
13. Hu YC, Wang YZ, Su R, Cao CR, Li F, Sun CW, Yang Y, Guan PF, Ding DW, Wang ZL et al (2016) A highly efficient and self-stabilizing metallic-glass catalyst for electrochemical hydrogen generation. Adv Mater 28(46):10293–10297
14. Guo H, Yan P, Wang Y, Tan J, Zhang Z, Sui M, Ma E (2007) Tensile ductility and necking of metallic glass. Nature Mater 6(10):735–739
15. Tian L, Cheng Y-Q, Shan Z-W, Li J, Wang C-C, Han X-D, Sun J, Ma E (2012) Approaching the ideal elastic limit of metallic glasses. Nature Commun 3(1):1–6
16. Ramachandramoorthy R, Schwiedrzik J, Petho L, Guerra-Nunez C, Frey D, Breguet J-M, Michler J (2019) Dynamic plasticity and failure of microscale glass: rate-dependent ductile–brittle–ductile transition. Nano Lett 19(4):2350–2359
17. Liu M, Cao C, Lu Y, Wang W, Bai H (2017) Flexible amorphous metal films with high stability. Appl Phys Lett 110(3):031901
18. Lin Y-C, Tsai Y-C, Ono T, Liu P, Esashi M, Gessner T, Chen M (2015) Metallic glass as a mechanical material for microscanners. Adv Funct Mater 25(35):5677–5682
19. An BW, Gwak E-J, Kim K, Kim Y-C, Jang J, Kim J-Y, Park J-U (2016) Stretchable, transparent electrodes as wearable heaters using nanotrough networks of metallic glasses with superior mechanical properties and thermal stability. Nano Lett 16(1):471–478
20. Valant M, Luin U, Fanetti M, Mavrič A, Vyshniakova K, Siketić Z, Kalin M (2016) Fully transparent nanocomposite coating with an amorphous alumina matrix and exceptional wear and scratch resistance. Adv Funct Mater 26(24):4362–4369

21. Mieszala M, Hasegawa M, Guillonneau G, Bauer J, Raghavan R, Frantz C, Kraft O, Mischler S, Michler J, Philippe L (2017) Micromechanics of amorphous metal/polymer hybrid structures with 3D cellular architectures: size effects, buckling behavior, and energy absorption capability. Small 13(8):1602514
22. Schlegel A, Wachter P, Ackermann K, Liard M, Güntherodt H-J (1979) Optical properties of metallic glasses. Solid State Commun 31(5):373–376
23. Yang C, Zhang C, Liu L (2017) Tuning colors in Zr-based thin film metallic glasses. J Alloys Compd 728:289–294
24. Liu X, Shao Y, Han Z, Yao K (2015) Morphology and structure evolution of metallic nanowire arrays prepared by die nanoimprinting. Sci Bull 60(6):629–633
25. Wang H, Li J, Li K, Lin Y, Chen J, Gao L, Nicolosi V, Xiao X, Lee J-M (2021) Transition metal nitrides for electrochemical energy applications. Chem Soc Rev 50(2):1354–1390
26. Na JH, Han KH, Garrett GR, Launey ME, Demetriou MD, Johnson WL (2019) Designing color in metallic glass. Sci Rep 9(1):1–8
27. Hua N, Huang L, Chen W, He W, Zhang T (2014) Biocompatible Ni-free Zr-based bulk metallic glasses with high-Zr-content: compositional optimization for potential biomedical applications. Mater Sci Eng C 44:400–410
28. Meagher P, O'Cearbhaill ED, Byrne JH, Browne DJ (2016) Bulk metallic glasses for implantable medical devices and surgical tools. Adv Mater 28(27):5755–5762
29. Cai C-N, Zhang C, Sun Y-S, Huang H-H, Yang C, Liu L (2017) ZrCuFeAlAg thin film metallic glass for potential dental applications. Intermetallics 86:80–87
30. Wang J, Wang L, Feng J, Tang C, Sun X, Peng H (2021) Long-term in vivo monitoring of chemicals with fiber sensors. Adv Fiber Mater 3(1):47–58
31. Ming WY, Guo XD, Xu YJ, Zhang GJ, Jiang ZW, Li YZ, Li XK (2023) Progress in non-traditional machining of amorphous alloys. Ceram Int 49(2):1585–1604
32. Jun WK, Willens RH, Duwez P (1960) Non-crystalline structure in solidified gold-silicon alloys. Nature 187:869–870
33. Zhang L, Huang H (2019) Micro machining of bulk metallic glasses: a review. Int J Adv Manu Technol 100(1–4):637–661
34. Wang WH, Dong C, Shek CH (2004) Bulk metallic glasses. Mater Sci Eng R Rep 44:45–89
35. Shen Y, Li Y, Chen C, Tsai H-L (2017) 3D printing of large, complex metallic glass structures. Mater Des 117:213–222
36. Zhang C, Wang W, Li Y-C, Yang Y-G, Wu Y, Liu L (2018) 3D printing of Fe-based bulk metallic glasses and composites with large dimensions and enhanced toughness by thermal spraying. J Mater Chem A 6:6800–6805
37. Pan C-T, Wu T-T, Liu Y-T, Yamagata Y, Huang JC (2009) Fabrication of aspheric surface using ultraprecision cutting and BMG molding. J Mater Process Technol 209:5014–5023
38. Lee CJ, Lin HK, Sun SY, Huang JC (2010) Characteristic difference between ITO/ZrCu and ITO/Ag bi-layer films as transparent electrodes deposited on PET substrate. Appl Surf Sci 257:239–243
39. Chu JP, Jang JSC, Huang JC, Chou HS, Yang Y, Ye JC, Wang YC, Lee JW, Liu FX, Liaw PK et al (2012) Thin film metallic glasses: Unique properties and potential applications. Thin Solid Films 520:5097–5122
40. Fujita K, Morishita Y, Nishiyama N, Kimura H, Inoue A (2005) Cutting characteristics of bulk metallic glass. Mater Trans 46:2856–2863
41. Han DX, Wang G, Li J, Chan KC, To S, Wu FF, Gao YL, Zhai QJ (2015) Cutting characteristics of Zr-based bulk metallic glass. J Mater Sci Technol 31:153–158
42. Nishiyama N, Horino M, Haruyama O, Inoue A (2001) Abrupt change in heat capacity of supercooled Pd–Cu–Ni–P melt during continuous cooling. Mater Sci Eng A 304–306:683–686
43. Drehman AJ, Greer AL, Turnbull D (1982) Bulk formation of a metallic glass: Pd40Ni40P20. Appl Phys Lett 41:716–717
44. Hu X, Ng SC, Feng YP, Li Y (2001) Cooling-rate dependence of the density of Pd40Ni10Cu30P20 bulk metallic glass. Phys Rev B 64

45. Schroers J, Johnson WL, Busch R (2000) Crystallization kinetics of the bulk-glass-forming Pd43Ni10Cu27P20 melt. Appl Phys Lett 77:1158–1160
46. Schroers J, Johnson WL (2004) Dispersion technique studies on Pd43Ni10Cu27P20. Mater Sci Eng A 375–377:781–784
47. Dongol R, Wang L, Cormack AN, Sundaram SK (2018) Molecular dynamics simulation of sodium aluminosilicate glass structures and glass surface-water reactions using the reactive force field (ReaxFF). Appl Surf Sci 439:1103–1110
48. Lee W (2013) An overview of ultra-precision diamond machining of microstructured freeform surfaces. J Mech Eng 49
49. Taniguchi N (1983) Current status in, and future trends of, ultraprecision machining and ultrafine materials processing. CIRP Ann 32:573–582
50. Bryan JB (1979) Design and construction of an ultraprecision 84 inch diamond turning machine. Precis Eng 1:13–17
51. Nishiyama N, Inoue A (1999) Supercooling investigation and critical cooling rate for glass formation in Pd–Cu–Ni–P alloy. Acta Mater 5:1487–1495
52. Kumar G, Schroers J (2008) Write and erase mechanisms for bulk metallic glass. Appl Phys Lett 92:031901
53. Lewandowski JJ, Wang WH, Greer AL (2005) Intrinsic plasticity or brittleness of metallic glasses. Philos Mag Lett 85:77–87
54. Narasimhan R, Tandaiya P, Singh I, Narayan RL, Ramamurty U (2015) Fracture in metallic glasses: mechanics and mechanisms. Int J Fract 191:53–75
55. Narayan RL, Tandaiya P, Narasimhan R, Ramamurty U (2014) Wallner lines, crack velocity and mechanisms of crack nucleation and growth in a brittle bulk metallic glass. Acta Mater 80:407–420
56. Madge S (2015) Toughness of bulk metallic glasses. Metals 5:1279–1305
57. Wang C, Cao QP, Wang XD, Zhang DX, Ramamurty U, Narayan RL, Jiang J-Z (2017) Intermediate temperature brittleness in metallic glasses. Adv Mater 29
58. Raut D, Narayan RL, Tandaiya P, Ramamurty U (2018) Temperature-dependence of mode I fracture toughness of a bulk metallic glass. Acta Mater 144:325–336
59. Komanduri R (1993) Machining and grinding: A historical review of the classical papers. Appl Mech Rev 46:80–132
60. Usui E, Shirakashi T, Kitagawa T (1978) Analytical prediction of three-dimensional cutting process—part 3: cutting temperature and crater wear of carbide tool. J Eng Ind 100:236–243
61. Shan C, Zhang X, Shen B, Zhang D (2019) An improved analytical model of cutting temperature in orthogonal cutting of Ti6Al4V. Chin J Aeronaut 32:759–769
62. Hu C, Zhuang K, Weng J, Pu D (2019) Three-dimensional analytical modeling of cutting temperature for round insert considering semi-infinite boundary and non-uniform heat partition. Int J Mech Sci 155:536–553
63. Plimpton S (1995) Fast parallel algorithms for short-range molecular dynamics. J Comput Phys 117:1–19
64. Wu CD, Fang TH, Su JK (2017) Nanometric mechanical cutting of metallic glass investigated using atomistic simulation. Appl Surf Sci 396:319–326
65. Zhu PZ, Qiu C, Fang FZ, Yuan DD, Shen X-C (2014) Molecular dynamics simulations of nanometric cutting mechanisms of amorphous alloy. Appl Surf Sci 317:432–442
66. Qiu C, Zhu P, Fang F, Yuan D, Shen X (2014) Study of nanoindentation behavior of amorphous alloy using molecular dynamics. Appl Surf Sci 305:101–110

# Chapter 5
# Tool Wear Mechanism and Suppression in Machining Ferrous Materials

**Abstract** Ferrous materials diamond cutting poses significant challenges due to the high wear rates of the tools. Understanding the underlying mechanisms of tool wear and developing effective suppression methods are critical for enhancing tool life and machining efficiency. This chapter examines the electrochemical mechanisms responsible for tool wear and explores various strategies to mitigate these effects. By understanding the electrochemical interactions and leveraging advanced tool wear suppression techniques, it is possible to enhance the performance and longevity of diamond tools, ultimately leading to more efficient and cost-effective machining processes.

**Keywords** Tool wear mechanism · Single-point diamond turning · Tool wear suppression · Ferrous materials

## 5.1 Electrochemical Tool Wear Mechanism

### *5.1.1 Electrochemical Tool Wear Rate*

Ferrous materials encompass a range of alloys such as chromium, manganese, and iron, among others, which are widely utilized in diverse industrial sectors [1, 2]. In the production of ferrous material parts, achieving complex shapes and high surface quality typically necessitates grinding and polishing, which are associated with prolonged time consuming [3]. Therefore, the need to explore more efficient alternatives, specifically SPDT, to enhance machining precision and efficiency becomes imperative [4].

The tools employed in SPDT are typically SCD that can be honed to achieve nano-level cutting edges, offering a distinct advantage over other tool materials by enabling the creation of nanometer-scale cutting layers [5]. However, it should be noted that once the diamond tool loses its sharp cutting edge, it becomes incapable of achieving nano-level surface roughness. Consequently, diamond tool wear (DTW)

 
G. Zhang and J. Wang, *Material-Oriented Cutting Processes in Precision Machining*, Springer Tracts in Mechanical Engineering,
https://doi.org/10.1007/978-981-96-2504-8_5

emerges as a critical challenge in SPDT. Despite the substantial mechanical resistance, brittleness, and chemical stability of diamond tools, wear is induced by the electrochemical reaction between diamond and Fe, which softens the diamond's hardness and accelerates tool wear.

Figure 5.1 displays the DTW of SCD tools in turning silicon, with a turning distance of 24 km, and steel materials, with a turning distance below 1 km [6, 7]. Notably, tool wear cannot be disregarded. High phosphorus electroless nickel, with a phosphorous concentration of 10–11%, represents a classic diamond-turnable material. Even after turning distances exceeding 20 km, the diamond tools retain the ability to machine surfaces with Ra below 10 nm, exhibiting minimal DTW. Following 200 km of turning, the flank wear width remains less than 5 μm, and the machined surface's Ra approximately reaches 25 nm [8, 9]. A comparison of tool wear and surface roughness for these materials is presented in Table 5.1, which elucidates the catastrophic nature of SCD tool wear when turning ferrous materials.

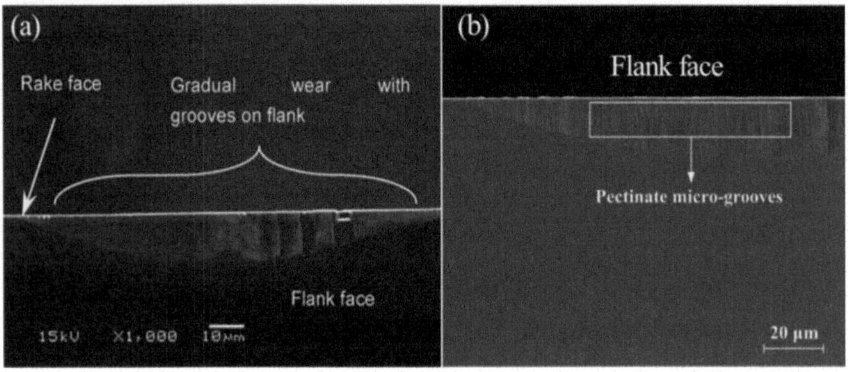


**Fig. 5.1** SEM images of diamond tools wear after turning **a** silicon materials for 24 km and **b** stainless steel for less than 1 km [6, 7]

**Table 5.1** Cutting performance comparison of diamond turning various materials [6, 9–11]

| Materials | Diamond turning properties | Turning distance | Maximum flank wear (μm) |
|---|---|---|---|
| High phosphorus electroless nickel | Diamond-turnable | 200 km | < 5 |
| Silicon | Diamond-turnable material, whereas the tool wear should not be ignored | 24 km | ≈ 20 |
| Stainless steel | Non-diamond-turnable | 250 m | > 20 |

In the context of ductile turning of silicon and non-diamond-turnable steel, SEM images in Fig. 5.1 illustrate the tool wear behavior of SCD tools [6, 7]. Even when dealing with diamond-turnable materials like silicon, tool wear is a significant factor. High phosphorus electroless nickel, like copper and aluminum alloy, known for its diamond-turnable properties, exhibits minimal wear even after extensive use. The diamond tools can maintain precision machining capabilities for mirror-level surfaces with very low roughness [8, 9]. The comparison presented in Table 5.1 highlights the substantial tool wear observed when turning ferrous materials using SCD tools.

Since the 1960s, significant attention has been focused on developing DTW suppression schemes employed in machining ferrous materials. However, due to cost limitations and methodological constraints, there is still ample room for progress in the field of ferrous material machining. Therefore, conducting in-depth research on the DTW mechanism when turning ferrous materials and advancing the development of effective wear suppression methods remain crucial aspects of SPDT.

Diamond tools used in turning undergo various types of mechanical wear, including adhesive, fatigue, and abrasive wear [8, 12–14]. However, due to the inherent hardness of diamonds, these physical wear factors do not significantly contribute to the overall DTW when machining ferrous materials. Instead, the main wear mechanism observed in this context is electrochemical wear.

### *5.1.2 Graphitization Wear*

Quantitative analysis to determine the proportion of different wear categories in overall diamond tool wear during turning is currently lacking. However, extensive experimental data and analysis indicate that the catastrophic DTW when turning ferrous materials is primarily attributable to chemical wear [15, 16].

Researchers Komanduri and Shaw first found that when grinding low carbon steel with diamonds, they observed a comb-like groove wear pattern on the diamond surface, as illustrated in Fig. 5.1b [13]. This wear phenomenon was attributed to diamond graphitization, as the hardness of diamonds prevents low carbon steel from leaving marks on the surface. Despite diamonds being stable at room temperature, they can undergo a transformation into graphite under specific conditions. The increase in temperature in the cutting zone during the grinding process, coupled with shear stress, promotes the graphitization process. Abrasive wear then removes the graphite, further accelerating this transformation.

Thornton and Wilks expanded upon the hypothesis of graphitization, stating that the temperature required for diamond graphitization in a vacuum was around 1800 K [17]. However, this temperature decreases to 1000 and 1100 K when an iron catalyst is present [18]. The authors also introduced the concept of "Flash Temperature" to explain the localized high temperature in the turning area. This high temperature induces diamond graphitization. They further observed that changes in turning parameters can lead to graphitization even at lower temperatures. Additionally, the appearance of a clean surface further facilitates wear [15, 18]. Zhang et al. successfully reduced diamond tool wear by implementing ultrasonic vibration to achieve rapid oxidation of the workpiece's clean surface during the machining process [19]. This experiment further confirmed the catalytic role played by iron in diamond tool wear.

In addition, Narulkar et al. employed MD to investigate the diamond-to-graphite transformation. They found the time consume (nano or picoseconds) of diamond graphitization with the iron exiting. Subsequently, graphite infiltrates the surrounding iron [20]. Guo et al. suggested that the Fe atom's suspended *d*-orbital is oriented at an angle of 30°–36° relative to the nucleus surface causing the diamond graphitization [21, 22].

Currently, two explanations exist to describe the mechanism of diamond graphitization. The first explanation suggests that the presence of iron decreases the graphitization temperature of diamond. As a result, the diamond directly undergoes graphitization under the catalytic action of iron; the second explanation proposes that iron and carbon atoms combine to form unstable FeC, which subsequently transforms into graphite.

### *5.1.3 Oxidative Wear*

Hitchiner and Wilks proposed that when machining ferrous materials using diamond tools, the interaction between iron and carbon results in the formation of iron-carbon compounds. The elevated temperatures and graphitization process further facilitate the creation of these compounds. X-ray photoelectron spectroscopy analyses of iron phase changes reveal that diamond shows a stronger preference for oxygen. As a result, there is a tendency for the formation of Fe–C–O compounds [23].

Tanaka et al. investigated the wear mechanism of diamond and identified a dual wear mechanism. Through experiments to simulate wear processes, they found two distinct wear mechanisms in DTW. Above 1000 K, a catalytic reaction is triggered, leading to an increase in wear rate with temperature elevation. Graphitization and diffusion are the primary wear mechanisms under these conditions. Below or close to 900 K, the diamond wear rate is considered to be lower. Oxidation processes, on the other hand, become the predominant wear mechanism when the oxygen partial pressure is increased [24].

A predictive model for chemical reactions was developed by Thornton and Wilks, they proposed that the wear process involves the participation of at least five different types of chemical bonding, as illustrated in Fig. 5.2 [25].

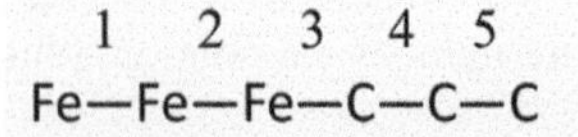


**Fig. 5.2** Various forms of chemical bonding in the diamond wear process

Atomic bonding within the iron lattice corresponds to bonding type 1, while the interfacial contact layer is associated with bonding type 3, and the diamond lattice is characterized by bonding type 5. Conversely, active chemical bonding within the interfacial contact layer is represented by bonding types 2 and 4. The strength and nature of these bonds (types 2, 3, and 4) play a vital role in regulating any chemical reactions that take place at the boundary layer.

In oxidative wear, the involved chemical reactions are complex, and they occur within a small span of time and space. Therefore, it becomes difficult to accurately determine the precise chemical formula of the compounds involved. Significant research was carried out by Zou et al. in this area, where they explained the fundamental process using MD simulations [7]. Using the Eq. (5.1), they were able to uncover the possible chemical reactions that may take place during oxidative wear. These reactions varied depending on the temperature at which the turning was taking place. According to the information shown in Table 5.2, the Gibbs free energy criterion exhibits a variety of behaviors at different temperatures.

Therefore, the DTW is influenced by the combined action of multiple wear mechanisms. Among these mechanisms, chemical wear, particularly graphitization, plays a significant role. Once diamond undergoes graphitization, the high affinity between iron and carbon, along with elevated interface temperatures, facilitates the rapid diffusion of the graphite formed into the iron under the influence of shear stress.

**Table 5.2** Evolution of Gibbs free energy (kJ/mol) following temperatures [7]

| *T/K* | | | | | | | | |
|---|---|---|---|---|---|---|---|---|
| No | 298 | 400 | 600 | 800 | 1000 | 1200 | 1400 | 1600 |
| A1 | – 396.06 | – 396.94 | – 398.66 | – 400.38 | – 402.10 | – 403.82 | – 405.54 | – 407.26 |
| B1 | – 277.68 | – 297.08 | – 335.12 | – 373.16 | – 411.20 | – 449.24 | – 487.28 | – 525.32 |
| C1 | – 501.98 | – 487.60 | – 459.40 | – 431.20 | – 403.00 | – 374.80 | – 346.60 | – 318.40 |
| D1 | – 1484.50 | – 1428.40 | – 1318.40 | – 1208.10 | – 1098.40 | – 988.40 | – 878.40 | – 768.40 |
| E1 | – 1015.32 | – 980.04 | – 910.86 | – 841.68 | – 772.50 | – 703.32 | – 634.14 | – 564.96 |
| A2 | 105.92 | 90.66 | 60.74 | 30.82 | 0.90 | – 29.02 | – 58.94 | – 88.86 |
| B2 | 296.31 | 237.58 | 122.42 | 7.26 | – 107.90 | – 233.06 | – 338.22 | – 453.38 |
| C2 | 223.20 | 186.16 | 113.54 | 40.92 | – 31.7 | – 104.32 | – 176.94 | – 249.56 |
| D2 | – 6.23 | – 4.60 | – 1.40 | 1.80 | 5.00 | 8.20 | 11.40 | 14.60 |
| E2 | – 29.42 | – 32.55 | – 34.10 | – 37.20 | – 40.30 | – 43.40 | – 46.50 | – 49.60 |
| F2 | – 13.57 | – 13.56 | – 13.54 | – 13.52 | – 13.50 | – 13.48 | – 13.46 | – 13.44 |

$$
\begin{cases}
\textit{Equation } A1: \mathrm{C} + \mathrm{O_2} = \mathrm{CO_2} \\
\textit{Equation } B1: \mathrm{2C} + \mathrm{O_2} = \mathrm{2CO} \\
\textit{Equation } C1: \mathrm{2Fe} + \mathrm{O_2} = \mathrm{2FeO} \\
\textit{Equation } D1: \mathrm{4Fe} + \mathrm{3O_2} = \mathrm{2Fe_2O_3} \\
\textit{Equation } E1: \mathrm{3Fe} + \mathrm{2O_2} = \mathrm{Fe_3}O_4 \\
\textit{Equation } A2: \mathrm{C} + \mathrm{2FeO} = \mathrm{2Fe} + \mathrm{CO_2} \\
\textit{Equation } B2: \mathrm{3C} + \mathrm{2Fe_2O_3} = \mathrm{4Fe} + \mathrm{3CO_2} \\
\textit{Equation } C2: \mathrm{2C} + \mathrm{Fe_3O_4} = \mathrm{3Fe} + \mathrm{2CO_2} \\
\textit{Equation } D2: \mathrm{CO} + \mathrm{FeO} = \mathrm{Fe} + \mathrm{CO_2} \\
\textit{Equation } E2: \mathrm{3CO} + \mathrm{Fe_2O_3} = \mathrm{2Fe} + \mathrm{3CO_2} \\
\textit{Equation } F2: \mathrm{4CO} + \mathrm{Fe_3O_4} = \mathrm{3Fe} + \mathrm{4CO_2}
\end{cases}
\tag{5.1}
$$

Experimental findings have demonstrated that the DTW is more severe when turning soft, low-carbon steel compared to hard, high-carbon steel. This difference in wear severity may be attributed to the varying diffusion forces influenced by the carbon content in the materials. Indirectly, this shows that oxidative wear is not the major cause of catastrophic diamond wear, which is an important proposition to consider. There are several reasons supporting this conclusion:

Reason 1: The close contact between the flank face of the diamond tool and the workpiece during machining hinders the penetration of the external gas atmosphere into the cutting area, thereby impeding most oxygen-dependent chemical reactions.

Reason 2: Eq. (5.1) reveals that all oxidative wear reactions involving oxygen. Researchers found that conducting turning experiments under $N_2$ or He gases actually increased diamond tool wear, indicating that oxidative wear is not the primary factor driving diamond tool deterioration.

As a result, oxygen plays a dual role in the graphitization process of diamonds. It can lead to diamond deterioration by reacting with the diamond at specific temperatures, and it can also reduce the catalytic effect of iron by reacting with it, thus inhibiting diamond graphitization. Table 5.3 summarizes the common forms of DTW.

Table 5.3 shows that mechanical wear is represented by types I to IV, while chemical wear is represented by types V and VI. In the realm of SPDT of ferrous materials, type 3 wear is particularly significant. Interestingly, the hardness of diamond prevents the formation of grooved "comb" wear on the diamond surface when machining ferrous materials. Therefore, prior to diamond tool wear, there must be certain alterations that decrease the hardness of diamond. This alteration is mainly attributed to type V wear, specifically graphitization.

Diamond graphitization significantly reduces the hardness of diamond. From an energetic perspective, under the influence of external factors, the stable diamond structure tends to transform into the more stable graphite structure. MD simulations and Raman spectroscopy confirm the theory and experimental evidence of diamond graphitization [7].

Type IV wear is consistently observed in conjunction with diamond tool wear, although it is not the primary factor contributing to the DTW. Type VI wear represents

**Table 5.3** The details of DTW mechanisms [7, 26, 27]

| Wear forms | Type I: Adhesion | Type II: Fatigue | Type III: Abrasive wear |
|---|---|---|---|
| Tool wear for details | 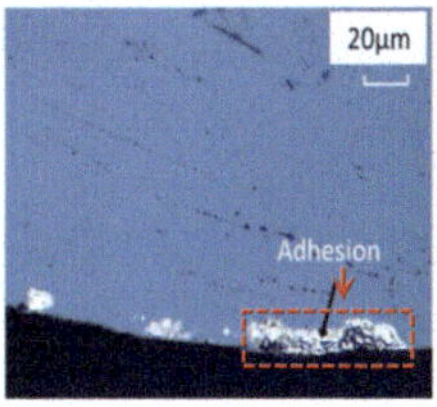 | 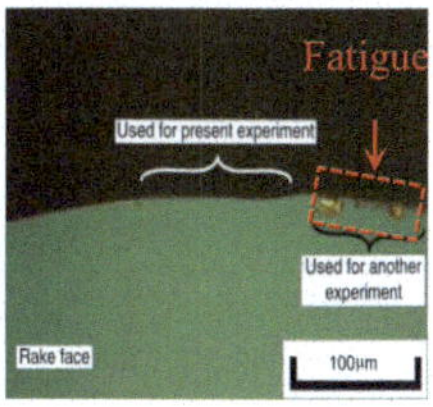 | 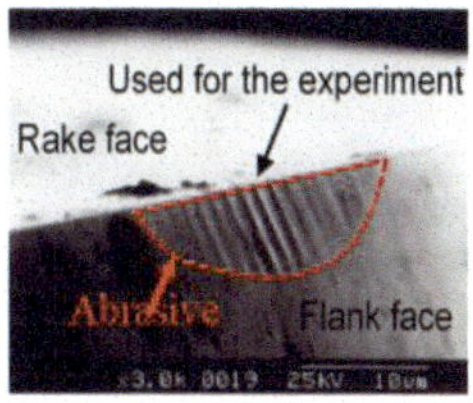 |
| **Wear forms** | **Type IV: Diffusion** | **Type V: Graphitization** | **Type VI: Oxidative wear** |
| Tool wear for details | 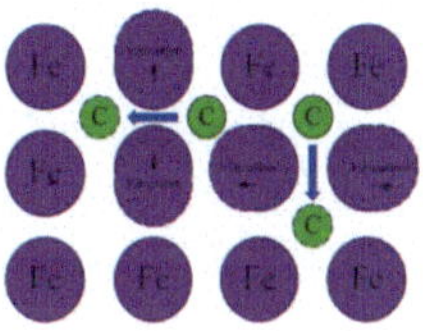 | 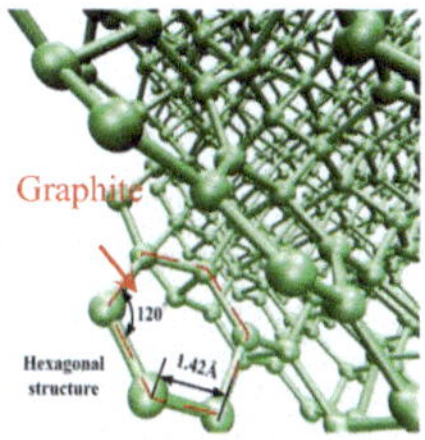 | $C_xO + Fe_aO_b \rightarrow Fe_aO_\beta C_\gamma$ |

the most intricate form due to the unique tribochemical reaction products involved, which differ from standard chemical reactions and encompass a diverse range of reactants and chemical processes in the context of DTW.

In terms of wear types, type I wear usually occurs when machining soft materials such as titanium alloys [28] and aluminum [29]. Type II wear receives more attention when dealing with hard materials like germanium [30] and silicon [31].

## 5.2 Tool Wear Suppression Methods

Since the 1960s, researchers have developed a range of strategies to suppress the phenomenon of diamond graphitization in DTW. This section will review these methods.

### *5.2.1 Effect of Tools, Workpieces and Cutting Parameters*

#### 5.2.1.1 Diamond Tools

The primary crystal planes used in SCD tools are (111), (110), and (100), and their various physical parameters are detailed in Table 5.4. Among these, (111) demonstrates the more higher mechanical properties, as indicated in Table 5.5. However, during actual machining operations, (111) faces challenges due to its minimal friction coefficient with cast iron surfaces. Moreover, the (111) plane has the largest spacing at 0.154 nm, making it more prone to microscopic cleavage and reducing the tool life. As a result, (110) and (100) crystal planes are commonly selected for the tool face [32].

Zong et al. conducted detailed investigations into the crystal plane selection for SCD tools, as depicted in Fig. 5.3. Through experiments involving four different cases of diamond tools, they found that (110) for the rake face, (110) for the flank

**Table 5.4** Physical parameters of different diamond's crystal planes (hexahedron) [32, 34, 35]

| Diamond crystal plane | (111) | (110) | (100) |
|---|---|---|---|
| Young's modulus | 1207 Gpa | 1165 Gpa | 1050 Gpa |
| Fracture surface energy | 5.3 J $m^{-2}$ | | |
| Cleavage angle | 109.5° | 144.73° | 125.27° |
| Relative density of C-atoms | 2.308 | 1.414 | 1 |
| Density of covalent bonds* | $4 \times \frac{3}{\sqrt{3}D^2}$ | $4 \times \frac{1}{\sqrt{2}D^2}$ | $2 \times \frac{2}{D^2}$ |
| Plane spacing | Bigger: 0.154 nm<br>Smaller: 0.051 nm | 0.126 nm | 0.089 nm |
| Relative density of covalent bonds | 3 | 1.22 | 1.73 |

* means that D is the side length of diamond unit cell (hexahedron)

**Table 5.5** Physical properties of different diamond's crystal planes [32, 33]

Micro tensile strength

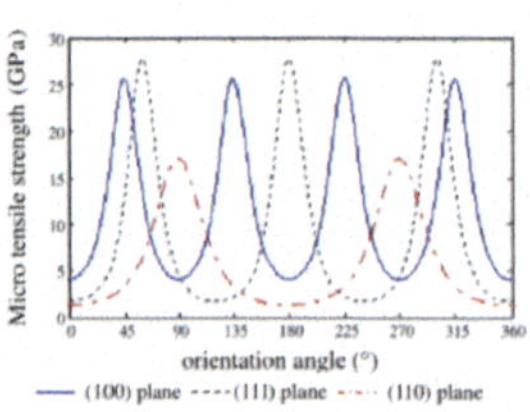


Micro shearing strength

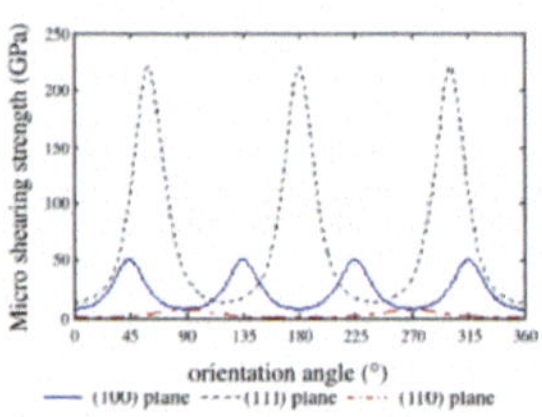


Micro compressive strength

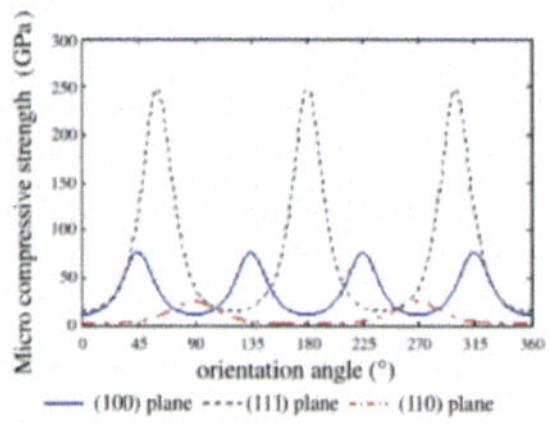


Knoop hardness

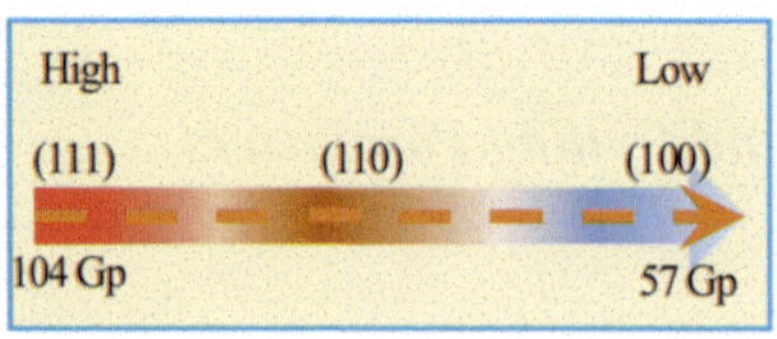


Friction coefficient*

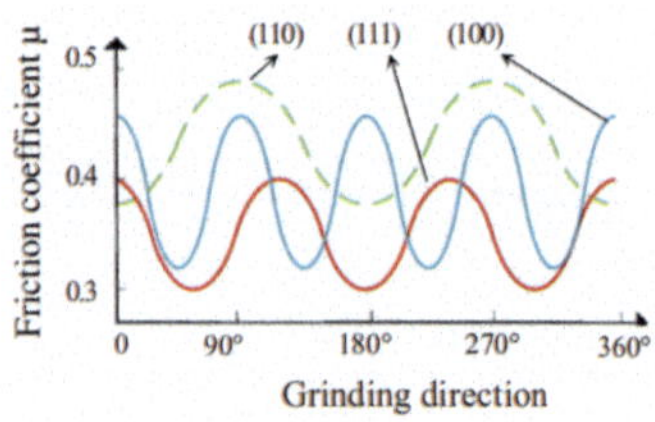


*Friction coefficient between hexahedral diamond crystals and a cast iron plate during the grinding process

face, and (110) for both rake and flank face configurations exhibited good durability [33].

Due to the distinct wear mechanisms observed when using diamond tools for turning ferrous and nonferrous materials, it is essential for the crystal planes of the diamond tool to possess the ability to inhibit diamond graphitization. In their study,

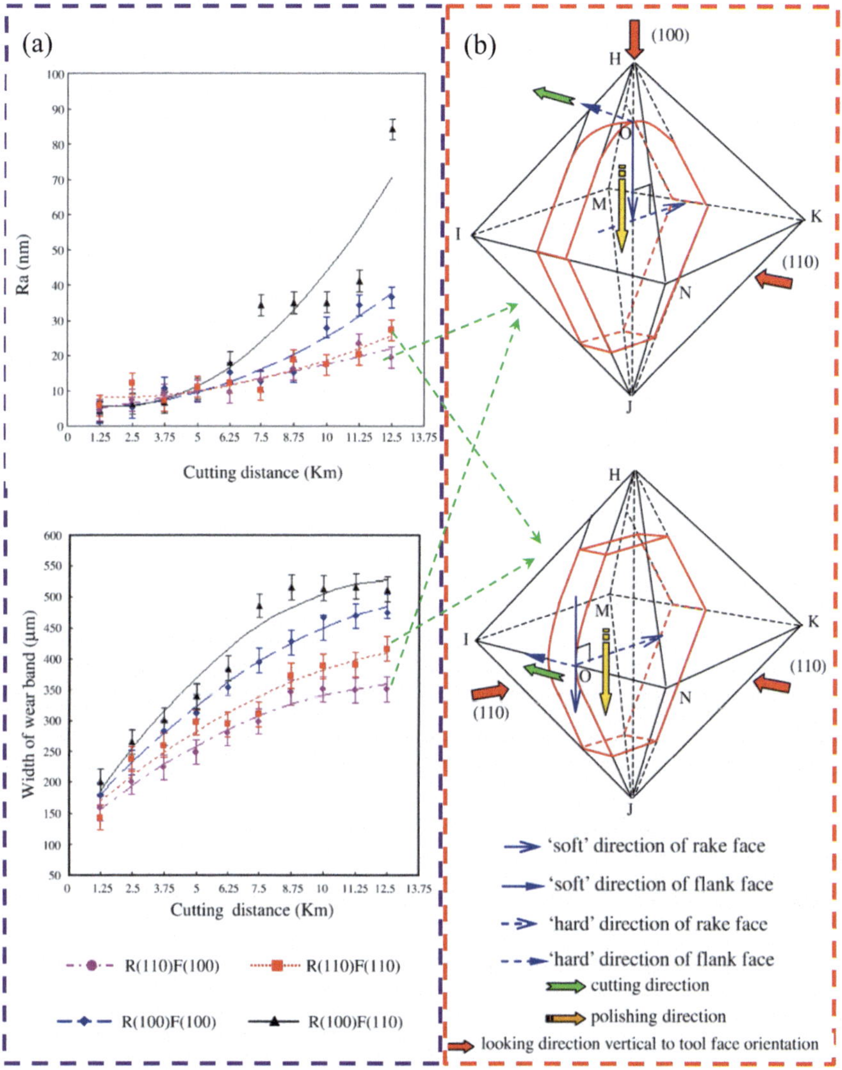


**Fig. 5.3** Evaluation of cutting performance of SCD tools with different crystal plane combinations: **a** variation of Ra changes with increasing cutting distance; **b** variation of wear band width variations as cutting distance increases; **c**, **d** polishing operation applied to diamond tools with varying rake and flank face orientations [33, 34, 40]

Seal et al. investigated the graphitization tendency of diamond crystal planes under high temperatures (1700 °C) in a vacuum [36]. The experimental findings revealed that the (100) crystal plane exhibits the highest ability to inhibit high-temperature graphitization, while the (110) crystal plane demonstrates the least inhibitory effect [20, 36].

However, it should be noted that the actual workpiece during machining is seldom composed of pure iron, and the carbon content tends to significantly influence the machining outcome. As a result, the Tersoff potential function exhibits relatively poor accuracy in modeling covalent structures. Jin et al. performed MD simulations on die steel using the high-precision modified embedding atomic method [37]. The simulation results indicated that as the turning length increases, the (111) exhibits the poorest resistance to diamond graphitization, whereas the (110) outperforms the (100) in terms of graphitization resistance [22, 38]. Nevertheless, the simulation results do not show a significant distinction between the inhibitory effects of the (110) and (100) crystal planes on graphitization, thus warranting further investigation. Bai et al. conducted MD simulations using the Tersoff potential function, employing the sixfold ring method. In their research, they also concluded that the (111) crystal plane is the most prone to graphitization. However, their findings on the (100) crystal plane differ from those of Jin et al., suggesting that the (100) crystal plane possesses the most pronounced inhibitory effect on diamond graphitization [39].

The discrepancy between the results presented in Refs. [29, 30] can be attributed to the utilization of different potential functions within molecular dynamics simulations. Nevertheless, MD simulations did confirm that under the catalysis of iron atoms, graphitization is more likely to occur on the (111) crystal plane of diamond.

The alignment principle [22] provides an explanation for why the diamond (111) crystal plane exhibits the poorest graphitization resistance. As depicted in Fig. 5.4a, the spacing ($2.46 \times 10^{-10}$ nm) between Fe atoms (a, b, c) and that between diamond atoms (1, 3, 5) ($2.51 \times 10^{-10}$ nm) is remarkably close, satisfying the alignment principle. Due to the presence of unpaired electrons, atoms a, b, and c form bonds with atoms 1, 3, and 5, respectively. Simultaneously, the downward movement of carbon atoms causes atoms 1 through 6 to converge into a flat surface, resulting in a graphite structure, as displayed in Fig. 4b, c.

Compared to the previous case, the utilization of diamond tools with the (110) crystal plane as the rake face and the (100) crystal plane as the flank face offers improved resistance to damage, preventing flaking and ensuring microchip integrity [18, 41, 42]. However, the wear rate of the (111) crystal surface in diamond lapping is still lower compared to (110) and (100) due to friction factors and forces. Therefore, despite being the most susceptible to graphitization in MD simulations, the (111) crystal plane still holds potential for use as both the rake and flank faces of diamond too.

The machining behaviors of SCD vary significantly among different crystal planes and orientations due to its strong anisotropy. Wear rates of (100), (110), and (111) crystal planes show notable discrepancies based on orientation, with each plane having its own high and low grinding rate directions that change with orientation. There are also substantial differences in grinding rates even within the high grinding

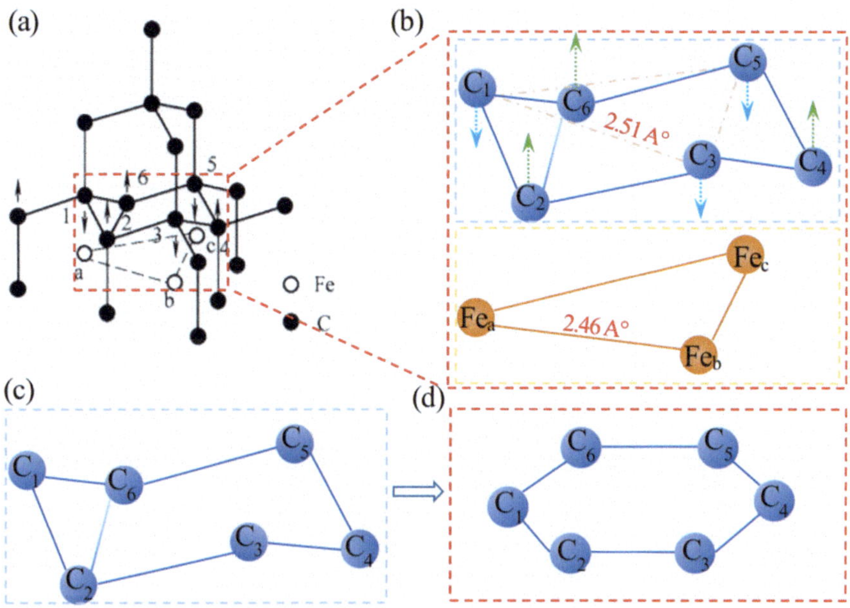


**Fig. 5.4** Schematic diagrams of **a**, **b** the diamond (111) crystal plane under alignment principle; **c**, **d** the generation of graphite structure [22]

rate directions [43]. For example, the grinding rate ratios for (100):(111):(110) are 5.8:1:12.8 [32]. Identifying the "soft" direction during diamond lapping and polishing can improve machining efficiency, which is crucial for diamond product manufacturing. Choosing the appropriate crystal orientation's "hard" direction as the wear direction for diamond tool flank face is essential to minimize tool wear in ultra-precision turning, especially when machining ferrous materials. Consideration can even be given to using the "hard" direction of the (111) crystal plane as the wear direction for the tool flank face if machining precision allows [44, 45]. Figure 5.5 illustrates the "soft" and "hard" directions of the three crystal planes.

#### 5.2.1.2 Effect of Cutting Parameters and Workpieces

Thornton and Wilks conducted a study on the turning of low-carbon steel using diamond tools. The experimental results revealed that as the turning speed increased to a critical value, the tool wear became more severe. At a turning speed of 30 m/s, the tool wear was 50 times greater compared to that at 11 m/s. Additionally, the tool wear increased with growing turning temperature [18]. That means the change of cutting parameters which tend to increase the cutting temperature will help to accelerate the diamond tool wear [46, 47]. Zou developed a prediction model for DTW that incorporates variables such as cutting force, cutting temperature, cutting

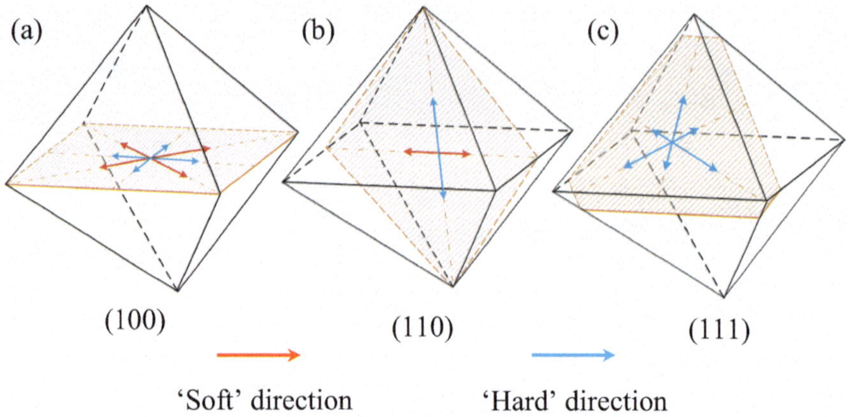


**Fig. 5.5** 'Hard' and 'Soft' direction of **a** (100), **b** (110) and **c** (111) plane of single crystal diamond

speed, and carbon content of workpieces. The model is based on the principles of friction and chemistry, as illustrated in Eq. (5.2).

$$W = 3.334 \times 10^{-8} F^{0.162} T^{2.204} V^{0.396} C^{-0.018} \tag{5.2}$$

where $W$ represents the degree of graphitization of diamonds, $F$ denotes the positive interface pressure (N), $T$ stands for the interfacial friction temperature (K), $V$ indicates the interfacial friction velocity ($ms^{-1}$), and $C$ is the carbon content of workpiece material (%).

This predictive model illustrates that as these cutting parameters increase, the DTW will also significantly accelerate. Conversely, increasing the carbon content of the workpiece material has minimal impact on diamond graphitization, as supported by grinding experiments conducted on diamond wheels [48, 49]. Moreover, according to the diffusion theory, diamond tools are more likely to undergo graphitization during the turning of low-carbon steels, compared to high-carbon steels. This is due to the higher concentration gradient of carbon atoms, which amplifies the diffusion force and facilitates the release of carbon atoms from the diamond matrix. The result is that carbon atoms have a tendency to build a structure that is more stable in graphite.

## 5.2.2 *Effect of External Energy Field Assisted*

### 5.2.2.1 Air Field Assisted Methods

Thornton and Wilks conducted a study on diamond wear impact at different air pressure levels, which are shown in Table 5.6 [18]. Tanaka et al. observed a correlation between oxygen partial pressure and the wear of diamond tools. The wear

**Table 5.6** Details of air field influence of diamond graphitization

| Variables | Degreed of diamond graphitization | | |
|---|---|---|---|
| Air pressure | | Cutting speed | |
| | Decrease | 0.16 m/s | Decrease |
| | | 11 m/s | Increase |
| | | 30 m/s | No effect |
| Oxygen partial pressure | Temperature | | |
| | 1000 K | Change | Reverse change |
| | 900 K | | Direction change |
| Atmosphere | $CO_2$ | | Decrease |
| | $CH_4$ | | Decrease or increase |
| | Inert gas | | Increase |

rate increased at temperatures exceeding 1000 K, but decreased with higher atmospheric oxygen partial pressure. Conversely, at approximately 900 K, the tool wear rate decreased initially, but then rose as the oxygen partial pressure increased [24].

Cassteven suggested that diamond synthesis could potentially occur in a low-pressure environment with carbon saturation, as the temperature in the cutting zone mirrors conditions ideal for diamond formation [50]. To verify this hypothesis, he replicated the low-pressure diamond synthesis conditions using a carbon-saturated atmosphere. However, the introduction of $CO_2$ did not have a significant impact on the longevity of diamond tools. Interestingly, the wear rate of low- and high-carbon steel samples decreased in the presence of $CH_4$ gas, indicating a potential strengthening of the bond between diamond and metal in the absence of water vapor or oxygen. This suggests that the effectiveness of DTW may be heightened under the influence of inert gases like He or $N_2$ [8, 50, 51]. In a separate study, Hitchiner and Wilks explored gas-assisted machining methods and discovered that employing $CH_4$ and $H_2$ gas protection during turning could enhance DTW when compared to using air. Nevertheless, due to the complexities introduced by the tool, machine, and environment, they arrived at a contradictory conclusion [23].

Hence, the impact of diamond graphitization by air pressure and gas-assisted machining does not show a clear reduction in tool wear as observed in the experiments mentioned. This lack of effect can be attributed to the minimal influence of air pressure and atmosphere on the cutting zone, where the tool's cutting edge maintains close contact with the material during the cutting process. In Table 5.6, an overview of the impact that the air field has on the graphitization of diamonds is shown.

#### 5.2.2.2 Coolant Assisted Methods

Viewed from the perspective that the chemical reactions can be influenced by the composition of cutting fluid, it is plausible that alternative elements present in the cutting fluid might initially interact with iron, thereby impeding the catalytic role of iron in diamond graphitization. Considering the aspect of thermally induced diamond graphitization, the utilization of cutting fluid as a coolant has the potential to reduce the cutting temperature in the machining process, thus partly restraining diamond graphitization [46, 52].

Inada et al. used carbon ions in a coolant injection cooling technique. This method creates a protective film on diamond tool surfaces and forms a 200 nm thick softened layer on workpiece surfaces, ideal for plastic cutting and reducing cutting temperatures. Experimental findings indicate that cooling the cutting process with carbon ions can enhance tool longevity and improve the workpiece's surface machining integrity, although DTW remains a challenge for practical production [53–55]. Zhou et al. observed that wet cutting of NAK80 with diamond tools led to increased tool wear, attributing this effect to the lubrication provided by the cutting fluid during the cutting process [56].

Li et al. used cryogenic coolants like liquid nitrogen ($N_2$) and liquid carbon dioxide ($CO_2$) for cooling the machining zone by employing the siphon principle to deliver ultra-low temperature fluid directly onto the machining area through a specialized nozzle. Experimental findings indicated a notable decrease in the DTW rate, ensuring a stable turning process. The surface roughness of the treated 45 steel remained consistent even after 20 min of cutting, leading to a substantial extension in tool life. Despite external factors affecting machining accuracy, the workpiece's surface roughness only achieved 0.4 μm, falling short of the desired mirror-level finish [58, 58]. Similarly, Bryan et al. also employed liquid nitrogen to cool the machining area during experiments involving diamond turning of stainless steel, which resulted in a significant reduction in DTW, enhancing the workpiece's surface roughness to below 25 nm [14].

#### 5.2.2.3 Electric Field and Multi-physics Coupling Assisted Methods

According to Zhang et al. [59], an electric field assisted machining method was proposed to address diamond graphitization as a chemical reaction. Through the use of inductive charging or conducting charge, this method entails increasing the electrical potential on the surface of the workpiece. By strengthening the workpiece surface, chemical reactions can be inhibited, leading to an extended lifespan of diamond tools and a reduction in the rate of chemical wear.

In a similar way, Fang et al. [60] presented a method of multi-physical field coupling to mitigate DTW. This technique utilizes various physical fields, including magnetic field, electric field, or laser radiation, around the diamond tool to induce the formation of low-energy dislocations. Changes in the material crystal's interface energy and structure can cause the solid-phase shift to happen during ultra-precision cutting. These micro-changes enable the improvement of cutting performance for diamond tools and workpieces, ultimately leading to a decrease in DTW.

Although these two auxiliary machining schemes are theoretically viable, it is essential to consider the impact of various factors on the workpiece's performance. This includes factors like the carbon content, temperature fluctuations during machining, and surface cleaning, as these elements may potentially disrupt the energy field. Furthermore, the specific microscopic action mechanisms and the practical implementation of these techniques pose significant challenges, and further research is required to elucidate them in detail.

#### 5.2.2.4 Workpiece Surface Modification Assisted Methods

According to the information that was stated in the introduction, the degree to which diamond graphitization occurs can be affected by particular elements that are present in materials. One example of such an element is carbon. During the early phases of the employment of diamond tools, natural diamonds that included impurities were utilized extensively in the process of machining. Researchers have made the observation that certain foreign elements have the ability to reduce the activity of carbon and iron atoms, which in turn effectively inhibits the graphitization of diamonds [15, 49, 61, 62]. After considering these points of view, two distinct types of auxiliary machining solutions have come into existence.

Research by Aronld et al. [63] and Vennemann et al. [64] revealed that the presence of phosphorus and nitrogen in the workpiece effectively inhibits diamond tool graphitization. Brinksmeier et al. [65–68] employed a thermochemical method to modify the sub-surface chemical composition of workpieces, establishing a chemical bond between carbon atoms of diamond tools and iron atoms of workpieces through ion nitriding, as shown in Fig. 5.6a. This method successfully reduces graphitization wear of diamond tools, decreasing flank wear width from 36 μm to less than 4 μm after a turning distance of 500 m, while achieving a surface roughness of 8–12 nm, as illustrated in Fig. 6b, c. Wang et al. employed plasma nitriding to machine mold steel, resulting in significant tool wear reduction after a turning distance of 5.4 km with a surface roughness of 20 nm [69]. Fang et al. compared the effects of ion nitriding and gas nitriding on tool wear inhibition in nitriding steel, finding that certain elements such as Al, Si, Cu, and Ni have positive effects, while Mo, Cr, and Mn have negative effects [70]. Saito et al. introduced a new nitriding method called electron-beam-excited-plasma nitriding, comparing its effects with ion nitriding and gas nitrocarburizing [71] (Fig. 5.7).

However, nitrogen ion implantation did not significantly improve the wear resistance of diamond tools, while the ion implantation of metal elements proved effective in enhancing tool wear resistance [72–74]. Lee et al. proposed that the induction of diamond tool graphitization becomes easier with a greater number of unpaired d-electrons. They utilized gallium ion implantation in diamond tools to machine ferrous metal, resulting in a workpiece surface roughness of 9.3 nm after turning 50 m, as depicted in Fig. 5.8. Furthermore, after a turning distance of 350 m, diamond tools displayed minimal signs of wear.

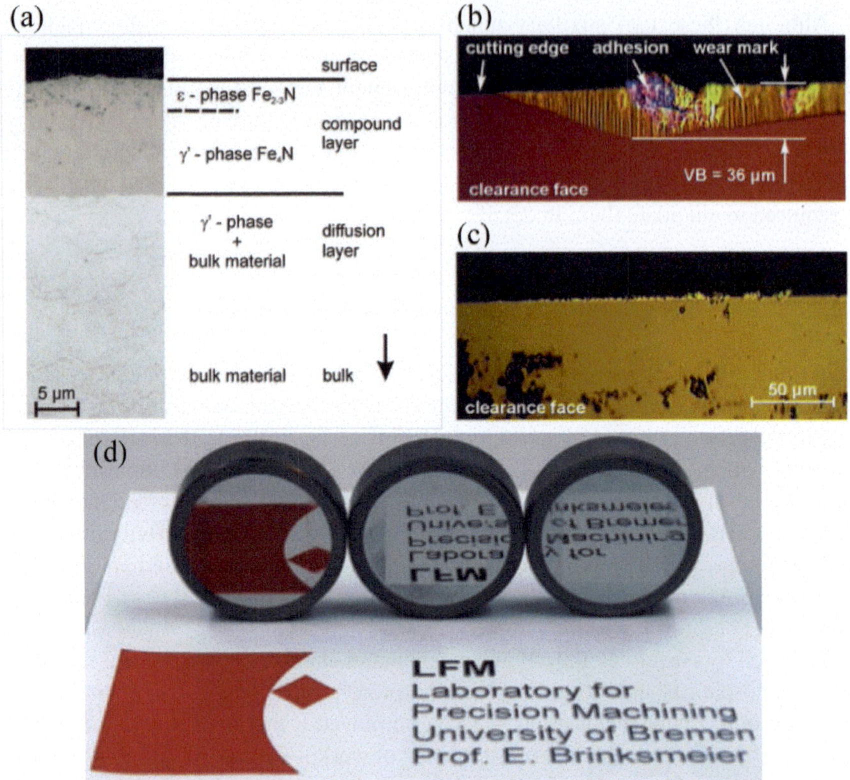


**Fig. 5.6** Diagram of **a** the nitriding layer, **b** the normal DTW in turning ferrous materials, **c** the DTW in turning nitriding materials and **d** the workpiece machining surface of mirror-level [65]

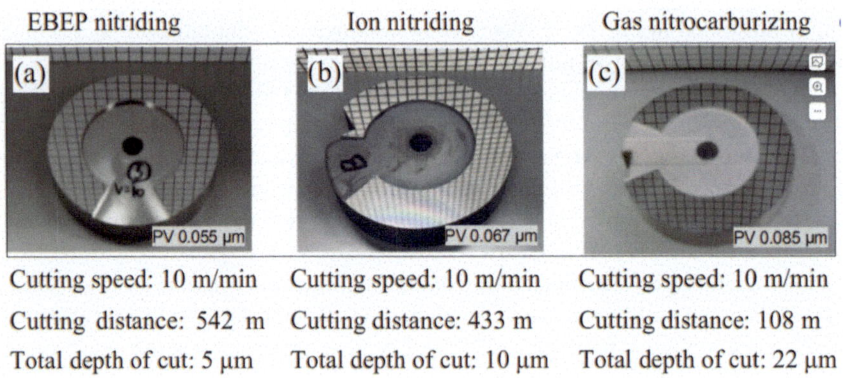


**Fig. 5.7** Photographs of finished surfaces of AISI 420 stainless steel in different surface treatment process [71]

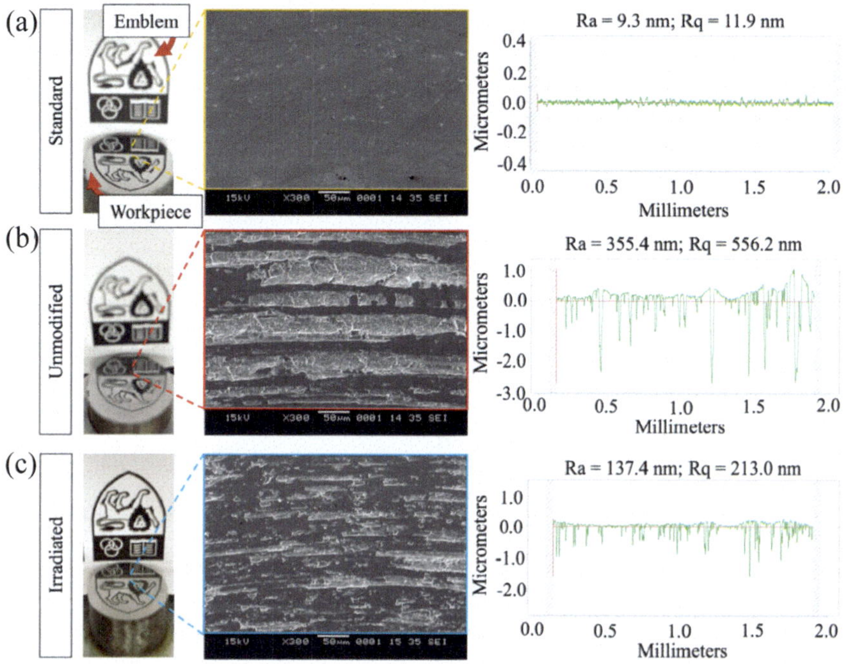


**Fig. 5.8** The machined surface quality of ferrous materials using a scanning electron microscope (SEM) and corresponding measurements of surface roughness: **a** 50 m turning of standard condition, **b** post-machining with a diamond tool over a distance of 350 m, and **c** post-machining with a Ga ion irradiated diamond tool over a distance of 350 m [77]

Therefore, the modification-assisted schemes operate by favoring the binding of foreign elements to iron atoms, subsequently suppressing the catalytic effect of iron atoms and mitigating the graphitization wear of diamond tools. It is important to note that while surface nitriding and ion implantation significantly suppress graphitization wear and facilitate the attainment of mirror-like surface roughness, the instability of the nitride layer and potential disruption of the diamond lattice structure can impact machining accuracy and long-term stability, necessitating further research [75]. On an alternative note, workpiece surface modification schemes fill in electron holes, thereby inhibiting the catalytic role of iron. Hence, multiple implementation approaches exist, such as Lee et al.'s utilization of marker ink absorbed on the workpiece surface, resulting in a 56% reduction in PCD tool wear [76].

#### 5.2.2.5 Coating Assisted Methods

Coating technology has long been recognized as an effective strategy to enhance tool wear resistance. In particular, metal nitride coatings have shown promising prospects for the improvement of diamond tool performance in the machining of ferrous materials [78].

Aronld et al. were among the pioneers in applying coating technology to diamond tools. They developed chemical nickel-phosphorus alloy coatings that effectively reduced tool wear. However, these coatings had limitations in terms of hardness and stability, rendering them unsuitable for practical applications [63]. Brinksmeier et al. conducted an experiment using TiN coatings, which demonstrated a reduction in flank wear of diamond tools. It is worth noting that the machining distance in their study was limited to 1130 mm, and the workpiece surface roughness was not specified [75]. Dong et al. compared the performance of three coatings: AlN, TiAlN, and TiN. Among them, the TiAlN coating exhibited a 50% reduction in diamond tool wear [79]. Meanwhile, Vennemann et al. developed a novel ternary coating with a nitrogen content ranging from 17 to 22%. This coating effectively prevented the adhesion and wear of diamond tools [64].

While coating technology has proven capable of inhibiting diamond tool graphitization, it may also lead to a decrease in the cutting edge sharpness of tools and the machining accuracy of workpieces. Moreover, the hardness of the coating is significantly lower than that of diamond, diminishing the hardness advantage of diamond tools, especially in the machining of high-hardness materials. At present, coating technology has not yet reached practical levels, but it still possesses great potential for further development.

#### 5.2.2.6 Alternative Tool Materials

Table 5.7 demonstrates the physical and mechanical properties of diamonds. These properties include ultra-high hardness, high thermal conductivity, high oxidation temperature, extremely sharp cutting edges (with edge radius as small as 10 nm or even 1 nm), and minimal long-term wear when turning diamond-turnable materials. Such attributes are essential for tools in ultra-precision machining. However, SCD present challenges due to their high cost and vulnerability to catastrophic tool wear when turning ferrous materials. As a result, researchers have explored alternative high-quality materials to replace SCD tools.

Table 5.8 presents various high-hardness materials used in tool manufacturing. Polycrystalline diamond (PCD) tools and diamond tools produced via chemical vapor deposition (CVD) or physical vapor deposition (PVD) do not possess physical properties superior to SCD and may also suffer from catastrophic chemical wear when used for turning ferrous materials [16]. Therefore, they are not suitable substitutes for diamond tools [16, 80–82]. In contrast, cubic boron nitride (CBN) tools, ceramic tools, and carbide tools exhibit no graphitization wear when employed for turning ferrous materials. As a result, extensive research has been conducted on these tool materials.

Liew et al. [83] found that PVD-coated carbide tools outperformed CVD coatings when turning stainless steel, achieving a surface roughness of about 100 nm after 440 m of turning, although significant tool wear was noted. In comparisons of wear

**Table 5.7** Physical and mechanical properties of diamonds

| Properties of diamonds | Values |
|---|---|
| Hardness HV | 6000– 10,000 (varies with crystal orientation and temperature) |
| Flexural strength | 210–490 MPa |
| Compressive strength | 1500–2500 MPa |
| Elastic modulus | (9–10.5) × $10^{11}$ N/$m^2$ |
| Thermal conductivity | (2–4)× 418.68 W/(m K) |
| Specific heat capacity | 0.516 J/(°C g) (Normal temperature) |
| Oxidation temperature | 900–1000 K |
| Graphitization temperature | 1800 K (In inert gas) |

**Table 5.8** Different high hardness materials in cutting tools

| Materials | Hardness HV | Main application (addition) with tools |
|---|---|---|
| Diamond | 6000–10,000 | SCD tools, PCD tools,<br>C (P) VD diamond tools |
| CBN | 6000–8500 | CBN tools |
| SiC | 3500 | Ceramic tools |
| TiC | 3200 | Carbide tools, ceramic tools |
| WC | 2400 | Carbide tools, ceramic tools |
| $Al_2O_3$ | 2200 | Ceramic tools |
| High carbon martensite | 1000 | High-quality stainless-steel tools |

performance among various tools for turning ferrous materials, Brinksmeier et al. [75] observed that alumina ceramic tools had the highest wear rate. Ghani et al. [84] reported 200 μm of flank wear and a surface roughness of approximately 1 μm after one minute of turning with alumina ceramic tools. Aslan et al. [85] optimized parameters for ceramic tools, achieving a surface roughness of 100 nm but also encountering tool wear up to 100 μm. Zhao et al. [86] found that alumina-based composite ceramic tools with WC and TiC additives showed less wear compared to carbide and standard alumina tools when turning AISI 1045 steel.

CBN tools, known for their hardness, have shown superior turning performance compared to ceramic or carbide tools. Fujisaki et al. [87] developed ultra-fine CBN tools, while Neo et al. [88] used pure CBN tools to achieve mirror surfaces with less than 20 nm roughness. Nishiguchi [89] and Moriwaki [90] developed single crystal and sintered CBN tools, respectively, which also achieved mirror-quality surfaces. However, despite these advancements, CBN tools still suffer from significant wear, making it difficult to maintain stable mirror finishes over extended use. Chou et al. utilized CBN tools with different grain sizes (ranging from 0.5 to 3 μm) for machining hardened steel, and Knuefermann et al. used nano-scale grain CBN tools for steel machining, both achieving tool performance comparable to that of SCD tools [91, 92]. Dosbaeva et al. [93] compared CVD-coated tools with PCBN tools and found

**Table 5.9** Comparison of various tool's cutting performance without assisted methods

| Tools | The surface roughness of workpieces is less than 20 nm? | The tool wear is serious? |
|---|---|---|
| | (Yes or No) | (Yes or No) |
| PCD tools | No | Yes |
| CBN tools | Yes | Yes |
| Ceramic tools | No | Yes |
| Carbide tools | No | Yes |

that PCBN exhibited a lower wear rate in the initial stages, with the tool flank width remaining below 0.1 mm. The study highlighted the tendency of diamond-coated tools to undergo graphitization when cutting ferrous materials, whereas CBN tools, despite not experiencing graphitization, have inferior sharpness and wear resistance compared to diamond tools, leading to less optimal performance over long distances.

A comparison of various tool's cutting performance without assisted methods is presented in Table 5.9.

#### 5.2.2.7 Ultrasonic Vibration Assisted Methods

Ultrasonic vibration cutting (UVC) is an advanced machining process that involves reciprocating vibration of the cutting tool or workpiece at a specific frequency and direction. UVC has emerged as a technological advancement in traditional machining, particularly for machining hard and brittle materials. It was discovered that UVC can reduce tool wear and achieve superior surface quality compared to traditional machining [94–97]. The application of ultrasonic vibration in ultra-precision cutting has greatly expanded its potential and various aspects of UVC research have been explored, including cutting force modeling, composite cutting, hard brittle cutting, mirror surface finishing, mechanical polishing, and micro/nano-machining [98–104]. UVC, as a highly popular cutting method, is continuously progressing toward intelligent machining [105]. Moreover, UVC also offers unique advantages in the field of SCD turning of ferrous materials.

Starting from the 1990s, Moriwaki and Shamoto et al. utilized one-dimensional ultrasonic vibration (traditional or linear) in ultra-precision turning [106–110]. By employing diamond tools and operating at a frequency of 40 kHz, they achieved a surface roughness of 26 nm in stainless-steel machining. With this method, the surface roughness remained below 100 nm, and the flank wear was less than 4 μm after turning 1600 m. Further advancements were made with elliptical vibration cutting (EVC or two-dimensional vibration cutting) that improved upon the performance of one-dimensional UVC (1D-UVC). EVC demonstrated a surface roughness of less than 20 nm and extended the tool life of diamond tools. The comparison of 1D-UVC and EVC for machining die steel is illustrated in Fig. 5.9. By employing EVC, tools

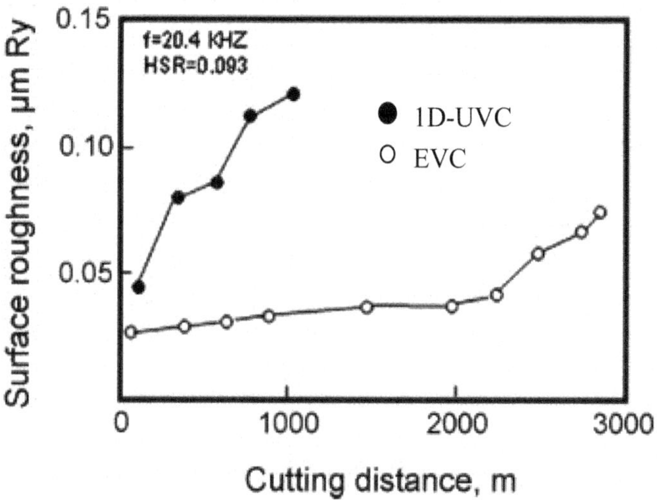


**Fig. 5.9** Comparison of workpiece surface roughness with 1D-UVC and EVC [110]

can generate smooth bevel wear, effectively suppressing diamond tool graphitization. All the tool trajectory diagrams for UVC are summarized in Table 5.10.

The key components of UVC include the ultrasonic generator, ultrasonic transducer, and ultrasonic horn, each fulfilling specific functions as outlined in Table 5.11. The working principle of UVC involves transmitting micron-level ultrasonic frequency to the tool, inducing tool vibration, and controlling the vibration direction, amplitude, and frequency. This enables the creation of periodic high-frequency separation between the workpiece and the tool, significantly enhancing workability. Table 5.12 presents the microscopic surface appearances of the workpiece under different waveforms. However, UVC is confined by its critical vibration cutting speed, limiting its efficiency and use primarily in low-speed cutting applications [111]. Researchers such as Sui et al. [112] and Zhang et al. [113] have explored high-speed ultrasonic cutting to improve the efficiency of UVC, as demonstrated in Table 5.10.

Ultrasonic vibration has emerged as a crucial method for extending the tool life in turning ferrous materials and has been considered by numerous scholars as a highly promising approach to mitigate diamond tool graphitization. In order to further advance the application of UVC in diamond turning of ferrous materials, extensive research has been carried out on various aspects such as UVC mechanism, device development, and parameter optimization [114].

Masahiko et al. [115] and other researchers have developed various ultrasonic vibration devices to improve machining efficiency and precision in ferrous materials. These advancements include programmable vibrating cutting tools [116], adjustable frequency devices [117], and ultra-precision 2D-UVC systems [118]. Klocke et al.

**Table 5.10** Three kinds of UVC [111]

| UVC | Speed | Schematics of tools trajectory |
|---|---|---|
| 1D-UVC | $v < v_c$ | 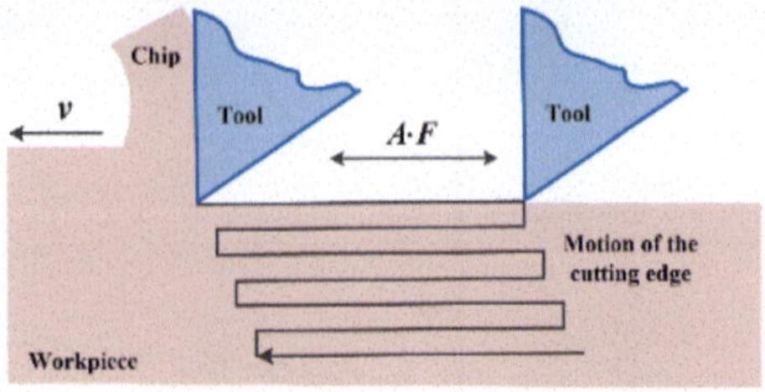  |
| EUVC | $v < v_c$ | 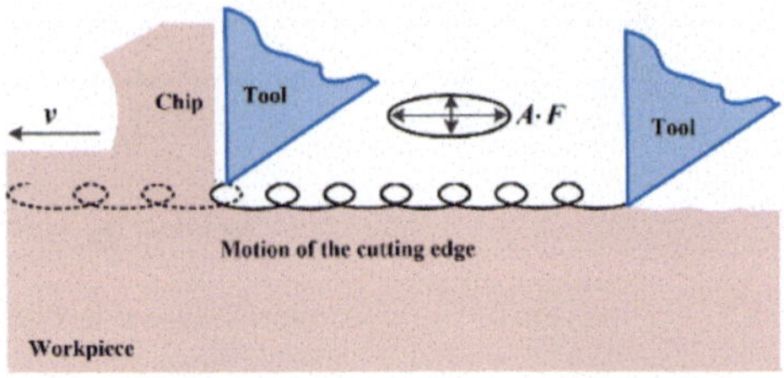  |
| HUVC | $v >> v_c$ | 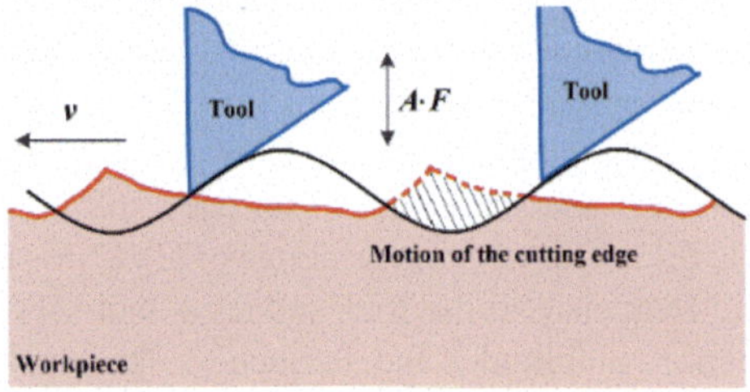  |

* The $v$ is cutting speed, and the $v_c$ is vibration critical cutting speed

**Table 5.11** The related device of ultrasonic vibration

| Devices | Functions |
|---|---|
| Ultrasonic generator | Converts current into a high-frequency AC signal that matches the ultrasonic transducer |
| Ultrasonic transducer | Converting incoming electrical power into mechanical power (ultrasonic) |
| Ultrasonic horn | Change the amplitude of ultrasonic vibration with the ultrasonic transducer |

[119] and Ammouri and Hamade [120–123] explored the use of ultrasonic vibration in machining quenched steel and micromachining, achieving notable improvements in shape accuracy and surface quality. Despite these innovations, challenges such as diamond tool wear during ultrasonic vibration machining remain unresolved [124, 125].

**Table 5.12** The microscopic appearance of the workpiece surface under different waveforms [27]

| Waveform | Frequency (HZ) | Cutting speed (m/min) | Machined surface morphology |
|---|---|---|---|
| Sinusoidal | 100 | 0.2 |  |
| Ramp | 100 | 0.2 |  |
| Zigzag | 100 | 0.2 |  |

For the turning mechanism in UVC and EVC, Moriwaki and Shamoto [106, 126] used mathematical modeling to demonstrate that EVC can achieve ultra-precision machining. Ma et al. developed a theoretical model of thrust in EVC [127, 128]. Researchers like Babitsky et al. [129] and Zhang et al. [130] explored the effects of vibration parameters, stress, and cutting forces on surface roughness and tool wear through finite element models. Mitrofanov et al. analyzed the effects of lubrication and cutting parameters on ultrasonic vibration-assisted cutting through finite element simulation [131]. Studies by Lu and Yang [132], and Ahmed et al. [133], further investigated cutting forces and friction in UVC, providing theoretical insights. Additionally, Fang and Wang [134, 135], and Zhou [136] analyzed the impact of UVC on tool wear, surface roughness, and diamond tool life in machining processes. Of course, there are some similar valuable research, like reference [137–141]. These investigations collectively contribute to understanding the mechanisms of UVC and its impact on machining efficiency and tool longevity.

The aforementioned studies have successfully elucidated the mechanisms of ultrasonic vibration-assisted machining (UVAM) in mitigating diamond tool wear and have contributed to the advancement of UVAM toward industrialization. Notably, thrust reduction, cutting deformation, average machining stress, and average processing temperature have been achieved through the intermittent tool-workpiece contact and tool-chip separation methods in ultrasonic-assisted machining. Moreover, it has been observed that higher vibration frequencies tend to result in less tool wear [142]. The formation of an intermediate oxide layer and increased air pressure on the contact surface promote a reduction in tool wear. However, the introduction of an external energy field through ultrasonic vibration leads to increased cutting tip temperature and chip temperature, with cutting temperature and energy consumption comparable to traditional cutting.

UVAM has proven to be highly effective in improving the quality of difficult-to-machine materials, such as diamond turning and ferrous materials [143]. Various UVAM methods based on UVC have been developed to enhance machining performance. For instance, Zhang et al. conducted experiments combining vibration assistance and $CO_2$ protection assistance in cutting ferrous materials. The results indicated that with a cutting distance of up to 2000 m, the workpiece achieved a surface roughness of less than 0.15 μm and a flank wear width of less than 5 μm [144]. In another study, Li introduced the combination of surface nitriding and ultrasonic vibration in turning WC materials, resulting in a surface roughness below 10 nm [145].

Researchers have also explored the integration of plasma jet and UVC to optimize machining outcomes [146]. Huang et al. successfully machined mold steel by combining cold plasma jet cutting with ultrasonic vibration, as depicted in Fig. 5.10. This process achieved a workpiece surface roughness of less than 2 μm and a 50% reduction in tool wear compared to traditional turning processes. However, it should be noted that this study utilized PCD tools and did not achieve mirror cutting [147]. Zhang et al. employed electrochemical treatment along with ultrasonic vibration to machine WC materials, as shown in Fig. 5.11. This method enabled the attainment of a mirror-level (Sa 23 nm) [26]. The utilization of ultrasonic elliptical vibration combined with varying air blow pressures in cutting carbon steel and stainless steel with PCD tools was investigated by Zhang et al. The experimental results demonstrated the significant influence of air blow pressure on diamond tool wear, as illustrated in Fig. 5.12 [148–150].

Another study by Zhang et al. focused on UVAM under oxygen-shielded conditions. It was found that the use of PCD tools and SCD tools, in combination with UVC, resulted in a significant reduction in flank face wear compared to using UVC alone. This can be attributed to the weakened catalytic effect of iron in the presence of oxygen, which inhibits graphitization [151]. Additionally, the application of oxidizing cutting fluid during UVC has shown the potential for reducing tool wear. Saito et al. conducted cutting experiments on hard die steel using coated tools with elliptical vibration. The experiments revealed that TiN-coated tools outperformed diamond-coated tools in terms of tool life, although the surface roughness (Rt) exceeded 100 μm [152]. Wang et al. explored ion-implanted tungsten carbide using ultrasonic vibration diamond cutting. However, the final processed workpiece exhibited a surface roughness at the micron level, which did not meet the requirements of ultra-precision machining [152, 153].

Thus, ultrasonic vibration-assisted methods have emerged as highly promising machining techniques with significant advantages, such as suppressing the graphitization of diamond tools and enabling the attainment of mirror-level surface roughness [154]. As a result, researchers have been conducting extensive studies and implementing various composite-assisted methods, primarily based on UVC. However, the introduction of vibration and energy fields, the utilization of ultrasonic vibration

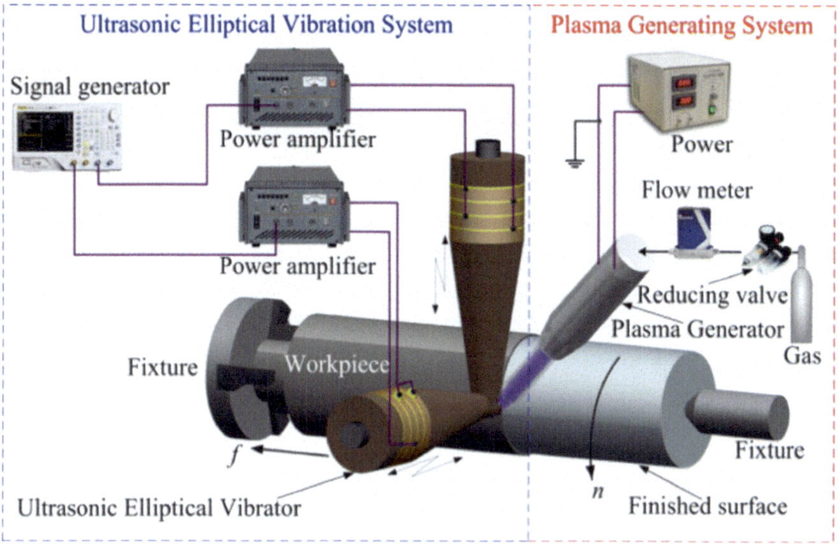


**Fig. 5.10** Schematic diagram of diamond cutting device assisted by cold plasma and ultrasonic vibration [147]

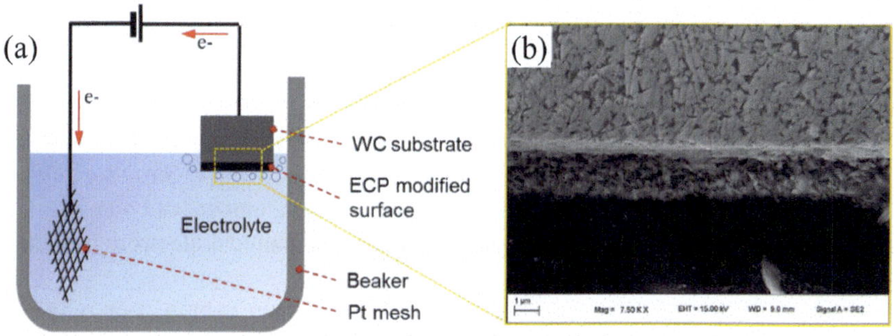


**Fig. 5.11** Schematic diagram of surface modification of electrochemically treated workpiece [26]

devices, workpiece workable diameter, machining interference, and the incorporation of corrugations can potentially impact the surface quality of the workpiece negatively. It is crucial to carefully consider these factors during the machining process to mitigate any potential harm to the surface quality.

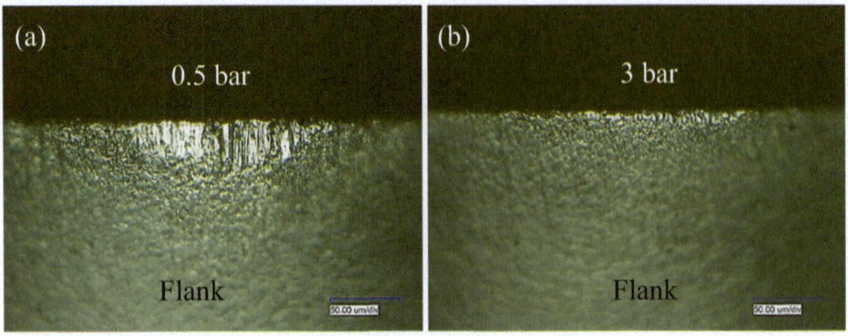


**Fig. 5.12** Microscope photographs of tool flank wear taken after EVC of steel using PCD tools under different air blowing pressure: **a** 0.5 bar and **b** 0.3 bar [19]

#### 5.2.2.8 Magnetic Field Assisted Machining

The magnetic field assisted method is based on the concept of unpaired d-electrons theory. Brinksmeier and Preuß conducted a study and concluded that certain materials can be categorized into easy-to-cut and difficult-to-cut diamond tools based on their wear characteristics [155]. Analyzing the distribution of these materials in the periodic table of elements, it can be observed that hard-to-cut materials are predominantly located in the left half of the table. Therefore, it becomes crucial to identify the primary cause of catastrophic wear in order to develop an effective method to suppress diamond tool wear.

Experimental findings by Paul et al. have demonstrated that unpaired *d*-electrons play a significant role in catastrophic wear [8]. These unpaired *d*-electrons exhibit a strong affinity towards diamond and iron. Hence, the development of an innovative approach to mitigate diamond tool wear by addressing unpaired d-electrons becomes imperative (Table 5.13).

Magnetic field (MF) has gained popularity as a cleaner energy source and is commonly used in various assisted machining techniques to mitigate tool wear and machining vibrations [156]. In the context of traditional turning, Nakano conducted a study to investigate the impact of MF on tool wear during steel turning. The experimental findings revealed that MF-assisted turning reduced flank wear and improved the surface roughness of the machined workpiece [157]. Patwari et al. also explored MF-assisted turning of preheated mild steel and observed a 20% improvement in tool life and surface roughness [158]. In another study, Zhang et al. found that MF-assisted turning reduced cutting force, tool wear, and enhanced surface roughness [159]. Over the past two decades, several research studies have demonstrated the feasibility of MF-assisted turning for machining ferrous materials.

In traditional turning, MF is typically applied as a pretreatment method to optimize tool performance. Various forms of MF pretreatment, including alternating and pulse MFs, have been employed [160, 161]. However, it is important to note that ultraprecision turning differs qualitatively from traditional turning, mainly utilizing diamond

**Table 5.13** Relation between the number of unpaired *d*-electrons and easy/difficult-to-cut type for diamond [8]

| Element | No. of unpaired *d*-electrons | Easy/ difficult-to-cut | Element | No. of unpaired *d*-electrons | Easy/ difficult-to-cut |
|---|---|---|---|---|---|
| Pu | 0 | E | U | 1 | D |
| Sn | 0 | E | Ni | 2 | D |
| Mg | 0 | E | Ti | 2 | D |
| Ag | 0 | E | Rh | 2 | D |
| Pb | 0 | E | Co | 3 | D |
| Ge | 0 | E | V | 3 | D |
| Au | 0 | E | Ru | 3 | D |
| Zn | 0 | E | Ta | 3 | D |
| Cu | 0 | E | Fe | 4 | D |
| In | 0 | E | Nb | 4 | D |
| Al | 0 | E | W | 4 | D |
| Be | 0 | E/D | Mn | 5 | D |
| Si | 0 | E/D | Cr | 5 | D |
| | | | Mo | 5 | D |
| | | | Re | 5 | D |

tools [162, 163]. Diamond, being a diamagnetic material, is not affected by MF. Therefore, further experimentation is required to investigate the influence of MF on diamond tool wear during ultraprecision turning of ferrous materials. Unfortunately, limited theoretical and experimental studies have focused on this aspect.

Yip et al. [156, 164–166] investigated the application of a magnetic field in SPDT when working with titanium alloys, as shown in Fig. 5.13a. When subject to a constant magnetic field during the cutting process, the titanium alloy generates eddy currents as depicted in Fig. 5.13b, accompanied by the Lorentz force, which contributes to damping vibrations at the tool's tip. In addition, utilizing the magnetic field's expansive influence on the substance, they were successful in diminishing the extent of scratches, thereby enhancing the surface finish as illustrated in Fig. 14a, b. This magnetic influence also augments the thermal dissipation properties of the titanium alloy, which correlates with a significant decrease in adhesion wear on the diamond tool. Specifically, there was a reported wear reduction between 57.2 and 68.99% when comparing dry magnetic field samples and dry non-magnetic field samples. This suggests that the lifetime of diamond cutting tools can effectively be extended through the implementation of a magnetic field.

Although titanium alloys fall under the category of materials that are challenging to machine with diamond tools, their machining difficulties differ from those of ferrous-based materials. Hence, the effect of magnetic fields on the wear suppression of diamond cutting tools during the machining of ferrous metals still requires further exploration.

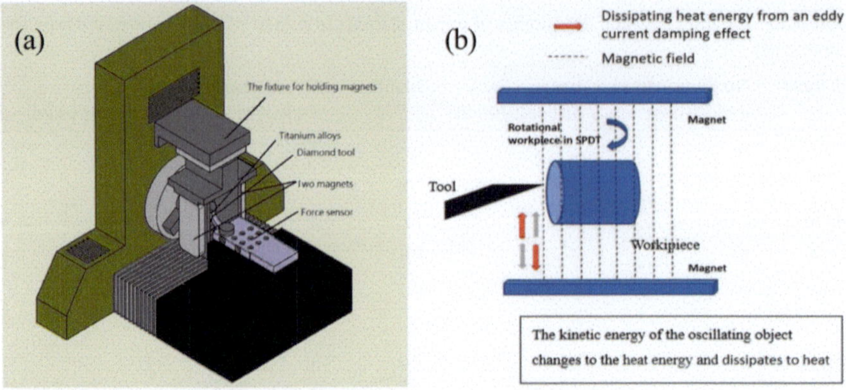


**Fig. 5.13** **a** Experimental setup of magnetic assisted diamond turning, **b** experimental setup of SPDT under an eddy damping effect

**Fig. 5.14** The micrographs of cutting groove of **a** dry magnetic field samples and **b** dry non-magnetic field samples

Wang et al. [167] pioneered the application of magnetic fields in single-point diamond turning of ferrous-based materials, achieving a reduction of over 30% in the flank wear of PCD tools. For specific details on their methodology and findings, one can refer to the study documented in reference [167].

The application of MF-assisted turning has demonstrated a significant reduction in flank wear on PCD tools. Graphitization, a well-known chemical wear mechanism in diamond tools during the ultraprecision turning of ferrous materials, can be effectively suppressed by MF-assisted turning. The chemical reactions involved in graphitization can be inhibited through this method.

Although the interaction energy between MF and chemical particles such as ions, molecules, and atoms is typically only around $10^{-4}$ to $10^{-6}$ times the thermal motion energy of these particles, the MF can still influence the electron spin state. The spin properties of d-electrons, which are affected by the MF, play a crucial role in chemical reactions. Understanding the spin of electrons within an MF is a challenging task, but quantum mechanics provides suitable calculations for individual electrons.

For instance, in the case where an electron in a specific domain experiences a uniform MF along the $x$ direction and its spin is initially upward at $t = 0$, the state wave function of the system at $t > 0$ can be defined using appropriate quantum mechanical calculations [168].

$$\psi(t) = \frac{1}{\sqrt{2}}[\psi_{x,+1}(0)e^{-\frac{i}{\hbar}E_1 t} + \psi_{x,-1}(0)e^{-\frac{i}{\hbar}E_2 t}] = \begin{bmatrix} \cos wt \\ -i \sin wt \end{bmatrix} \tag{5.3}$$

where $\psi_{x,+1}(0)$ and $\psi_{x,-1}(0)$ are a complete orthogonal system formed by the eigenstates of $\hat{\sigma}_x$ in the representation of $\left[\hat{\sigma}^2, \hat{\sigma}_z\right] = 0$, $E_1$ and $E_2$ are the energies of electrons in different states, and $h$ is the Planck constant.

According to Eq. (5.4), the measured average values of each component of the electron spin operator can be obtained at $t > 0$.

$$\begin{cases} \langle s_x \rangle = 0 \\ \langle s_y \rangle = -\frac{\hbar}{2} \sin 2wt \\ \langle s_z \rangle = \frac{\hbar}{2} \cos 2wt \end{cases} \tag{5.4}$$

Correct, by considering the $z$ direction as an example, the initial state can be represented as $\langle s_z \rangle = \frac{\hbar}{2}$, indicating that the MF alters the electron spin state, thereby directly influencing the chemical reaction. The change in electron spin state plays a crucial role in chemical reactions. According to principles of quantum chemistry, achieving electron pairing in chemical reactions necessitates different electron spin states. If the electron spin states are identical, a chemical reaction cannot occur.

Furthermore, MFs possess the capability to enhance electron spin polarization. Consequently, they reduce the number of electrons with different spin states that can be paired with other electrons. As graphitization is a chemical reaction influenced by the MF, the MF can inhibit graphitization by affecting electron spin states. By promoting electron spin polarization, the MF has an impact on the entropy change of chemical reactions.

In the turning region, it is reasonable to assume constant pressure and temperature. The Gibbs function, which is defined as the difference between enthalpy and temperature multiplied by entropy, plays a vital role in determining the feasibility of a chemical reaction.

$$\Delta G = \Delta H - T\Delta S \tag{5.5}$$

where $\Delta G$ represents the Gibbs free energy, $\Delta H$ denotes the enthalpy change, $T$ corresponds to the temperature, and $\Delta S$ is the entropy change.

Indeed, the Gibbs free energy can be utilized to determine the direction of a chemical reaction. The thermodynamics of the chemical reaction between diamond and graphite suggests that graphitization of diamond can be suppressed due to an increase in Gibbs free energy (see Eq. 5.5).

$$\Delta G \begin{cases} < 0 \, \text{Diamond} \rightarrow \text{Graphite} \\ = 0 \, \text{Chemical reaction does not proceed} \\ > 0 \, \text{Graphite} \rightarrow \text{Diamond} \end{cases} \tag{5.6}$$

Since chemical reactions also rely on entropy, the MF can promote the spin polarization of *d*-electrons, resulting in an ordered system. Notably, in ferromagnetic materials, unpaired *d*-electrons spontaneously magnetize, forming magnetic domains. These magnetic domains serve as the minimum unit for analyzing electron spin polarization in pure iron, for example. The polarization of magnetic domains can shift the system from a disordered state to an ordered state. An ordered state can contribute to a decrease in entropy of the system. Consequently, $\Delta S < 0$ results in an increase in $\Delta G$, which effectively suppresses the wear caused by diamond graphitization.

## References

1. Davies MA, Evans CJ, Patterson SR, Vohra R, Bergner BC (2003) Application of precision diamond machining to the manufacture of micro-photonics components. Proc SPIE 5(183):94–108
2. Fu Y (2007) Research on precision diamond cutting of ferrous metals. Harbin Institute of Technology, pp 45–50 (in Chinese)
3. Zhou F, Yu AB, Wang CC (2003) The research development of metal-bonded diamond wheels dressing technology. Precis Manuf Autom 2:12 (in Chinese)
4. Ikawa N, Donaldson RR, Komanduri R, König W, Aachen TH, McKeown PA, Moriwaki T, Stowers IF (1991) Ultraprecision metal cutting—the past, the present and the future. CIRP Ann 40(2):587–594
5. Xiao XX, Zhong ZM (1984) Diamond tools for ultra-precision cutting. Aeronaut Manuf Technol 02:9–14 (in Chinese)
6. Uddin AS, Seah KHW, Rahman M, Li XP, Liu K (2007) Performance of single crystal diamond tools in ductile mode cutting of silicon. J Mater Process Technol 185(1–3):24–30
7. Zou L, Yin JC, Huang Y, Zhou M (2018) Essential causes for tool wear of single crystal diamond in ultra-precision cutting of ferrous metals. Diam Relat Mater 86:29–40
8. Paul E, Evans CJ, Mangamelli A, McGlauflin ML, Polavani RS (1996) Chemical aspects of tool wear in single point diamond turning. J Int Soc Precis Eng Nanotechnol 18(1):4–19
9. Pramanik A, Neo KS, Rahman M, Li XP, Sawa M, Maeda Y (2008) Ultraprecision turning of electroless nickel: effects of crystal orientation and origin of diamond tools. Int J Adv Manuf Technol 43(7–8):681–689
10. Pramanik A, Neo KS, Rahman A, Li XP, Sawa M, Maeda Y (2003) Cutting performance of diamond tools during ultra-precision turning of electroless-nickel plated die materials. J Mater Process Technol 140(1–3):308–313
11. Moriwaki T (1995) Experimental analysis of ultraprecision machining. Int J Jpn Soc Precis Eng 29(4):287–290

12. Lu HY (2003) Discussion on wear mechanism of diamond tool. Mach Manuf 1:38–39 (in Chinese)
13. Komanduri R, Shaw MC (1975) Wear of synthetic diamond when grinding ferrous-metals. Nature 255(5505):211–213
14. Evans C, Bryan JB (1991) Cryogenic diamond turning of stainless steel. CIRP Ann 40(1):571–575
15. Thornton AG, Wilks J (1980) The wear of diamond tools turning mild-steel. Wear 65(1):67–74
16. Gimenez S, Van der Biest O, Vleugels J (2007) The role of chemical wear in machining iron based materials by PCD and PCBN super-hard tool materials. Diam Relat Mater 16(3):435–445
17. Zou L (2015) Study on wear and inhibition of black metal diamond cutting tools. Harbin Institute of Technology Doctor, pp 25–27 (in Chinese)
18. Thornton AG, Wilks J (1978) Clean surface-reactions between diamond and steel. Nature 274(5673):792–793
19. Zhang XQ, Liu K, Kumar AS, Rahman M (2014) A study of the diamond tool wear suppression mechanism in vibration-assisted machining of steel. J Mater Process Technol 214(2):496–506
20. Narulkar R, Bukkapatnam S, Raff LM, Komanduri R (2009) Graphitization as a precursor to wear of diamond in machining pure iron: A molecular dynamics investigation. Comput Mater Sci 45(2):358–366
21. Guo XG, Zhai CH, Jin ZJ, Guo DM (2015) The study of diamond graphitization under the action of iron-based catalyst. J Mechan Eng 51(17):162
22. Wang JP, Zhang GQ, Chen N, Zhou MH, Chen YB (2021) A review of tool wear mechanism and suppression method in diamond turning of ferrous materials. Int J Adv Manuf Technol 113:3027–3055
23. Hitchiner MP, Wilks J (1984) Factors Affecting Chemical Wear during Machining. Wear 93(1):63–80
24. Tanaka H, Shimada S, Ikawa N, Yoshinaga M (2001) Wear mechanism of diamond cutting tool in machining of steel. Precis Mach Adv Mater 196:69–78
25. Thornton AG, Wilks J (1979) Tool wear and solid-state reactions during machining. Wear 53(1):165–187
26. Zhang XQ, Huang R, Liu K, Kumar AS, Deng H (2018) Suppression of diamond tool wear in machining of tungsten carbide by combining ultrasonic vibration and electrochemical processing. Ceram Int 44(4):4142–4153
27. Shamoto E, Suzuki N (2014) Ultrasonic vibration diamond cutting and ultrasonic elliptical vibration cutting, pp 405–454
28. Yip WS, To S (2018) Sustainable manufacturing of ultra-precision machining of titanium alloys using a magnetic field and its sustainability assessment. Sustain Mater Technol 16:38–46
29. Ge YF, Xu JH, Yang H (2010) Diamond tools wear and their applicability when ultra-precision turning of SiCp/2009Al matrix composite. Wear 269(11–12):699–708
30. Ohta T, Yan JW, Yajima S, Takahashi Y, Horikawa N, Kuriyagawa T (2007) High-efficiency machining of single-crystal germanium using large-radius diamond tools. Int J Surf Sci Eng 1(4):374–392
31. Durazo-Cardenas I, Shore P, Luo X, Jacklin T, Impey SA, Cox A (2007) 3D characterisation of tool wear whilst diamond turning silicon. Wear 262(3–4):340–349
32. Yuan ZJ, Yao YX, Zhou M, Bai QS (2003) Lapping of single crystal diamond tools. CIRP Ann Manuf Technol 52(1):285–288
33. Zong WJ, Sun T, Li D, Cheng K (2009) Design criterion for crystal orientation of diamond cutting tool. Diam Relat Mater 18(4):642–650
34. Zong WJ, Li ZQ, Sun T, Cheng K, Li D, Dong S (2010) The basic issues in design and fabrication of diamond-cutting tools for ultra-precision and nanometric machining. Int J Mach Tools Manuf 50(4):411–419
35. Hurt HH, Decker DL (1984) Tribological considerations of the diamond single-point tool 0508:126

36. Seal M (1963) The effect of surface orientation on the graphitization of diamond. Phys Status Solid 3(4):658–664
37. Jin ZJ, Xie F, Guo XG, Shi SJ (2016) Wear mechanism of single crystal diamond tool against mold steel by molecular dynamics simulation. Nanotechnol Precis Eng 14(6):410–415
38. Bai Q, Wang Z, Guo Y, Chen J, Shang Y (2018) Graphitization behavior of single crystal diamond for the application in nano-metric cutting. Curr Nanosci 14(5):377–383
39. Yuan ZW, Jin ZJ, Kang RK, Wen Q (2012) Tribochemical polishing CVD diamond film with FeNiCr alloy polishing plate prepared by MA-HPS technique. Diam Relat Mater 21:50–57
40. Zong WJ, Li ZQ, Sun T, Li D, Cheng K (2010) Analysis for the wear resistance anisotropy of diamond cutting tools in theory and experiment. J Mater Process Technol 210(6–7):858–867
41. Yuan JZ, He JC (1992) Natural diamond tool crystal selection on cutting deformation and machining surface quality. Tool Technol 26(11):28–31 (in Chinese)
42. Zhou M, Yuan JZ (1999) The selection of crystal surface of diamond tool affects the cutting process and durability. J Harb Inst Technol 31(4):78–79 (in Chinese)
43. Zhang JH (1985) Precision cutting theory and technology. Shanghai Science and Technology Press, pp 139–151 (in Chinese)
44. Grodzinski P (1953) Diamond technology: production methods for diamond and gem stones. N A G Press, New York
45. Li ZQ, Han JC, Sun T, Zong WJ, Dong S (2011) Periodic bond chain model for estimation of the anisotropy of diamond cutting tools. Nanotechnol Precis Eng 9(2):176–179 (in Chinese)
46. Hitchiner MP, Wilks J (1987) Some remarks on the chemical wear of diamond and cubic Bn during turning and grinding. Wear 114(3):327–338
47. Lata S, Rana R, Hitesh (2018) Investigation of chip-tool interface temperature: effect of machining parameters and tool material on ferrous and non-ferrous metal. Mater Today Proc 5(2):4250–4257
48. Graham W, Nee A (1974) The grinding of tool steels with a diamond abrasive wheel. Int J Mach Tool Des Res 14(2):175–185
49. Graham W, Nee A (1974) The grinding of tool steels—a comparison between diamond and cubic boron nitride. Prod Eng 53(6):186
50. Casstevens JM (1982) Diamond turning of steel in a carbon-saturated atmosphere. NASA STI/Recon Tech Rep N 5(1):9–15
51. Brinksmeier E (1999) Single point diamond turning of steel. Proc EUSPEN 628
52. Zhao QL, Chen MJ, Liang YC, Dong S (2002) Wear analysis of single crystal diamond turning tool in ultra-precise single point cutting. J Tribol 5:2–8 (in Chinese)
53. Inada A, Min S, Ohmori H (2011) Micro cutting of ferrous materials using diamond tool under ionized coolant with carbon particles. CIRP Ann Manuf Technol 60(1):97–100
54. Inada A, Min S, Dornfeld D, Ohmori H (2008) Effects of ion-shot coolant system on cutting of the ferrous material with a diamond tool. Proc IMETI2008 Orlando 2:13–17
55. Inada A, Ohmori H, Min S, Dornfeld D (2010) Investigation of the effects of an electrolytic coolant with a nano carbon additive in diamond micro cutting on ferrous materials. J Adv Mechan Des Syst Manuf 4(5):1076–1083
56. Zhou MH, Zhang GQ, Chen N (2020) Effects of lubricant on cutting performance in single-point diamond turning of ferrous metal NAK 80. Int J Adv Manuf Technol 109(9–12):2549–2558
57. Li JN, Yuan JZ (1987) Study on ultrafine cutting property of diamond in ferrous metal. J Harbin Inst Technol 01:138–139 (in Chinese)
58. Li JN, Yuan JZ (1989) Ultra-low temperature diamond ultra-precision cutting of ferrous metal. J Mechan Eng 25(1):69–72 (in Chinese)
59. Zhang G (2007) Method for extending diamond tool life in diamond machining of materials that chemically react with diamond. U.S. Patent 7,198,043
60. Fang F, Zhang H, Qiu Z, Tang X (2010) A method for the suppression of single crystal diamond tool wear in ultra-precision machining by the action of multiple physical fields. China. Patent CN101817694A
61. Evans T, Sauter DH (1961) Etching of diamond surfaces with gases. Phil Mag 6(63):429

62. Simons EL, Cannon P (1966) A means of increasing the oxidation resistance of diamond. Nature 210(5031):90–91
63. Arnold JB, Morris TO, Sladky RE, Steger PJ (1977) Machinability studies of infrared window materials and metals. Opt Eng 16(4):164324
64. Vennemann A, Stock HR, Kohlscheen J, Vorgel E, Mayr P (2003) New diamond turnable coatings of ternary systems for ultraprecision cutting. Surf Coat Technol 174:973–978
65. Brinksmeier E, Glab R, Osmer J (2006) Ultra-precision diamond cutting of steel molds. CIRP Ann Manuf Technol 55(1):551–554
66. Osmer J, Glabe R, Riemer O, Brinksmeier E, Butepage S, Hoffmann F (2009) Diamond milling of nitrided steels for optical mold making. J Vac Sci Technol B 27(3):1238–1240
67. Brinksmeier E, Gläbe R, Osmer J (2010) Diamond cutting of FeN-layers on steel substrates for optical mould making. Key Eng Mater 438:31–34
68. Brinksmeier E, Glaebe R, Osmer J (2011) Surface integrity demands of high precision optical molds and realization by a new process chain. In: 1st Cirp conference on surface integrity (Csi), vol 19, pp 40–43
69. Wang YL, Zhao QL, Shang YJ, Lv PX, Guo B, Zhao LL (2011) Ultra-precision machining of Fresnel microstructure on die steel using single crystal diamond tool. J Mater Process Technol 211(12):2152–2159
70. Dai TF, Fang FZ, Hu XT (2009) Tool wear study in diamond turning of steels. J Vacuum Sci Technol B Microelectr Nanometer Struct 27(3):1335–1339
71. Saito H, Jung HJ, Shamoto E, Hara Y, Hara T (2020) Suppression of tool damage in ultra-precision diamond machining of stainless steel by applying electron-beam-excited plasma nitriding. J Int Soc Precis Eng Nanotechnol 63:126–136
72. Stock HR, Schlett V, Kohlscheen J, Mayr P (2001) Characterization and mechanical properties of ion-implanted diamond surfaces. Surf Coat Technol 146:425–429
73. Zhang GL (1985) Ion implantation of diamond dies. Wire industry 52(617):314–317
74. Picraux ST (1982 (by editor)) Metastable materials formation by ion implantation. North-Holland New York
75. Brinksmeier E, Glabe R (2001) Advances in precision machining of steel. CIRP Ann Manuf Technol 50(1):385–388
76. Lee YJ, Shen YK, Wang H (2020) Suppression of polycrystalline diamond tool wear with mechanochemical effects in micromachining of ferrous metal. J Manuf Mater Proces 4(3):81
77. Lee YJ, Hao L, Luder J, Chaudhari A, Wang SY, Manzhos S, Wang H (2019) Micromachining of ferrous metal with an ion implanted diamond cutting tool. Carbon 152:598–608
78. Yuan JL, Wang ZW, Wen DH, Lv BH, Dai Y (2007) Review of the status of ultra-precision machining. J Mechan Eng 43(1):35–48 (in Chinese)
79. Dong J, Mehner A, Gläbe R, Hoffmann F, Mayr P, Brinksmeier E (2004) Schutz von Diamantwerkzeugen vor chemischem Verschleiß. HTM Härtereitechnische Mitteilungen 59(4):284–290
80. Belmonte M, Ferro P, Fernandes AJS, Costa FM, Sacramento J, Silva RF (2003) Wear resistant CVD diamond tools for turning of sintered hardmetals. Diam Relat Mater 12(3–7):738–743
81. Ginting A, Skein R, Cuaca D, Herdianto P, Masyithah Z (2018) The characteristics of CVD- and PVD-coated carbide tools in hard turning of AISI 4340. Measurement 129:548–557
82. Yuan ZW, Jin ZJ, Ma XW, Dong BX (2009) Fabrication and characterization of FeNiCr Matrix-TiC composite for polishing CVD diamond film. J Mater Sci Technol 25(03):319–324
83. Liew WYH (2009) Experimental study on the performance of coated carbide tools in the ultra-precision machining of stainless steel. Tribol Trans 52(3):293–302
84. Ghani AK, Choudhury IA, Husni (2002) Study of tool life, surface roughness and vibration in machining nodular cast iron with ceramic tool. J Mater Process Technol 127(1):17–22
85. Aslan E, Camuscu N, Birgoren B (2007) Design optimization of cutting parameters when turning hardened AISI 4140 steel (63 HRC) with Al2O3+TiCN mixed ceramic tool. Mater Des 28(5):1618–1622
86. Zhao J, Yuan XL, Zhou YH (2010) Cutting performance and failure mechanisms of an Al2O3/WC/TiC micro-nano-composite ceramic tool. Int J Refract Metal Hard Mater 28(3):330–337

87. Fujisaki K, Yokota H, Furushiro N, Yamagata Y, Taniguchi T, Himeno R, Makinouchi A, Higuchi T (2009) Development of ultra-fine-grain binderless cBN tool for precision cutting of ferrous materials. J Mater Process Technol 209(15–16):5646–5652
88. Neo KS, Rahman M, Li XP, Khoo HH, Sawa M, Maeda Y (2003) Performance evaluation of pure CBN tools for machining of steel. J Mater Process Technol 140(1–3):326–331
89. Nishiguchi T, Masuda M (1988) Precision cutting of ferrous metals with single crystal CBN cutting tools. J JSPE 54(2):384
90. Moriwanki T, Mukai T (1990) Machinability of CBN tools in precision machining of stainless steel. Inst Eng Austr 95:129
91. Chou YK, Evans CJ (1997) Tool wear mechanism in continuous cutting of hardened tool steels. Wear 212(1):59–65
92. Knuefermann M, Read R, Nunn R, Clark I, Fleming A (2000) Ultraprazisionsbearbeitung geharteter Stahlbauteile mit Amborite DBN45. IDR 34(3):222–230
93. Dosbaeva GK, El Hakim MA, Shalaby MA, Krzanowski JE, Veldhuis SC (2015) Cutting temperature effect on PCBN and CVD coated carbide tools in hard turning of D2 tool steel. Int J Refract Metal Hard Mater 50:1–8
94. Hatefi S, Abou-El-Hossein K (2020) Review of non-conventional technologies for assisting ultra-precision single-point diamond turning. Int J Adv Manuf Technol 111(9–10):2667–2685
95. Gaidys R, Dambon O, Ostasevicius V, Dicke C, Narijauskaite B (2017) Ultrasonic tooling system design and development for single point diamond turning (SPDT) of ferrous metals. Int J Adv Manuf Technol 93(5–8):2841–2854
96. Jung HJ, Hayasaka T, Shamoto E (2018) Elliptical vibration cutting for difficult-to-cut materials, pp 1–43
97. Zhang JG, Suzuki N, Shamoto E (2019) Advanced applications of elliptical vibration cutting in micro/nanomachining of difficult-to-cut materials, pp 167–200
98. Kurniawan R, Kumaran ST, Ali S, Nurcahyaningsih DA, Kiswanto G, Ko TJ (2018) Experimental and analytical study of ultrasonic elliptical vibration cutting on AISI 1045 for sustainable machining of round-shaped microgroove pattern. Int J Adv Manuf Technol 98(5–8):2031–2055
99. Bai W, Sun RL, Gao Y, Leopold J (2015) Analysis and modeling of force in orthogonal elliptical vibration cutting. Int J Adv Manuf Technol 83(5–8):1025–1036
100. Geng D, Zhang D, Li Z, Liu D (2017) Feasibility study of ultrasonic elliptical vibration-assisted reaming of carbon fiber reinforced plastics/titanium alloy stacks. Ultrasonics 75:80–90
101. Chen JB, Fang QH, Wang CC, Du JK, Liu F (2016) Theoretical study on brittle–ductile transition behavior in elliptical ultrasonic assisted grinding of hard brittle materials. Precis Eng 46:104–117
102. Saito H, Jung H, Shamoto E, Wu TC, Chien JT (2017) Mirror surface machining of high-alloy steels by elliptical vibration cutting with single-crystalline diamond tools Influence of alloy elements on diamond tool damage. J Int Soc Precis Eng Nanotechnol 49:200–210
103. Liu DF, Yan RM, Chen T (2017) Material removal model of ultrasonic elliptical vibration-assisted chemical mechanical polishing for hard and brittle materials. Int J Adv Manuf Technol 92(1–4):81–99
104. Zhang JG, Cui T, Ge C, Sui YX, Yang HJ (2016) Review of micro/nano machining by utilizing elliptical vibration cutting. Int J Mach Tools Manuf 106:109–126
105. Jung HJ, Hayasaka T, Shamoto E (2018) Study on process monitoring of elliptical vibration cutting by utilizing internal data in ultrasonic elliptical vibration device. Int J Precis Eng Manuf-Green Technol 5(5):571–581
106. Shamoto E, Moriwaki T (1994) Study on elliptical vibration cutting. CIRP Ann 43(1):35–38
107. Shamoto E, Suzuki N, Tsuchiya E, Hori Y, Inagaki H, Yoshino K (2005) Development of 3 DOF ultrasonic vibration tool for elliptical vibration cutting of sculptured surfaces. CIRP Ann Manuf Technol 54(1):321–324
108. Moriwaki T, Shamoto E (1991) Ultraprecision diamond turning of stainless steel by applying ultrasonic vibration. CIRP Ann 40(1):559–562

109. Moriwaki T, Shamoto E (1995) Ultrasonic elliptical vibration cutting. CIRP Ann 44(1):31–34
110. Shamoto E, Moriwaki T (1999) Ultaprecision diamond cutting of hardened steel by applying elliptical vibration cutting. CIRP Ann 48(1):441–444
111. Zang DY, Liu YH, Geng DX, Jiang XG (2019) Research progress of ultrasonic processing technology. Electroform Mold 5:1–10 (in Chinese)
112. Sui H, Zhang XY, Zhang DY, Jiang XG, Wu RBA (2017) Feasibility study of high-speed ultrasonic vibration cutting titanium alloy. J Mater Process Technol 247:111–120
113. Zhang XY, Sui H, Zhang DY, Jiang XG (2018) An analytical transient cutting force model of high-speed ultrasonic vibration cutting. Int J Adv Manuf Technol 95(9–12):3929–3941
114. Cerniway MA (2002) Elliptical diamond milling: kinematics, force and tool wear
115. Jin M, Murakawa M (2001) Development of a practical ultrasonic vibration cutting tool system. J Mater Process Technol 113(1–3):342–347
116. Kim HS, Kim SI, Lee KI, Lee DH, Bang YB, Lee KI (2009) Development of a programmable vibration cutting tool for diamond turning of hardened mold steels. Int J Adv Manuf Technol 40(1–2):26–40
117. Overcash JL, Cuttino JF (2009) Design and experimental results of a tunable vibration turning device operating at ultrasonic frequencies. J Int Soc Precis Eng Nanotechnol 33(2):127–134
118. Moriwaki T (2010) Development of 2DOF ultrasonic vibration cutting device for ultraprecision elliptical vibration cutting. Adv Precis Eng 447–448:164–168
119. Klocke F, Olaf D, Benjamin B (2011) Tooling system for diamond turning of hardened steel moulds with apsheric or non-rotational symmetrical geometries. In: Proceedings of the 11th international Euspen conference, Como, Italy
120. Ammouri AH, Hamade RF (2011) BUEVA: a bi-directional ultrasonic elliptical vibration actuator for micromachining. Int J Adv Manuf Technol 58(9–12):991–1001
121. Tan RK, Zhao XS, Zou XC, Sun T (2018) A novel ultrasonic elliptical vibration cutting device based on a sandwiched and symmetrical structure. Int J Adv Manuf Technol 97(1–4):1397–1406
122. Yin Z, Fu YC, Xu JH, Li H, Cao ZY, Chen YR (2016) A novel single driven ultrasonic elliptical vibration cutting device. Int J Adv Manuf Technol 90(9–12):3289–3300
123. Xu WX, Zhang LC (2015) Ultrasonic vibration-assisted machining: principle, design and application. Adv Manuf 3(3):173–192
124. Pan Y, Kang R, Dong ZG, Du WH, Yin S, Bao Y (2020) On-line prediction of ultrasonic elliptical vibration cutting surface roughness of tungsten heavy alloy based on deep learning. J Intell Manuf
125. Lin J, Lu M, Zhou X (2016) Development of a non-resonant 3D elliptical vibration cutting apparatus for diamond turning. Exp Tech 40(1):173–183
126. Zhou M, Eow YT, Ngoi BKA, Lim EN (2003) Vibration-assisted precision machining of steel with PCD tools. Mater Manuf Processes 18(5):825–834
127. Babitsky VI, Mitrofanov AV, Silberschmidt VV (2004) Ultrasonically assisted turning of aviation materials: simulations and experimental study. Ultrasonics 42(1–9):81–86
128. Ma CX, Shamoto E, Moriwaki T, Wang LJ (2004) Study of machining accuracy in ultrasonic elliptical vibration cutting. Int J Mach Tools Manuf 44(12–13):1305–1310
129. Ma CX, Shamoto E, Moriwaki T (2004) Study on the thrust cutting force in ultrasonic elliptical vibration cutting. Mater Sci Forum 471–472:396–400
130. Zhang YL, Zhou ZM, Xia ZH (2005) Research on the influence of vibration parameters on diamond cutting stainless steel parts. J Dalian Univ Tech 45(6):819–822
131. Mitrofanov AV, Ahmed N, Babitsky VI, Silberschmidt VV (2005) Effect of lubrication and cutting parameters on ultrasonically assisted turning of Inconel 718. J Mater Process Technol 162:649–654
132. Lu ZS, Yang L (2006) Research and simulation of the influence of ultrasonic vibration cutting frequency on cutting force. Aviat Precis Manuf Technol 05:10–14 (in Chinese)
133. Ahmed N, Mitrofanov AV, Babitsky VI, Silberschmidt VV (2007) 3D finite element analysis of ultrasonically assisted turning. Comput Mater Sci 39(1):149–154

134. Fang FZ, Wang L (2008) Cutting of tool steels under ultrasonic vibration. Modern Manuf Eng 11:54–58 (in Chinese)
135. Wang L (2008) Study on ultrasonic assisted vibration diamond mirror cutting technology for carbon steel. Tianjin University, pp 40–50 (in Chinese)
136. Zhou ZM, Zhang YL, Dong J, Li XY (2009) Effect of cutting parameters on diamond tool life during cutting stainless steel. Mater Sci Forum 626–627:99–104
137. Overcash JL, Cuttino JF (2009) In-process modeling of dynamic tool-tip temperatures of a tunable vibration turning device operating at ultrasonic frequencies. J Int Soc Precis Eng Nanotechnol 33(4):505–515
138. Song YC, Nezu K, Park CH, Moriwaki T (2009) Tool wear control in single-crystal diamond cutting of steel by using the ultra-intermittent cutting method. Int J Mach Tools Manuf 49(3–4):339–343
139. Shi XH, Li Z, Shao H (2010) Analysis of ultrasonic vibration—assisted turning temperature field. Tool Eng 07:45–48 (in Chinese)
140. Zhang XQ, Senthil KA, Rahman M, Nath C, Liu K (2012) An analytical force model for orthogonal elliptical vibration cutting technique. J Manuf Process 14(3):378–387
141. Jiao F, Liang Z, Li Y, ChongYang Z, Ying N (2013) The tool wear mechanism of ultrasonic vibration cutting and its effect on the surface roughness of workpiece. J South Univ Sci Technol (Nat Sci Edn) 32(4):449–453 (in Chinese)
142. Zhang YL, Zhou ZM, Lv Y, Wang JL, Shao L, Lqbal A (2013) Wear behavior of natural diamond tool in cutting tungsten-based alloy. Int J Adv Manuf Technol 69(1–4):329–335
143. Hatefi S, Abou-El-Hossein K (2020) Review of hybrid methods and advanced technologies for in-process metrology in ultra-high-precision single-point diamond turning. Int J Adv Manuf Technol 111(1–2):427–447
144. Zhang YL, Zhou ZM, Hhuang CY, Xia ZH (2004) Study on vibration and gas - protected cutting technology of natural diamond. Mechan Sci Technol 3:339–340
145. li ZJ (2013) Study on the technology and mechanism of ultra-precision cutting black metal. Tianjin University
146. Tang QC, Yin SH, Chen FJ, Huang S, Luo H (2018) New technology for cutting ferrous metal with diamond tools. Diam Relat Mater 88:32–42
147. Huang S, Liu X, Chen FZ, Zheng HX, Yang XL, Wu LB, Song JL, Xu WJ (2016) Diamond-cutting ferrous metals assisted by cold plasma and ultrasonic elliptical vibration. Int J Adv Manuf Technol 85(1–4):673–681
148. Zhang XQ, Senthil KA, Rahman M, Nath C, Liu K (2011) Experimental study on ultrasonic elliptical vibration cutting of hardened steel using PCD tools. J Mater Process Technol 211(11):1701–1709
149. Li GG, Rahim MZ, Pan WC, Wen C, Ding SL (2020) The manufacturing and the application of polycrystalline diamond tools—a comprehensive review. J Manuf Process 56:400–416
150. Zhang XQ, Deng H, Liu K, Wang H (2017) Effect of air blow pressure in ultrasonic vibration cutting of steel using PCD tools. In: Proceedings of the 17th international conference of the European society for precision engineering and nanotechnology, EUSPEN, pp 123–124
151. Zhang XQ, Deng H, Liu K (2019) Oxygen-shielded ultrasonic vibration cutting to suppress the chemical wear of diamond tools. CIRP Ann 68(1):69–72
152. Saito H, Jung HJ, Shamoto E (2016) Elliptical vibration cutting of hardened die steel with coated carbide tools. Precis Eng 45:44–54
153. Wang JS, Fang FZ, Yan GP, Guo YB (2019) Study on diamond cutting of ion implanted Tungsten Carbide with and without ultrasonic vibration. Nanomanuf Metrol 2(3):177–185
154. Willert M, Zielinski T, Rickens K, Riemer O, Karpuschewski B (2020) Impact of ultrasonic assisted cutting of steel on surface integrity. Procedia CIRP 87:222–227
155. Brinksmeier E, Preuß W (2012) Micro-machining. Phil Trans R Soc A 370:3973–3992
156. Yip WS, To S (2017) Tool life enhancement in dry diamond turning of titanium alloys using an eddy current damping and a magnetic field for sustainable manufacturing. J Clean Prod 168:929–939

157. Nakano F, Yanagiharra K, Yamaguchi H et al (2000) Basic study on magnetic-field assisted cutting. II. Improvement of machinability in cutting austenitic stainless steel. Seisan-Kenkyu. J Ins Indus Sci Univ Tokyo (Japan) 52(9):42–45
158. Patwari M, Mahmood MN, Noor S (2013) Investigation of machinability responses during magnetic field assisted turning process of preheated mild steel. Procedia Eng. 56:713–718
159. Zhang JZ, Li XR, Wang ZL (2001) Test for cutting process of technology system with magnetic field. J Liaoning Tech Univ 20:89–92
160. Zhang L, Liu Z (2013) Study on effect of magnetization treatment on hardness of high speed steel. Hot Work Technol 16:85–87
161. Ma LP, Liang ZQ, Wang XB et al (2015) Influence of pulsed magnetic treatment on microstructures and mechanical properties on M42 high speed steel tool. J Metal 51(3)
162. Ikawa N, Donaldson RR, Komanduri R et al (1991) Ultraprecision metal cutting—the past, the present and the future. CIRP Ann 40(2):587–594
163. Wei W, Li J, Chen Y et al (2020) Analysis of characteristics of arc shape caused by intermittent cutting force and dynamic vibration in ultra-precision cutting processes. Proc Instit Mechan Eng Part B J Eng Manuf 234(1):095440542092873
164. Yip WS, To S (2017) Reduction of material swelling and recovery of titanium alloys in diamond cutting by magnetic field assistance. J Alloys Compd 722:525–531
165. Yip WS, To S (2017) An application of eddy current damping effect on single point diamond turning of titanium alloys. J Phys D-Appl Phys 50(43):435002
166. Yip WS, To S (2019) Reduction of tool tip vibration in single point diamond turning using an eddy current damping effect. Int J Adv Manuf Technol 103(5–8):1799–1809
167. Wang JP, Wen YF, Han JH, Ma S, Zhang GQ (2022) A novel tool wear suppression method for polycrystalline diamond by applying magnetic field in turning of ferrous materials. P I Mech Eng B-J Eng 236(13):1772–1781
168. Cui HY, Zhang HF, Guo QL (2005) Discussion on electronic spin in a localization and uniform magnetic field. J Jiamusi Univ 23(2):324–326

# Chapter 6
# Cutting Process in Machining Ferrous Materials

**Abstract** Intermittent cutting techniques, particularly EVC, have emerged as a promising solution for decreasing tool wear during SPDT of ferrous materials. This unique intermittent mechanism enhances the accessibility of cutting fluid to the tool-workpiece contact area and therefore influence the machined surface. This chapter aims to provide a comprehensive understanding of the cutting process in diamond cutting ferrous materials by examining the benefits of intermittent cutting technology and the critical role of lubricants.

**Keywords** Cutting process · Precision machining · Ferrous materials · Diamond tools

## 6.1 Intermittent Cutting Technology

Ferrous materials, known for their exceptional mechanical properties, find extensive applications in various industries. However, when it comes to SPDT, these materials pose a significant challenge due to their tendency to cause severe tool wear [1]. This severe tool wear hampers the efficient and high-precision application of ferrous materials in fields such as optics [2].

To address this issue, intermittent cutting techniques, particularly EVC, have emerged as a promising solution for minimizing tool wear during SPDT of ferrous materials. EVC involves exciting the tool tip with a single or double excitation, causing it to vibrate in an elliptical trajectory at a high frequency. This intermittent cutting method enhances the accessibility of cutting fluid to the tool-workpiece contact area, thereby allowing for improved surface quality of the workpiece and reduced tool wear.

The introduction of EVC technology by Shamoto in 1994 [3] significantly advanced the utilization of SPDT of ferrous materials.

In the field of SPDT of ferrous metals, EVC has a better tool wear suppression effect than traditional vibration (linear vibration) cutting. As the cutting distance increases, both the cutting forces and the surface roughness show an upward trend.



G. Zhang and J. Wang, *Material-Oriented Cutting Processes in Precision Machining*, Springer Tracts in Mechanical Engineering,
https://doi.org/10.1007/978-981-96-2504-8_6

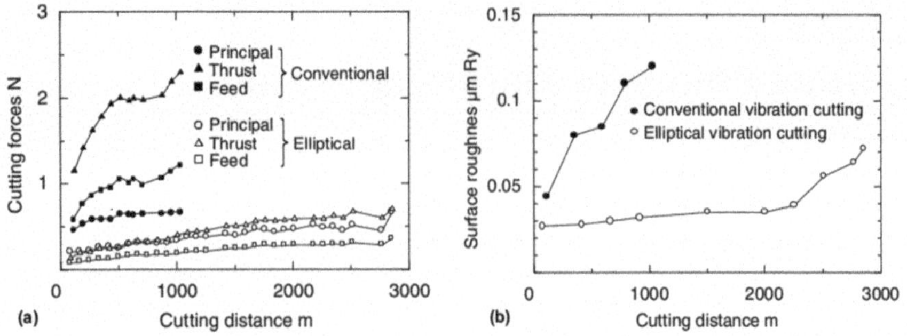


**Fig. 6.1** Turning performance of diamond tools: **a** relationship between cutting forces and cutting distances, **b** correlation between surface roughness and cutting distances

Figure 6.1 provides a summary of the cutting forces and surface roughness observed at different cutting distances [4].

Nonetheless, when employing EVC, notably lower cutting forces, especially the thrust force, are recorded compared to traditional vibration cutting. Moreover, optical quality surfaces with a maximum roughness of less than 0.05 mm Rz can be achieved up to a cutting distance of 2250 m with EVC. On the contrary, in conventional vibration cutting, surface roughness significantly escalates at around 300 m due to the chipping of the diamond cutting edge.

Compared to traditional non-vibration cutting, Wang et al. [5] conducted experiments to exhibit the benefits of EVC in enhancing surface quality and reducing tool wear when diamond cutting hardened steel. In conventional cutting, severe tool wear and a decline in surface quality were observed at a cutting distance of 150 m, as shown in Fig. 6.2a. In contrast, EVC resulted in minimal wear on the diamond tool's cutting edge and maintained a machined Rz below 40 nm even at an extended cutting distance of 190 m, as illustrated in Fig. 6.2b.

EVC is a machining process where the tool tip vibrates in an elliptical trajectory induced by a single or double excitation at high frequency. Material is intermittently removed as the tool initiates cutting at position B during each vibration cycle, resulting in the formation of chips. Position C marks the cutting end point as the tool gradually separates from the workpiece material. To ensure the tool remains detached from the workpiece material during each vibration cycle, the cutting feed rate is kept below the maximum vibration speed. This process is illustrated in Fig. 6.3 [6].

EVC provides several advantages over conventional cutting processes. This method employs intermittent cutting to efficiently minimize chip thickness and cutting force. This periodic cutting process enables effective cooling of the tool and workpiece material through the use of cutting fluid or the surrounding air in the cutting area. As a result, the temperature in the cutting zone decreases, preventing the workpiece material from sticking to the cutting tool's rake face [7–9]. This effective suppression of electrochemical wear in diamond tools enables the long-distance mirror surface machining of iron-based materials, as shown in Fig. 6.3 [4].

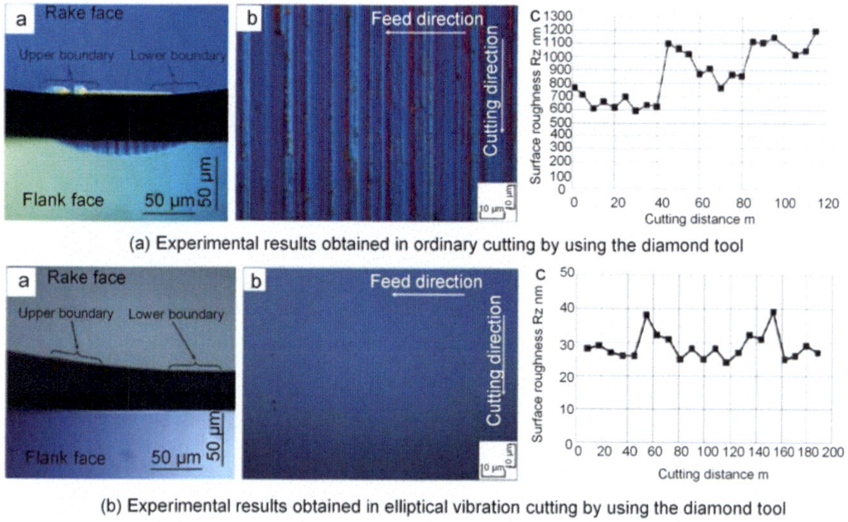


**Fig. 6.2** Comparison between traditional cutting and ultrasonic elliptical vibration machining quality and tool wear

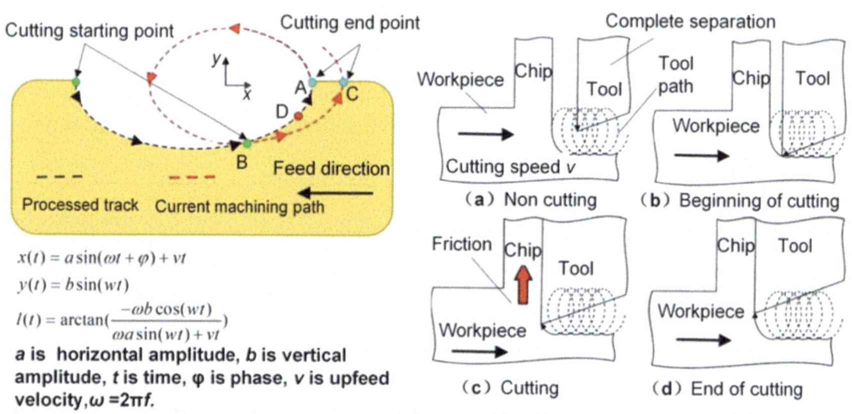


**Fig. 6.3** EVC principle [4]

Currently, UVC devices can be found not only in laboratory environments, but also in a variety of industrial and commercial products [10]. Leading manufacturers in this sector include TAGA Electric in Japan and Son-X from the Fraunhofer Institute for Production Technology (IPT) in Aachen, Germany. For example, TAGA Electric has introduced the Sonic Impulse EL-50 series of EVC devices, which offer vibration frequencies exceeding 30 kHz (Fig. 6.4).

Moreover, numerous scientific research institutes worldwide have also developed various ultrasonic vibration systems. As the technological society continues to

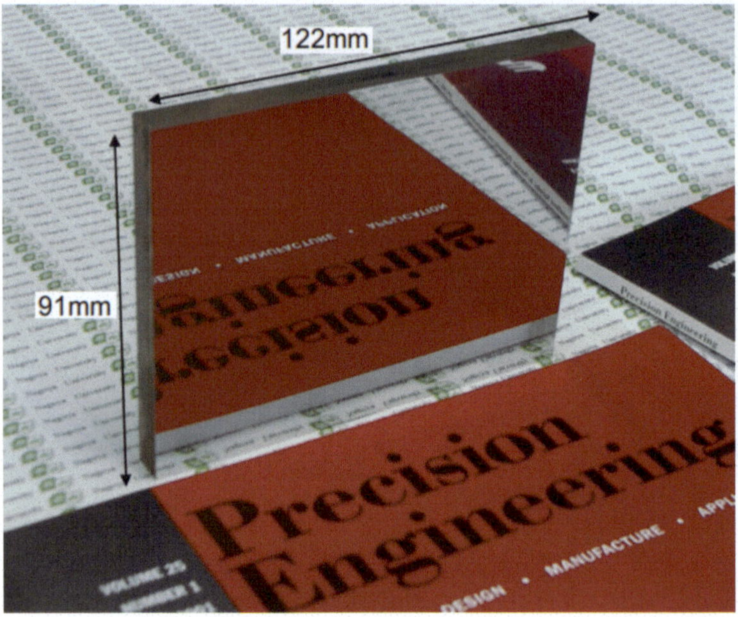


**Fig. 6.4** Photograph of large hardened-steel mirror surface finished by EVC [4]

progress, it is expected that more commercial ultrasonic vibration devices will be manufactured to meet the high standards required for SPDT of ferrous materials.

## 6.2 Lubricant Effects

In comparison to environments without lubricant (NL), SPDT utilizes various lubrication methods that rely on different lubrication media and techniques. These methods include high pressure gas machining (HPGM), minimum quantity lubrication machining with oil (MQLM-oil), and minimum quantity lubrication machining with water (MQLM-water).

MQLM conditions are widely considered to result in a slight reduction in tool wear during the ultra-precision machining of ferrous metals, as compared to HPGM conditions. In traditional turning, lubricating the tip flank face through the MQLM technique has been found to decrease tool wear and enhance tool life in comparison to HPGM conditions [11]. However, this conclusion may not be applicable to ultra-precision cutting, as the lubricant has limited cooling and lubricating effects on the tool tip [12]. Initial experiments have also disproven this notion, with MQLM conditions resulting in several times higher tool wear than HPGM conditions. Furthermore, several previous studies indicate that lubricants can be detrimental to both human health and the environment [13]. To investigate environmentally friendly Ti-6Al-4V cutting performance, cryogenic gas and the MQLM processing environment are utilized for single-point diamond tool cutting of Ti-6Al-4V, which is not typically regarded as "diamond turnable" materials. However, certain essential experimental aspects have been neglected thus far, specifically regarding the role of lubricant in the process of diamond-cutting ferrous metals.

Currently, the preferred material for optical molds is ferrous metal. Additionally, ultra-precision diamond cutting enables the direct machining of high-precision and intricate optical surfaces [15], making more efficiency. However, the use of single crystal diamond for machining ferrous metals is unsatisfactory due to poor surface quality and significant tool wear. Extensive tests have been undertaken under various cutting circumstances (MQLM-oil, MQLM-water, and HPGM) to get a comprehensive knowledge of the mechanism involved and examine the effects of environmentally friendly processing conditions on diamond-cutting ferrous metals. These experiments encompass various aspects, including tool wear patterns, chip characteristics, and workpiece surface performance. The findings of these experiments are of great significance for future applications of lubricants in intermittent cutting technology.

## *6.2.1 Machined Surface*

### 6.2.1.1 Experimental Design

In the cutting experiments, SCD tools were employed, each featuring a nose radius of 0.5 mm, a rake angle of 0°, a clearance angle of 10°, and a cutting edge radius of approximately 100 nm. To ensure consistency in the contrast experiments, the tool cutting edge was dispersed across an arc blade with an included angle of 110°, as shown in Fig. 6.5c. The cutting experiments utilized the tool's main edge and two side edges, with a 45° angle between them. It is worth noting that the main edge exhibits slightly better wear resistance than the side edge [16, 17].

The experimental cutting of NAK 80 workpieces was carried out using the Nanotech 450UPL. To maintain consistent cutting distances and theoretical material removal, a depth-of-cut ($D_c$) of 5 μm and a feed rate of 2.5 mm/min were set. Cylindrical workpieces with a diameter of 10 mm were used with spindle speeds set to 500 rpm. Prior to the cutting, the experiment area on the ultra-precision machine tool was preliminary turned using a common CBN tool, resulting in an initial surface roughness of 190 nm.

During the implementation of the MQLM method, oil was sprayed onto the cutting zone at a flow rate of 0.2 ml/min throughout the cutting process. Under both the MQLM and HPGM circumstances, pressurized gas was chosen and supplied to the cutting zone through a nozzle with a diameter of 2.3 mm. The pressure was set at 0.231 megapascals, and the flow rate was set at 32.7 L per minute. The oil used in this study is composed of isoparaffin.

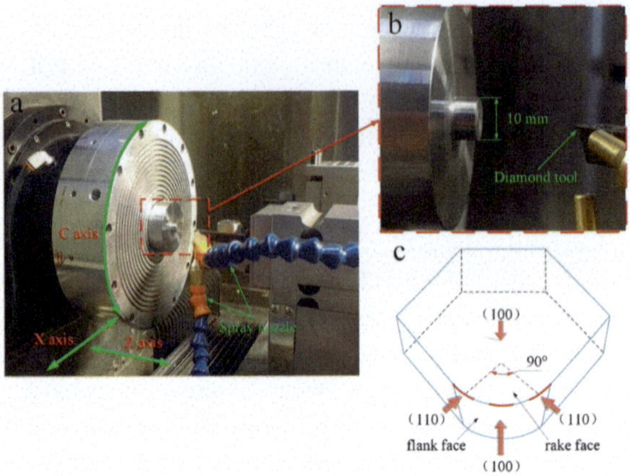


**Fig. 6.5** **a** Experimental set up and its **b** enlarged view, **c** Schematic diagram of diamond tool

**Table 6.1** Details of cutting experiments

| Diamond tool | Cutting condition | Cutting distance (m) | Tool wear |
|---|---|---|---|
| Side edge of tool A | MQLM-oil | 15 | Catastrophic |
| Side edge of tool A | HPGM | 15 | General |
| Main edge of tool A | MQLM-oil | 15 | Catastrophic |
| Side edge of tool B | MQLM-oil | 15 | Catastrophic |
| Side edge of tool B | MQLM-water | 15 | Catastrophic |
| Side edge of tool A | HPGM | 75 | General |
| Side edge of tool A | HPGM | 135 | General |

In order to thoroughly investigate the reasons behind the disparities between preliminary results and traditional viewpoints, factors such as lubricant composition, cutting edge direction, and cutting tool material were comprehensively examined. Detailed information regarding the cutting experiments, including tool wear results, is presented in Table 6.1. In this context, the term "catastrophic" refers to the flank face wear belt with a longitudinal width exceeding 20 μm. However, in general, it refers to that is limited to within 8 μm.

#### 6.2.1.2 Wear Characteristics of the Diamond Tool

Preliminary studies have yielded unexpected results in terms of tool wear when comparing different cutting environments, as depicted in Fig. 6.6. Figure 6.6a shows the flank face of a new diamond tool, while Fig. 6.6b displays an optical microphotograph of the side edge of the same diamond tool after being used in the MQLM-oil cutting environment for approximately 15 m of cutting distance. Figure 6.6c presents the main edge of the diamond tool in the MQLM-oil cutting environment. It is evident from the tool wear pattern that substantially less wear occurs in the HPGM conditions in comparison to MQLM conditions (refer to Fig. 6.6f for results under the HPGM conditions). The results shown in Fig. 6.6b, c indicate that the tool wear in the MQLM-oil cutting environment is not solely attributed to the crystal orientation selection of the tool cutting edge.

In addition to the aforementioned experiments, the same specimen was machined using the main edge of another diamond tool in the MQLM-oil cutting environment to further investigate the reason for the discrepancy in tool wear between the HPGM and MQLM environments. The tool wear pattern observed in Fig. 6.6d is almost identical, providing further evidence. Furthermore, Fig. 6.6e presents the cutting edge in the MQLM-water cutting environment, and the results demonstrate that tool wear depicted in Fig. 6.6b–e is not directly related to the lubricant composition.

Figure 6.6f showcases the tool wear under the HPGM conditions after turning 135 m. Tool wear under the HPGM conditions is significantly lower than in the MQLM cutting environment, even when considering that the cutting distance in the HPGM environment is ten times greater than in the MQLM environment.

Figure 6.6g illustrates the diamond tool wear outcomes under different turning conditions. The wear level is represented by the longitudinal width of the wear area on the tool's flank face. After turning of 15 m, the average tool wear under MQLM conditions exceeds 20 μm, compared to only 2 μm under HPGM conditions. As the cutting distance increases, the tool wear under HPGM conditions gradually rises. However, when the cumulative cutting distance reaches 135 m, the tool wear remains below 8 μm. This suggests that MQLM conditions lead to more extensive tool damage than HPGM conditions.

When observing the picture of tool wear under MQLM conditions in Fig. 6.6, pectinate micro-grooves can be seen on the flank surface, which is an indication of tool graphitization. This phenomenon has been reported in previous studies as well [18]. The tool wear observed under MQLM conditions is primarily attributed to chemical wear.

#### 6.2.1.3 Chip Analysis

The photographs captured by SEM, Fig. 6.7a, b, were used to examine the chips produced in the MQLM and HPGM after turning 15 m. For after turning 75 and 135 m in the HPGM environment, the chip shapes are shown in Fig. 6.7c, d. Under the current experimental conditions, it was observed that the chip shape obtained in the HPGM environment was more complete and stable. In Fig. 6.7a, the red-dotted box represents the ideal and complete contour chip shape, while the yellow-dotted box indicates the flaw area in cutting. Through SEM analysis, it was found that this flaw is common under the MQLM conditions and it affects the integrity of the chip as the cutting progresses. The experimental results in Fig. 6.7b–d under the HPGM conditions demonstrate that increasing the cutting distance gradually reveals chip flaws. These flaws can be attributed to tool wear, and the flaw area on the chip becomes larger as the tool wear increases. This is further supported by

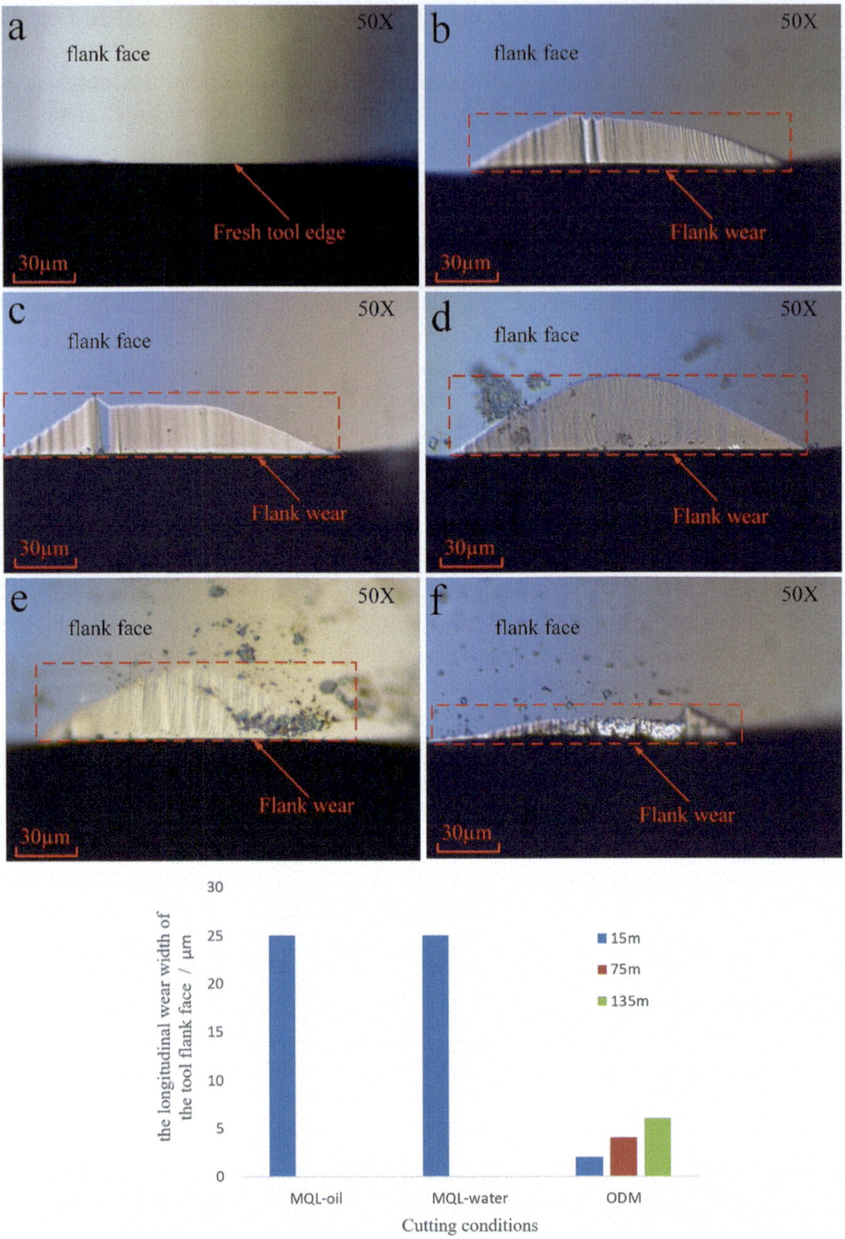


**Fig. 6.6** Tool wear photographs: **a** Fresh cutting edge, and **b–d** lubricated by MQLM-oil, **e** MQLM-water, and **f** HPGM

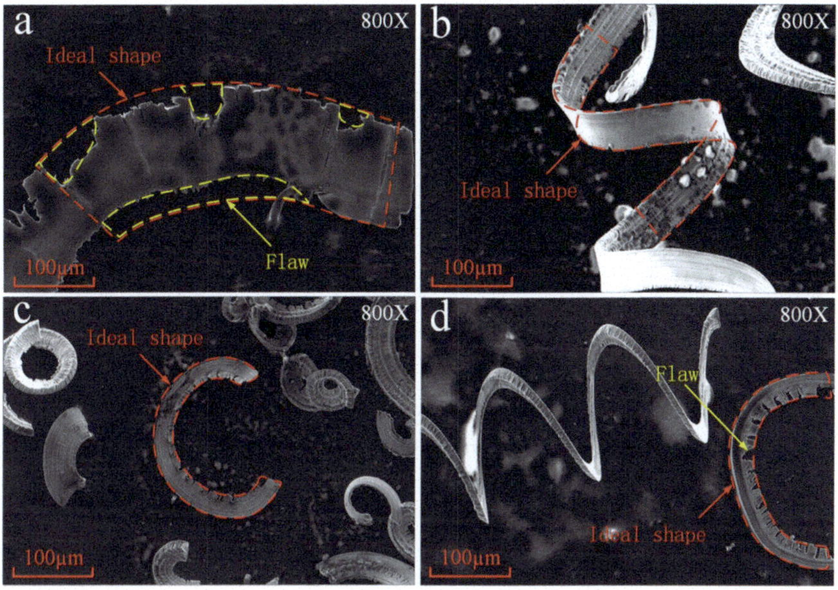


**Fig. 6.7** SEM photographs of various chip's morphology: **a** MQLM-oil cutting 15 m, **b** HPGM cutting 15 m, **c** HPGM cutting 75 m, and **d** HPGM cutting 135 m

the findings in Sect. 3.1.2 regarding tool wear. Additionally, the tool wear amount differs, causing variations in the chip width due to changes in the tool edge width after the tool wears during actual cutting. By measuring the chip widths in the SEM photos at ×800 magnification, it was observed that chip widths obtained in the HPGM environments are below 45 μm, whereas those in the MQLM environments are above 110 μm, exceeding the theoretical value (70 μm) based on the diamond geometry, depth-of-cut, and feed rate.

The different tool wear patterns observed in the cutting environment, as discussed in Sect. 6.2.1.2, are the primary cause of the variation. In order to analyze and explain this intriguing chip phenomenon, a schematic diagram illustrating different tool wear patterns when cutting with a diamond tool is presented in Fig. 6.8.

In Fig. 6.8a, the schematic diagram represents a fresh diamond tool cutting the workpiece surface. The theoretical chip widths are calculated based on the diamond tool geometry, depth-of-cut, and feed rate. Moving on to Fig. 6.8b, it depicts a scenario where the front edge of the diamond tool has worn off while cutting the workpiece surface. In this case, the chip widths are influenced by the diamond tool's geometry, the actual depth-of-cut, and the width of the tool wear.

Figure 6.8c illustrates a severe chemical wear on the diamond tool while cutting the workpiece surface. In this situation, the chip widths solely depend on the width of the tool wear. This indicates that the tool wear pattern directly contributes to the differences observed in chip widths.

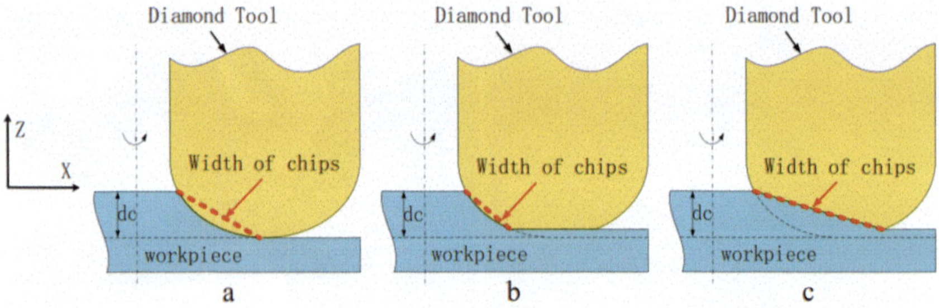


**Fig. 6.8** Schematic diagram of diamond cutting using: **a** fresh diamond tool, **b** sliver tool wear, **c** severe tool wear

The analysis results of the EDS chips, as seen in Fig. 6.7a, b, are presented in Fig. 6.9. The sampling areas were randomly distributed on the back surface of the chip, as depicted in Fig. 6.9a, b.

The results reveal that the carbon content in the chip under the MQLM condition is significantly higher compared to the HPGM condition. Specifically, the carbon content in the MQLM chip is measured to be 48.82%, while it is 16.25% in the HPGM chip. Additionally, the EDS analysis detected 5.86% oxygen in the chips under the MQLM cutting conditions, which is the most commonly identified element apart from iron and carbon. This observation suggests that oxygen from the air might contribute to the chemical wear, which has also been reported in other literature [19].

The EDS measurement results presented in Fig. 6.9c correspond to the free face of the chip and can be interpreted as the EDS data for the original material due to the limited depth of the cutting experiment. While the results do not explicitly mention the carbon content, it is evident in the peak. It is worth noting that the carbon content in this scenario is notably lower in comparison to the findings illustrated in Fig. 6.9a, b.

#### 6.2.1.4 Effects of Cutting Conditions on the Workpiece Surface

Figure 6.10 displays the results obtained under the MQLM condition, while Fig. 6.11 presents the results achieved under the HPGM condition. By comparing the photo simulations in Figs. 6.10b and 6.11b, it can be observed that pits are visible on the workpiece surface during HPGM. Furthermore, Figs. 6.10c and 6.11c show sections of the workpiece surface that are 25.5 μm long in both the MQLM and HPGM environments. The section positions are marked in Figs. 6.10a and 6.11a. It is notable that in the HPGM environment, a distinct outline of the tool path can be observed.

The micro-hardness values of the machined surfaces, as observed in Figs. 6.10 and 6.11, are presented in Fig. 6.12. The measurements were conducted using an FMARS9000 automatic micro-hardness tester, with sample points taken along the feed direction. Two loading conditions were employed, with forces of 25 gf and 10 gf.

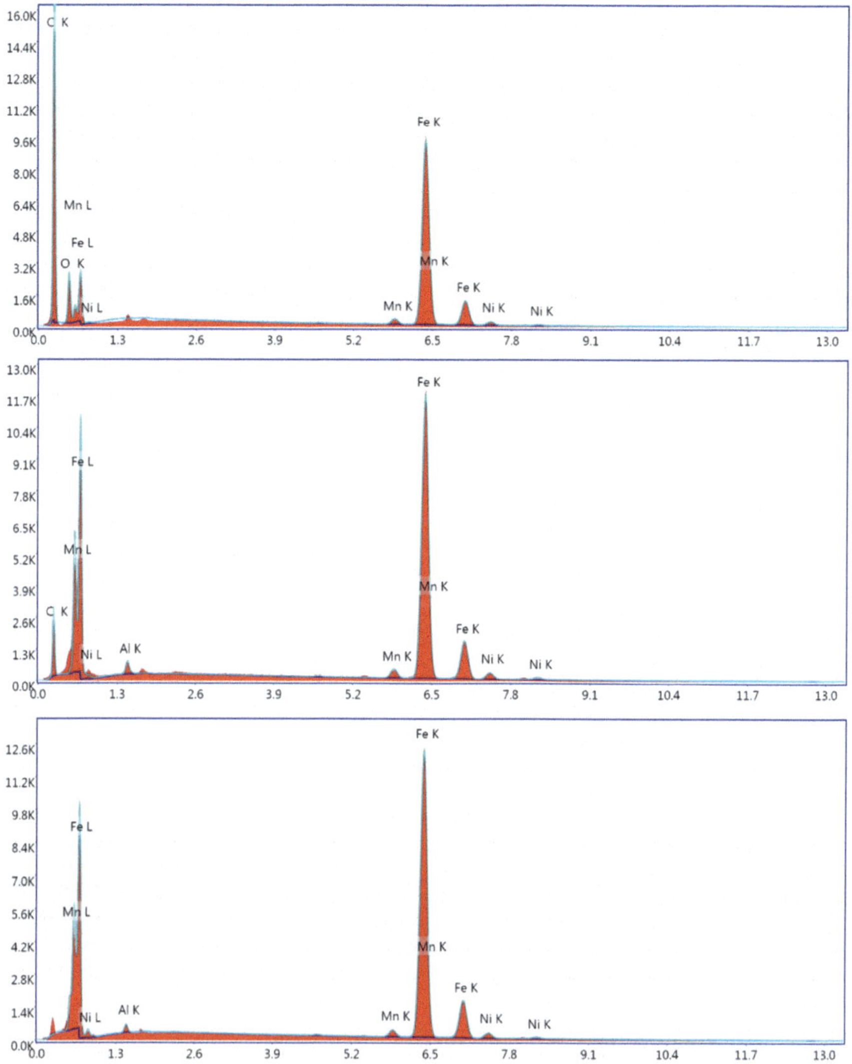


**Fig. 6.9** Chips of EDS analysis results: **a** back surface of chip under MQLM, **b** back surface of chip under HPGM and **c** free surface of chip

The micro-hardness of surfaces machined under the HPGM condition was notably higher than that under the MQLM condition, a common observation during the processing of various materials [20]. The difference in surface micro-hardness between the two conditions can be attributed to significant work hardening effects. Strain aging, recrystallization, and plastic deformation on the workpiece surface

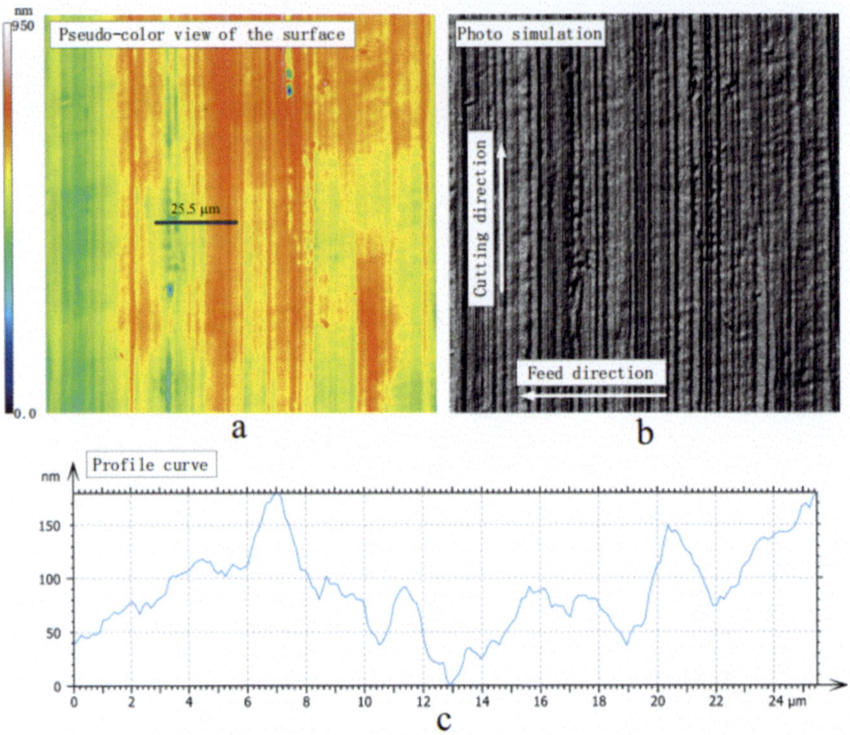


**Fig. 6.10** Measured surface morphology in the MQLM environment: **a** pseudo-color view of the surface, **b** photo simulation, **c** profile curve

occur due to thermal effects (e.g., high temperature and rapid quenching) and mechanical effects (e.g., high stress and strain), serving as the primary factors contributing to the variations in surface micro-hardness [21].

#### 6.2.1.5 Mechanisms of Wear and Surface Formation

The heat that is produced by the cutting tool during the process of material removal has an effect not only on the region that is being processed but also on the area that is around it that has not been treated. This is shown in Fig. 6.13a. Because of the cutting conditions (MQLM or HPGM) and the heat, the hardness of these places goes through a process of change. The hardness of the surface that has been cranked up is evaluated with the use of a hardness tester, and the consequences of this evaluation are explained in Sect. 6.2.2.4. Simultaneously, a change in hardness that is comparable should take place in the region that has not been handled. There is an increase in the presence of hard particles in the workpiece as a result of the hardness of the region that has not been treated and consequently rises. Additionally, this behavior has been reported in a number of additional research [22].

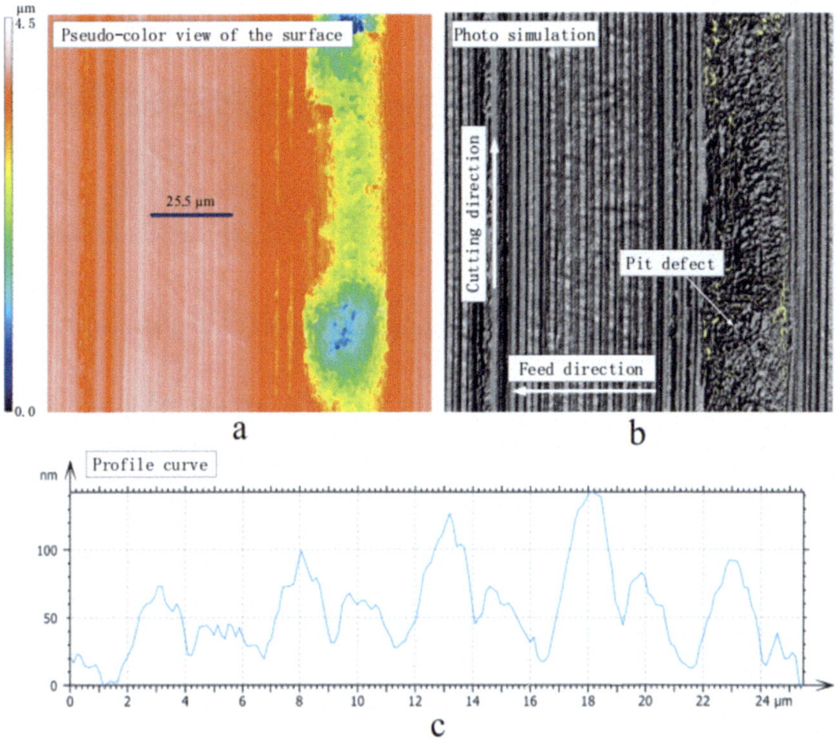


**Fig. 6.11** Measured surface morphology in the NL environment: **a** pseudo-color view of the surface, **b** photo simulation, **c** profile curve

**Fig. 6.12** Machined surface's micro-hardness in the MQLM and HPGM environments

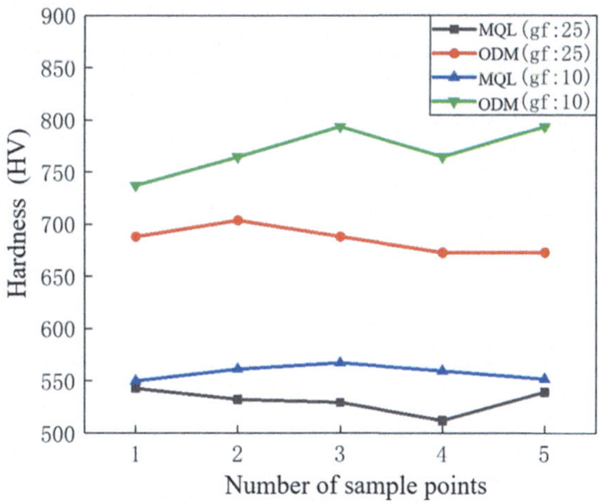

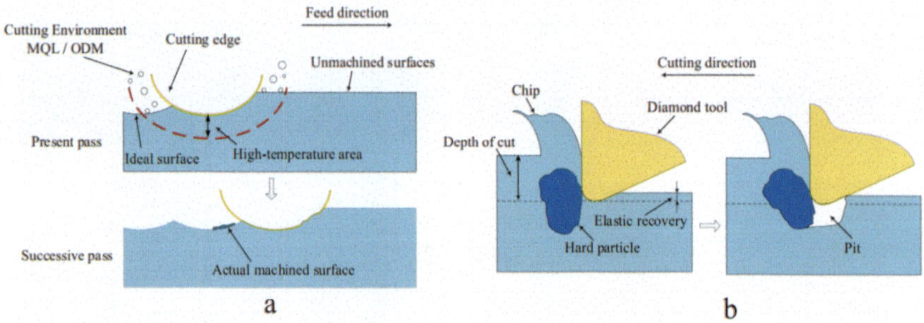


**Fig. 6.13** Schematic diagram of machined surface generation: **a** feed direction **b** cutting direction

Based on the results presented in Sect. 6.2.1.4, it is observed that there are pits on the workpiece surface under the HPGM conditions. These pits emerge on the workpiece surface during material removal when the cutting tool cannot shear hard particles and uproots them. This also explains why HPGM circumstances cause more pits.

Cutting area temperature directly affects tool chemical wear [12]. The tool wear data show that MQLM circumstances do not cool the cutting region, supporting literature [12]. This is because the cutting edge and workpiece are close throughout the turning operation, unlike vibration turning. The ultra-precision turning process is tiny, making it difficult for the lubricant to reach the micron level in the limited region. Thus, lubrication and higher cutting temperatures change surface qualities after cutting and the hardness and elastic modulus of surfaces to be machined. Changes in characteristics affect cutting friction and heat dissipation.

The tool wear rate is directly linked to the temperature within the cutting area [12]. Research findings suggest that under MQLM conditions, the cooling of the cutting area is insufficient, consistent with previous studies [12]. This inadequate cooling can be attributed to the close proximity of the cutting edge to the workpiece throughout the turning process, unlike in vibration turning. Furthermore, the ultra-precision turning process's small scale presents a challenge for the lubricant to penetrate the narrow area down to the micron level. As a result, the combination of lubricant and elevated temperatures in the cutting region not only modifies the surface properties post-cutting but also impacts the properties of the surfaces to be machined, such as hardness and elastic modulus. These property alterations influence the friction coefficient and heat dissipation coefficient during the cutting operation.

The results presented in Sect. 6.2.1.4 demonstrate that under the HPGM conditions, the hardness of the workpiece after turning is actually higher than that under the MQLM conditions. The higher hardness of the workpiece results in greater flank face forces acting on the cutter and increases the likelihood of mechanical wear due to the presence of hard particles. Simultaneously, the elastic recovery of the workpiece material causes the cutting edge involved in the machining to be tightly wrapped by the flank face, significantly influencing the temperature in the cutting area [23]. When the tool is turning soft materials, as shown in Fig. 6.13b, it more easily presses into the workpiece material, leading to a tighter wrapping of the cutting edges [24], which makes heat dissipation from the cutting edge more challenging. This raises the temperature in the cutting area and increases the possibility of chemical wear. Therefore, when turning NAK 80 under the MQLM conditions, the turning area temperature actually increases instead of decreases, resulting in significantly less tool chemical wear compared to the HPGM conditions.

## *6.2.2 Tool Marks*

During SPDT, it is not feasible to observe the instantaneous state of both the workpiece and tool by abruptly pausing the experiment due to the high-speed rotation of the workpiece. Nonetheless, it is crucial to observe and study the instantaneous state of the tool and workpiece materials during the experiment. This is particularly important for examining a complete tool mark, as it provides valuable insights into how lubricant influences tool wear and surface quality. Understanding these dynamics can offer significant benefits in terms of optimizing the cutting process and achieving improved machining outcomes.

### 6.2.2.1 Experimental Design

In order to investigate the impact of lubricant in SPDT of ferrous metals, this section focuses on two sets of single-groove scraping experiments and provides a detailed analysis of the experimental results. The white light interferometer is a commonly used tool for accurately measuring the surface morphology of workpieces in ultra-precision machining. In this section, the morphology of machined surfaces and grooves was captured using the white light interferometer in VSI mode without filtering. Specifically, experiments were conducted using low hardness die steel AISI 4140.

The cutting force was measured using an ultra-precision dynamometer. Prior to the single-groove scraping experiment, the workpieces were rough-machined using a PCD tool. Three machining environments were selected for the rough-machining of the workpiece surface: MQLM-oil, ordinary dry machining (ODM), and HPGM. The surface morphology and roughness obtained using HPGM and ODM were found to be similar. The surface morphology of the specific rough-machined surfaces was captured using the white light interferometer, as depicted in Fig. 6.14. The surface roughness value, Sa (as shown in Fig. 6.14), indicates that MQLM-oil exhibits improved machined quality compared to ODM and HPGM, highlighting the benefits of MQLM-oil in enhancing the machined surface quality.

The single-groove scraping tests are conducted in two phases using four natural single-crystal diamond tools (Yuhe, China) with varying radii: T1 (496.2 μm radius), T2 (490.9 μm radius), T3 (492.6 μm radius), and T4 (483 μm Every diamond tool has a 0° rake angle and 10° clearance angle.

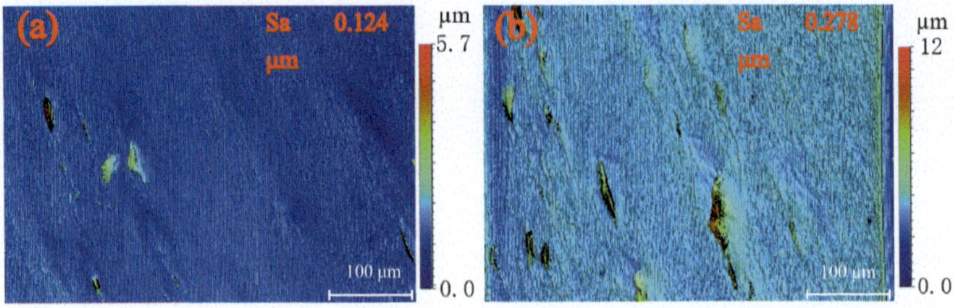


**Fig. 6.14** Machined surface morphology captured by means of the white light interferometer in **a** MQLM-oil and **b** HPGM environments

The experiment began with 360° single-groove scraping on rough-machined surfaces from the three machining environments. The theoretical depth of the single groove was 12 μm, with a *Z*-axis feed of 2 μm. The ultra-precision dynamometer measured cutting force, as illustrated in Table 6.2E-1 to E-3.

The second phase of the experiment was performed on the rough-machined surface from the three machining environments (Fig. 6.15). The circular groove included six uniform portions (N1-N6). The experiment used a 2 μm Z-axis feed and a 60° scraping angle for a single groove. The ultra-precision dynamometer measured cutting force.

**Table 6.2** Detailed parameters in the single-groove scraping experiments

| Group number | Rough-machining environment | Tool number | Turning environment | Single scarping depth * angle | Total scarping depth (μm) |
|---|---|---|---|---|---|
| E-1 | HPGM/ODM/MQLM-oil | T1 | MQLM-oil | 2 μm * 360° | 12 |
| E-2 | HPGM/ODM/MQLM-oil | T2 | HPGM | 2 μm * 360° | 12 |
| E-3 | HPGM/ODM/MQLM-oil | T3 | ODM | 2 μm * 360° | 12 |
| N-1 | HPGM/ODM/MQLM-oil | T4 | MQLM-oil | 2 μm * 360° | 2 |
| N-2 | HPGM/ODM/MQLM-oil | T4 | MQLM-oil | 2 μm * 300° | 4 |
| N-3 | HPGM/ODM/MQLM-oil | T4 | MQLM-oil | 2 μm * 240° | 6 |
| N-4 | HPGM/ODM/MQLM-oil | T4 | MQLM-oil | 2 μm * 180° | 8 |
| N-5 | HPGM/ODM/MQLM-oil | T4 | MQLM-oil | 2 μm * 120° | 10 |
| N-6 | HPGM/ODM/MQLM-oil | T4 | MQLM-oil | 2 μm * 60° | 12 |

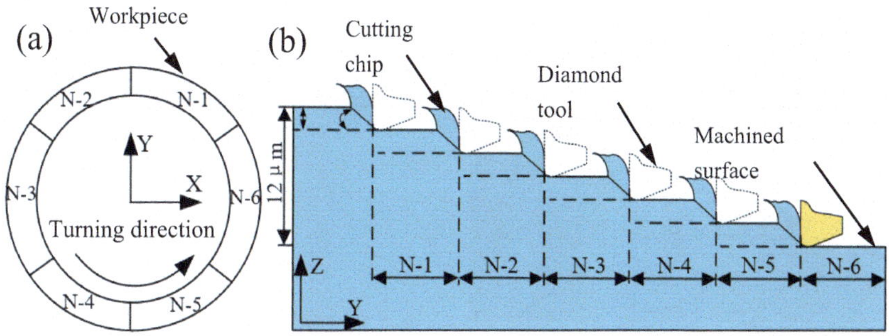


**Fig. 6.15** Diagram of **a** single-groove scarping experiments and **b** cutting depth changed from N-1 to N-6

After the experiments, the white light interferometer was used to capture the surface morphology of the workpiece. Additionally, the cutting edge of the diamond tool was observed under a metallographic microscope, revealing no noticeable signs of wear in any of the single-groove scraping experiments. Furthermore, due to the short duration and slow scraping speed compared to turning, there was minimal wear on the diamond tools during the scraping experiments. This implies that the wear status of the cutting tool used in these experiments can be considered negligible.

#### 6.2.2.2 Surface Morphology of Single-Groove

In the ODM environment, PCD tools were utilized for the rough machining of the AISI 4140 workpiece with a diameter of 10 mm. The resulting surface morphology, displaying a surface roughness value of 0.278 μm, can be observed in Fig. 6.15b. By implementing a scraping depth of 2 μm at a radius of 3 mm, a total scraping depth of 12 μm was achieved on the workpiece.

Figure 6.16a exhibits a section of the surface morphology obtained through HPGM, demonstrating similarities to the surface obtained in the ODM environment. As for Fig. 6.16c, it presents a portion of the surface morphology attained through MQLM-oil. Upon comparison of these experimental outcomes, it becomes evident that pits can be observed on the groove's surface in both the HPGM and ODM environments, while the MQLM-oil environment shows a notable absence of pits.

Upon conducting statistical analysis regarding the frequency and depth of pit occurrences, it was discovered that the HPGM environment exhibited the highest occurrence rate, with pits displaying the deepest depth. Evaluating the data from the single-groove surface, it was observed that the number of pits in a single-groove within the HPGM and ODM environments far surpassed those found in the MQLM-oil environment. Additionally, the depth of pits encountered in the MQLM-oil environment was comparatively smaller than that observed in the other two machining environments.

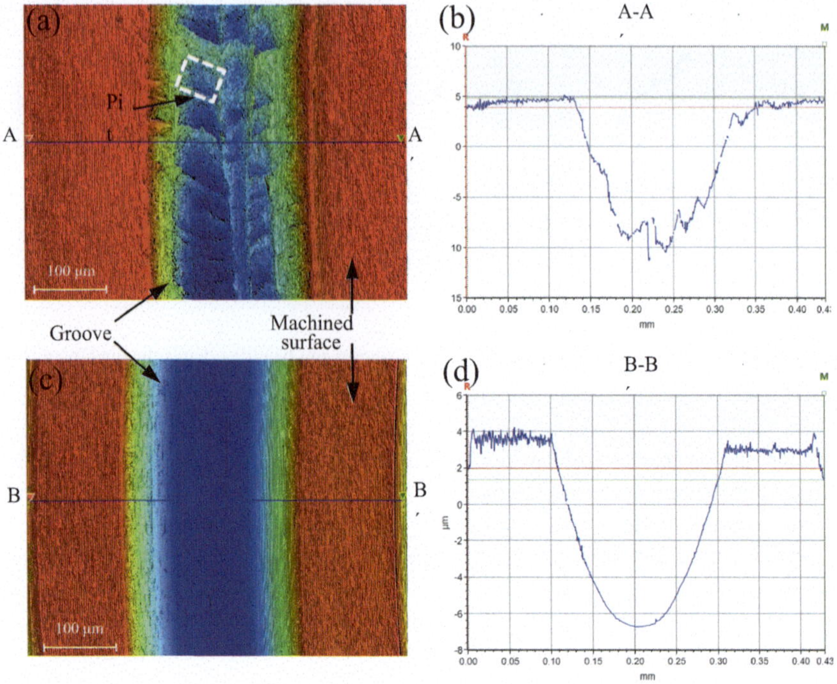


**Fig. 6.16** Single-groove scarping morphology captured by means of the white light interferometer in **a** HPGM environment and **c** MQLM-oil environment; **b** the cross section curve of A-A′ and **d** the cross section curve of B-B′

The presence of pits adversely affects the morphology of the single-groove, as indicated in Fig. 6.16b, d. Further insights into the pit depths in the three machining environments can be gleaned from Fig. 6.17.

The rough-machined surface was subjected to repeated experiments denoted as E-1, E-2, and E-3 in the MOLM-oil and ODM environments, respectively. It was observed that the experimental outcomes of rough-machining in the ODM environment closely resembled those obtained in the HPGM environment, while the experimental results of rough-machining in the MOLM-oil environment are illustrated in Fig. 6.17b.

Upon analysis of the single-groove surfaces in the HPGM and ODM environments, it became apparent that these surfaces exhibited a significant number of pits, ranging in depth from 12 to 16 μm. In contrast, the single-groove surface in the MQLM-oil environment displayed minimal presence of pits. A comparison between the experimental results in the HPGM rough-machining environment revealed that even in the MQLM-oil environment, a few pits were still observed during the single-groove scraping experiment conducted on the rough-machined surface.

In order to investigate the impact of the rough-machining environment on subsequent machining, a more comprehensive second part experiment focusing on scraping

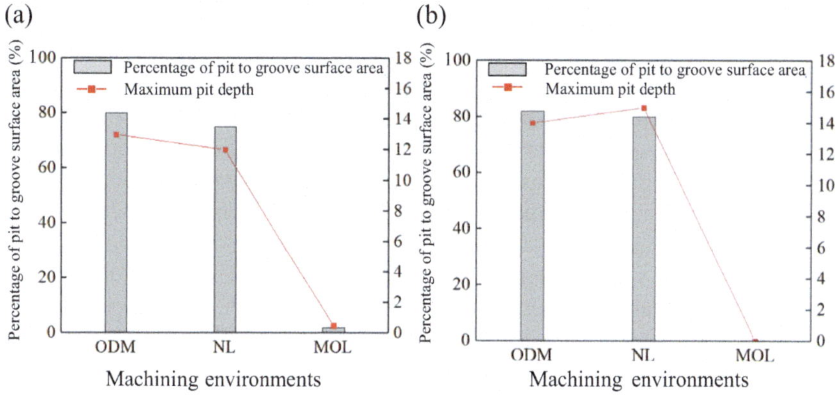


**Fig. 6.17** The statistical data of the pits in the groove scraping experiments of E-1, E-2 and E-3 in rough-machined surface machined in **a** HPGM and **b** MQLM-oil environments

depth was conducted. After rough-machining the AISI 4140 workpiece in the HPGM environment, experiments labeled as N-1 to N-6 were carried out (refer to Table 6.2), and the corresponding experimental results are displayed in Fig. 6.18a.

It was observed that in both the HPGM and ODM environments, the maximum pit depth did not exhibit significant variations with an increase in cumulative cutting depth. Conversely, in the MQLM-oil environment, the maximum pit depth decreased noticeably with an increase in cumulative cutting depth. Repeat experiments were conducted for rough-machining workpieces in both the HPGM and MQLM-oil environments. It is noteworthy that the rough-machining results obtained in the ODM environment closely resembled those obtained in the HPGM environment. Figure 6.18b provides a detailed view of the pits observed in the MQLM-oil environment.

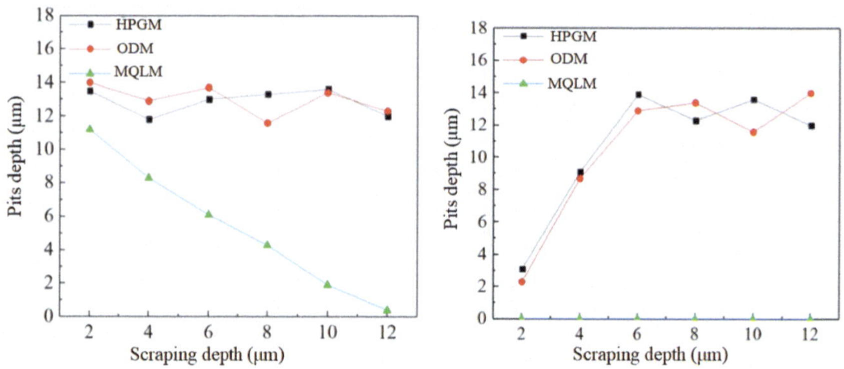


**Fig. 6.18** Relation of the scraping depth and the depth of the pits in rough-machined surface of **a** HPGM environment and **b** MQLM-oil environment

Comprehensively summarizing the surface pit data obtained from the rough-machining conducted in the MQLM-oil environment, it becomes evident that there were negligible occurrences of pits on the single-groove surface of the workpiece in this particular environment. It is worth mentioning that the results obtained from rough-machining in the other environments displayed similar characteristics, wherein the depth of pits initially increased and then remained largely unchanged.

Consequently, it can be deduced that the MQLM-oil environment significantly reduces the occurrence of pits. Additionally, it is evident that other machining environments result in a substantial number of pits in SPDT ferrous metals. Furthermore, once pits emerge, their maximum depth does not exhibit an indefinite increase with an increase in cumulative cutting depth. Instead, the maximum depth tends to stabilize within the range of 12–16 μm.

#### 6.2.2.3 Cutting Force Analysis of Single-Groove Scraping Experiments

The experiment involved capturing and processing the cutting force. The force magnitude during each scraping depth of 2 μm is illustrated in Fig. 6.19. In the case of rough-machining the workpiece surface in the MQLM-oil environment, the scraping experimental results in the three environments are depicted in Fig. 6.19a. Initially, the average cutting forces in the three environments are nearly identical. As the scraping depth increases, the turning force escalates in all three environments. Notably, the force increase is most pronounced in the HPGM environment, while it is less apparent in the MQLM-oil environment. Moreover, the average cutting force in the MQLM-oil and HPGM environments remains almost equal in the 2–8 μm scraping depth range, but in the HPGM environment, the average cutting force shows a significant surge after surpassing 8 μm.

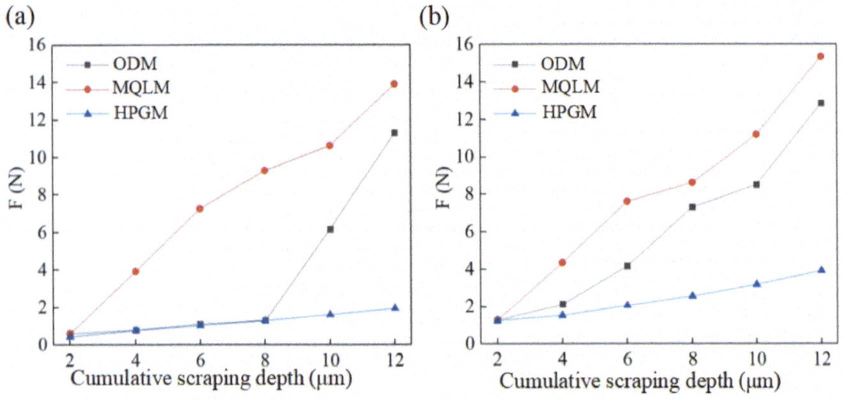


**Fig. 6.19** The average force of single-groove scraping experiments of E-1 (MQLM-oil), E-2 (HPGM) and E-3 (ODM) on the rough-machined surface of **a** MQLM-oil environment and **b** HPGM environment

When rough-machining the workpiece surface in the HPGM environment, the scraping experimental results in the three environments are illustrated in Fig. 6.19b, with the ODM results aligning closely to those obtained in the HPGM environment. Initially, the average cutting forces in all three environments are approximately the same. As the scraping depth increases, the turning force rises in the HPGM environment. Specifically, when the cumulative cutting depth is 2 μm, the surface cutting force from rough-machining in the MQLM environment measures only around 0.6 N, whereas the other environments yield approximately 1.1 N in surface cutting force.

It can be observed that the choice of rough-machining method influences the initial scraping experiment results, albeit to a lesser extent as the cumulative cutting depth increases. Analyzing the data from the machined surface, it becomes apparent that the captured surface morphology of the groove in different scraping environments is almost identical. This suggests that the rough-machining environment exerts a significant impact on subsequent finishing processes, but this influence does not continue indefinitely, with a critical depth of 12 μm.

#### 6.2.2.4 Physical Mechanism of the Lubricant Influence on Single-Groove Surface Morphology

The presence of hard carbide particles constitutes the primary physical mechanism behind the lubricant's impact on the morphology of the single-groove surface. In die steel AISI 4140, there are numerous hard carbide particles. As depicted in Fig. 6.20, traces of hard particles measuring approximately 2 μm can be observed on the SPDT ferrous metals' surface, exhibiting two orders of magnitude higher concentration compared to the workpiece [5, 25].

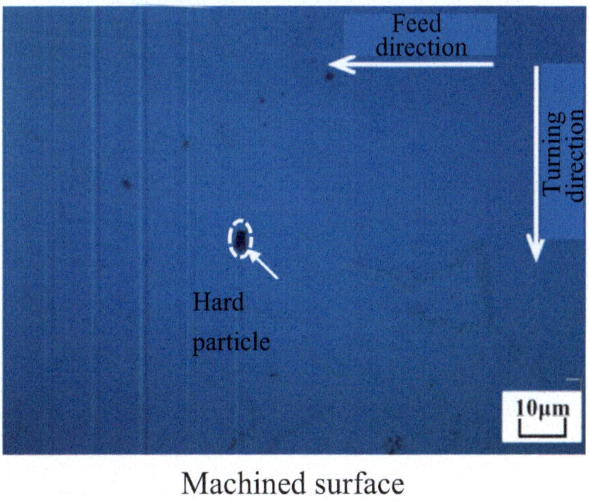


**Fig. 6.20** Hard particles on the machined surface in SPDT ferrous metals

Figure 6.21a provides an enlarged surface simulation image of a pit area, offering insight into the pit's morphology along the cutting direction. The pit generation mechanism is illustrated in Fig. 6.21b. It can be observed that the pit consistently displays a steep surface on one side and a gentle surface on the other. This phenomenon is clearly attributed to the displacement of hard particles by the diamond tool during the cutting process. In the high-precision servo lathe, the position of the diamond tool mounted on the *Z*-axis remains fixed, indicating that the pit within the material cannot be a result of plastic removal caused by the diamond tool.

The mechanism underlying pit formation can be succinctly illustrated in Fig. 6.21b. During the single-groove scraping process, the diamond tool operates at a low speed. Upon contact with the soft ferrous metal, the material can be plastically removed, giving rise to the formation of the single-groove morphology. However, when the tool encounters hard particles during the turning process, the hard particles cannot be plastically removed as their hardness is significantly lower than that of the diamond tool. Consequently, the diamond tool extracts the hard particles from the ferrous metal, leading to the formation of pits.

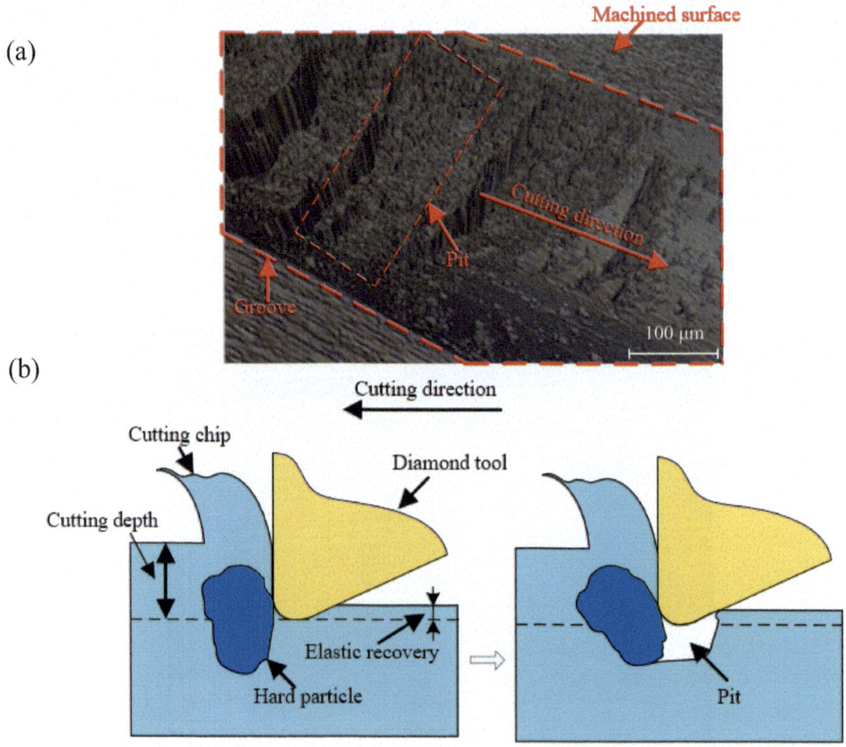


**Fig. 6.21** **a** an enlarged surface of a pit area on the groove surface and **b** the illustration of the pit generation mechanism on the ferrous metal surface in SPDT

Cutting force may be used to assess pit parameters in machining because hard carbide particles affect it. Figure 6.22 shows the cutting force and single-groove shape in the three conditions.

In the MQLM-oil environment, when the diamond tool initially cuts into the material, an abrupt change in the instantaneous cutting force can be observed. This phenomenon is also observed in the other two machining environments. However, only in the MQLM-oil environment does the cutting force remain relatively stable during subsequent processing. Conversely, in the HPGM and ODM environments, the cutting force displays significant fluctuations. This indicates that the material removal in these two environments is unstable, and the formation of pits inevitably leads to continuous variations in the documented cutting force.

Furthermore, there exists a direct relationship between the size and depth of the pits in the single-groove and the cutting force. In the HPGM environment, the cutting force fluctuates extensively over an extended period, resulting in a small number of fluctuations throughout the entire groove machining process. Consequently, this leads to the formation of larger pits characterized by continuous patterns. On the

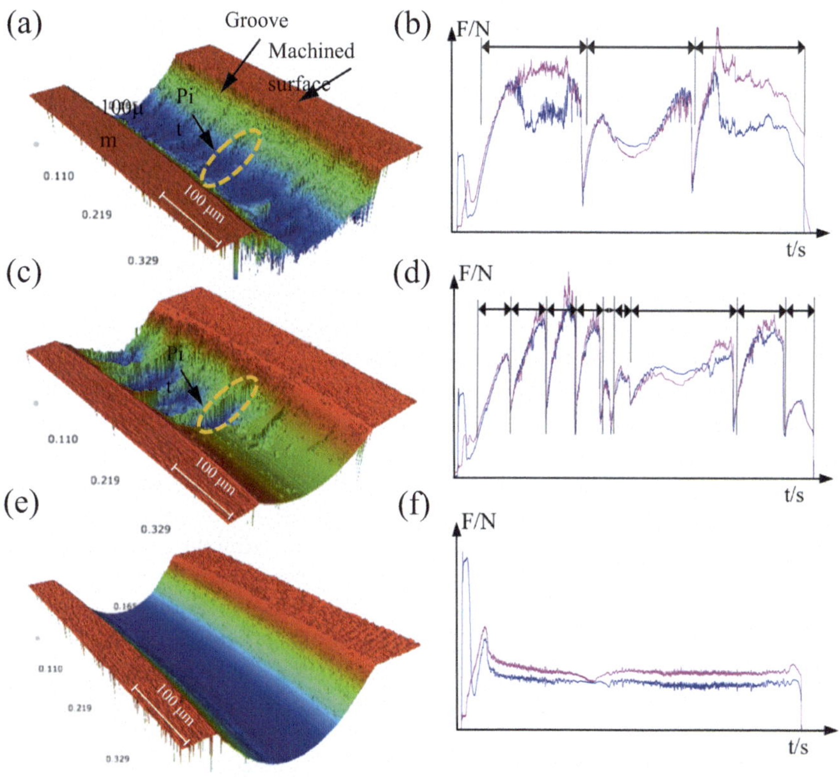


**Fig. 6.22** Groove surface morphology and cutting force in **a**, **b** HPGM environment, **c**, **d** ODM environment and **e**, **f** MQLM-oil environment

other hand, in the ODM environment, individual force signals exhibit short-lived significant fluctuations, and there are numerous fluctuations throughout the entire groove processing process, albeit with large variations in timespan. This gives rise to predominantly intermittent pits in the groove, with only a few displaying continuous patterns. In contrast, under the MQLM-oil environment, the cutting force exhibits smooth and continuous changes, indicating the absence of pits. Therefore, the cutting force can be effectively utilized to accurately characterize the properties of pits during the cutting process.

#### 6.2.2.5 Influence of Closed-Loop-Stiffness of the Lathe on Single-Groove Surface Morphology

Another potential explanation for the occurrence of pits is the insufficient closed-loop stiffness of the ultra-precision lathe, leading to spindle crawling during processing. In the experimental setup, particularly during the scraping of the single-groove, the diamond tool undergoes significant stress while removing the workpiece material. This force is ultimately transmitted to the *C*-axis. Considering that the *C*-axis is an aerostatic spindle, its stiffness is limited. Under high forces, it is prone to crawling, causing an unstable material removal process.

The machining process that takes place on the ultra-precision lathe is shown in Fig. 6.23a. This process involves the transmission of the interaction force that is caused by the interference between the tool and the workpiece via the spindle and the lathe bed, which finally results in the formation of a closed-loop force that causes the spindle to crawl. During the machining process, the tool is subjected to two principal forces in the *Z*-axis and *Y*-axis directions, as shown in Fig. 6.23b. The force acting in the direction of the *Z*-axis may cause crawling along the *Z*-axis, while the force acting in the direction of the *Y*-axis can cause crawling along the *C*-axis. During the relative sliding process between the main shaft and the guide rail, the main shaft may be prohibited from advancing when its power cannot overcome the encountered resistance. However, as the driving equipment continues to exert force on the main shaft, it eventually overcomes the resistance and makes a leap forward. This phenomenon is commonly referred to as crawling of the main shaft.

Evidently, the lubrication effects produced by different lubricant environments are inconsistent, and the MQLM-oil environment exhibits a lower equivalent friction coefficient compared to the ODM environment. A higher equivalent friction coefficient implies increased resistance to overcome, thereby increasing the likelihood of spindle crawling. As shown in Fig. 6.23b, the occurrence of crawling may result in an uneven tool path when the tool is subjected to the effect of the resulting force that is applied by the tool. Additionally, crawling in the direction of the *Z*-axis can cause the single-groove surface to sink lower.

Figure 6.24 illustrates the numerous types of pits that may be created by a single groove due to the presence of a variety of variables. As can be seen in Fig. 6.24a, pits that are formed by hard particles are limited by the size of the particles themselves and do not have the capability to considerably influence the cross-sectional profile of

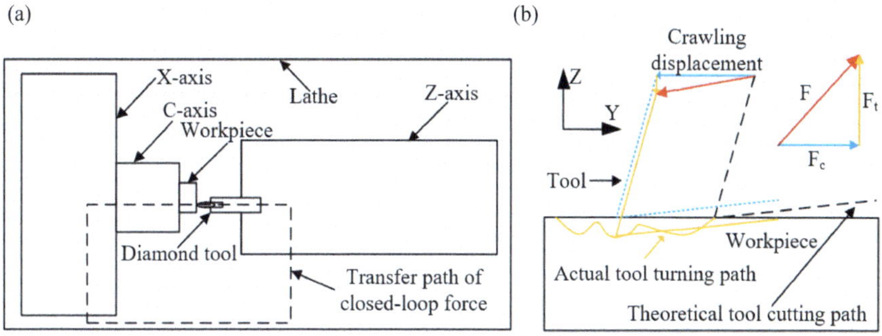


**Fig. 6.23** **a** is the transfer path of closed-loop force on the lathe and **b** is the transfer direction of the force

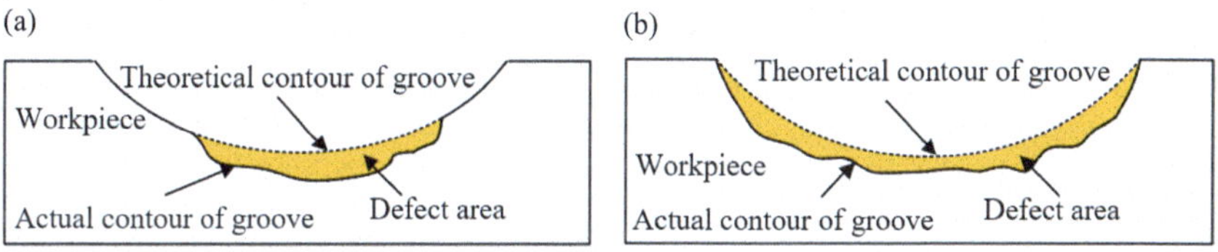


**Fig. 6.24** Different types of pit on the groove surface of **a** hard particles caused the defect area and **b** crawling caused the defect area

the complete single-groove. Conversely, pits generated by the crawling phenomenon result from irregular changes in the tool's trajectory. These pits are expected to affect the entire cross-sectional profile of the single-groove, as depicted in Fig. 6.24b.

## References


1. Davies MA, Evans CJ, Patterson SR, Vohra R, Bergner BC (2003) Application of precision diamond machining to the manufacture of micro-photonics components. Proc SPIE 5(183):94–108
2. Paul E, Evans CJ, Mangamelli A, McGlauflin ML, Polavani RS (1996) Chemical aspects of tool wear in single point diamond turning. J Int Soc Precis Eng Nanotechnol 18(1):4–19
3. Shamoto E, Moriwaki T (1994) Study on elliptical vibration cutting. CIRP Ann 43:35–38
4. Shamoto E, Suzuki N (2014) Ultrasonic vibration diamond cutting and ultrasonic elliptical vibration cutting. Compreh Mater Process 11:405–454
5. Wang Y, Suzuki N, Shamoto E, Zhao Q (2011) Investigation of tool wear suppression in ultraprecision diamond machining of die steel. Precis Eng 35:677–685
6. Jung H, Hayasaka T, Shamoto E (2018) Study on process monitoring of elliptical vibration cutting by utilizing internal data in ultrasonic elliptical vibration device. Int J Precise Eng Manuf-Green Technol 5: 571–81
7. Moriwaki T, Shamoto E (1995) Ultrasonic elliptical vibration cutting. CIRP Ann 44:31–34
8. Amini S, Hosseinabadi HN, Sajjady SA (2016) Experimental study on effect of micro textured surfaces generated by ultrasonic vibration assisted face turning on friction and wear performance. Appl Surf Sci 390:633–648

9. Lin JQ, Han JG, Zhou XQ, Hao ZP, Lu MM (2016) Study on predictive model of cutting force and geometry parameters for oblique elliptical vibration cutting. Int J Mech Sci 117:43–52
10. Song Y, Park CH, Moriwaki T (2010) Mirror finishing of Co-Cr-Mo alloy using elliptical vibration cutting. Precis Eng 34:784–789
11. Attanasio A, Gelfi M, Giardini C, Remino C (2006) Minimal quantity lubrication in turning: effect on tool wear. Wear 260(3):333–338
12. Aramcharoen A (2016) Influence of cryogenic cooling on tool wear and chip formation in turning of titanium alloy. In: Wertheim R, Ihlefeldt S, Hochmuth C, Putz M (eds) 7th HPC 2016—CIRP conference on high performance cutting, vol 46, Procedia CIRP, vol 46. Elsevier Science BV, Amsterdam, pp 83–86
13. Bennett EO (1983) Water based cutting fluids and human health. Tribol Int 16(3):133–136
14. Huang P, Li HC, Zhu WL, Wang HT, Zhang GQ, Wu XY, To SE, Zhu ZW (2020) Effects of eco-friendly cooling strategy on machining performance in micro-scale diamond turning of Ti-6Al-4V. J Clean Prod 243:9
15. Cheng K, Huo D (eds) (2013) Micro-cutting: fundamentals and applications. Wiley
16. Zong WJ, Wu D, Yao XL, Sun T (2016) Strength dependent evaluation method for the wear resistance of multifaceted diamond Berkovich indenter in scratch test. J Mater Process Technol 234:45–57
17. Zong WJ, Li ZQ, Sun T, Li D, Cheng K (2010) Analysis for the wear resistance anisotropy of diamond cutting tools in theory and experiment. J Mater Process Technol 210(6–7):858–867
18. Thornton AG, Wilks J (1978) Clean surface-reactions between diamond and steel. Nature 274(5673):792–793
19. Casstevens JM (1983) Diamond turning of steel in carbon-saturated atmospheres. Precis Eng-J Am Soc Precis Eng 5(1):9–15
20. Xavior MA, Manohar M, Madhukar PM, Jeyapandiarajan P (2017) Experimental investigation of work hardening, residual stress and microstructure during machining Inconel 718. J Mech Sci Technol 31(10):4789–4794
21. Ulutan D, Ozel T (2011) Machining induced surface integrity in titanium and nickel alloys: a review. Int J Mach Tools Manuf 51(3):250–280
22. Wang YL, Suzuki N, Shamoto E, Zhao QL (2011) Investigation of tool wear suppression in ultraprecision diamond machining of die steel. Precis Eng-J Int Soc Precis Eng Nanotechnol 35(4):677–685
23. Putz M, Dix M, Neubert M, Schmidt T (2016) Mechanism of cutting elastomers with cryogenic cooling. CIRP Ann-Manuf Technol 65(1):73–76
24. Zhu WL, Duan F, Zhang XD, Zhu ZW, Ju BF (2018) A new diamond machining approach for extendable fabrication of microfreeform lens array. Int J Mach Tools Manuf 124:134–148
25. Evans C, Bryan JB (1991) Cryogenic diamond turning of stainless steel. CIRP Ann 40(1):571–575

# Chapter 7
# Size Effect on Cutting Process in Precision Machining

**Abstract** Precision machining is a critical aspect of modern manufacturing, enabling the production of components with extremely tight tolerances and high surface quality. One of the key factors influencing the performance and outcomes of precision machining is the size effect, which refers to the varying behavior of materials when the size of a workpiece or feature is reduced to a certain scale. Two key factors that contribute to dimensional effects in precision machining are tool radius and cutting depth. This chapter will focus on these two factors contributing to size effects in precision machining. It will discuss the impact of tool radius and cutting depth on size effects in micro-milling and ultra-precision fly cutting processes, respectively.

**Keywords** Size effect · Cutting process · Precision machining · Micro milling · Ultra-precision fly cutting

## 7.1 Size Effect of Cutting Edge

Micro-milling technology finds extensive application in the aerospace, automobile, medical, precision mold, and other industries [1]. In contrast to traditional milling, micro-milling involves cutting edge arc radius ($r_0$) and tool diameter ($D$) that are of comparable magnitude [2]. Consequently, various size effects resulting from factors such as cutting edge arc radius, tool deflection, and material grain size need to be considered [3–6]. It is important to note that in micro-milling, smaller cutting parameters do not always yield better surface quality.

In micro-milling processes, material removal mainly takes place on the cutting edge arc surface (AC section) rather than the front cutting surface (CD section) due to the small ratio ($f_z/r_0$) of the feed rate per tooth ($f_z$) to the cutting edge arc radius ($r_0$). Additionally, the enhanced interaction between the back cutting surface (AE section) and the workpiece surface induces elastic deformation, resulting in significant scraping, squeezing, and ploughing of the cutting edge arc surface on the workpiece, as shown in Fig. 7.1. This phenomenon, known as the size effect in this



G. Zhang and J. Wang, *Material-Oriented Cutting Processes in Precision Machining*, Springer Tracts in Mechanical Engineering,
https://doi.org/10.1007/978-981-96-2504-8_7

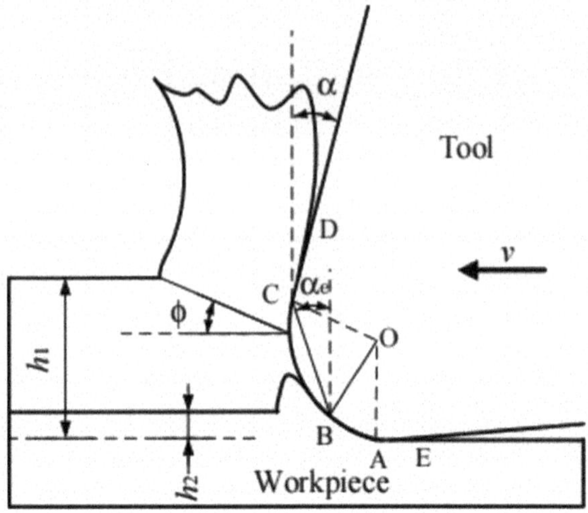


**Fig. 7.1** Chip formation schematic at different cutting thicknesses ($h_1$ and $h_2$) during micro-milling

chapter, occurs when the cutting parameters approach the size of $r_0$, leading to issues such as poor surface quality, increased burrs, and elevated tool wear in micro-milling operations [7–9]. Conversely, when $f_z$ surpasses $r_0$ in micro-milling, the tool's rake face becomes actively involved in cutting, leading to a cutting mechanism resembling that of macro cutting as the rake face's participation proportion continues to rise [10].

In the micro-milling process, when $r_0$ approaches or equals the minimum cutting chip thickness ($h_{min}$), the positive rake angle of the cutting edge shifts to a negative rake angle [19]. This transition leads to elastic deformation between the tool and the workpiece, causing ploughing without chip formation. To ensure a consistent and stable chip formation during cutting, it is essential to increase the cutting thickness ($h_2$) (see Fig. 7.1), enabling $h_{min}$ to shift from elastic–plastic deformation to shear deformation [20, 21].

In the realm of micro-milling research, researchers typically start by determining the range of $h_{min}$ using experiments or simulations. Following this, they employed qualitative analysis techniques to explore the impact of micro-milling on surface quality and devise optimization strategies for cutting parameters. However, there has been limited exploration into quantitatively analyzing the size effect on the cutting edge to minimize burrs and surface roughness. The key aspect of the size effect lies in the progressive shift of the material removal mechanism from ploughing and friction to shear with an increasing $f_z/r_0$ ratio. This transition enables the conversion of elastic deformation to plastic deformation, reducing frictional resistance between the tool and the workpiece material and facilitating chip generation. The presence of residual chips at the base and sides of the micro-slot significantly influences surface quality and burr formation during micro-milling operations.

The determination of $h_{min}$ range in this section is based on analyzing the milling force variation in the ploughing and shearing regions during micro-milling. Subsequently, the impact of $f_z/r_0$ on the top burr size of the micro-slot, surface roughness (Ra) at the bottom of the micro-slot, and micro-milling force is examined, leading to the identification of strong size effect zones, transition size effect zones, and optimal size effect zones. As a result, a size effect band for micro-milling is established. Additionally, a comparison of the size effect zones for different $f_z/r_0$ values is conducted to optimize cutting parameters, avoiding the utilization of parameters within the strong size effect zone. This section presents a theoretical framework for setting micro-milling cutting parameters and serves as a technical guide for enhancing surface quality in micro-milling.

## 7.1.1 Micro-milling Experiments

### 7.1.1.1 Determination of $h_{min}$

In micro-milling, Yuan et al. [22] suggested the existence of a separation point between plough cutting and shear cutting areas on the cutting edge arc, leading to indirect determination and estimation of $h_{min}$. Their study revealed $h_{min}$ values around 20–40% of $r_0$ by analyzing the relationship between the force ratio of $F_x/F_y$ and $h_{min}$. Additionally, Liu et al. [23] developed a model for $h_{min}$ considering thermal softening and strain strengthening effects in workpiece materials, which indicated $h_{min}$ values of 20–35% of $r_0$ for 1040 steel and 35–40% of $r_0$ for Al6082-T6 aluminum.

The tool morphology used in the test is shown in Fig. 7.2, featuring a $r_0$ of approximately 2.5 μm. Previous studies [22, 23] have demonstrated that different workpiece materials lead to varying $h_{min}$ values. Theoretically, the $h_{min}$ can be around 0.2–0.4 times the $r_0$ value. Figure 7.3 illustrates the changes in milling force measured during micro-milling in the feed direction ($F_y$) and perpendicular to it ($F_x$). Figure 7.3a presents the original milling force data from the measurements, while Fig. 7.3b gives a closer look at the measured milling force. It is clear from Fig. 7.3b that the filtered and measured milling forces ($F_x$ and $F_y$) exhibit similar variations and closely match in magnitude. By averaging the absolute amplitude values from 10 peaks and 10 valleys in five sets of cutting force cycles, the $F_x$ and $F_y$ values for different $F_z$ can be determined. The measured cutting force amplitude remains relatively low, and the use of an 800 Hz low-pass filter effectively removes the impact of high-frequency noise caused by machine tool vibrations and other interfering factors on micro-milling force measurements.

Figure 7.4 depicts the milling force curves ($F_x$ and $F_y$) as a function of $f_z$ during the micro-milling of Al7075. The micro-milling force initially increases and then decreases in a non-linear manner as $f_z$ varies between 0.2 and 1 μm/tooth, indicating the presence of the size effect in the micro-milling process [24, 25]. At an $f_z$ value of 0.65 μm/tooth, both $F_x$ and $F_y$ reach their peaks, denoted as A and B respectively. Hence, the corresponding value of $h_{min}$ can be inferred from the milling force to be approximately 0.65 μm (as shown in Fig. 7.4). This falls within the range of 0.2–0.4 $r_0$, in agreement with the theoretical estimate of $h_{min}$, thereby establishing consistency between the theoretical and experimental results. Consequently, $h_{min}$ in this section is determined to be within the range of 0.5–1 μm, providing a theoretical foundation for

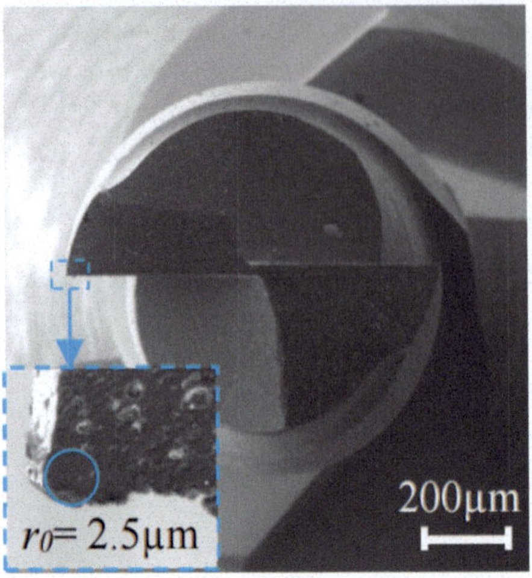


**Fig. 7.2** The edge morphology of a flat end milling tool with a diameter of 1 mm

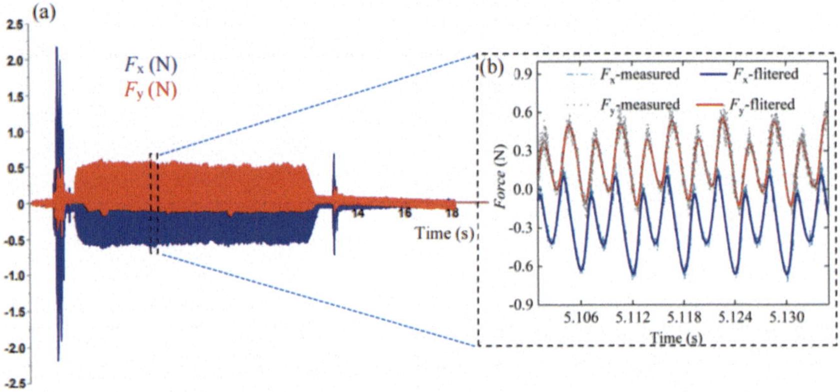


**Fig. 7.3** Forces during milling in the direction of the feed ($F_y$) and in the direction of the step ($F_x$): **a** initial milling force, **b** comparison between the measured and filtered milling force

subsequent tests investigating the effects of size effects and the selection of cutting parameters within different size effect bands.

#### 7.1.1.2 Single Factor Micro-milling Test

A single-factor test on the micro-milling slot was conducted using a five-axis linked mill machining center (DMG MORI). The machine tool features a power of 1.5 kW, with a maximum spindle speed of 18,000 rpm and a positioning error of 3 μm. The experimental setup for micro-milling is illustrated in Fig. 7.5.

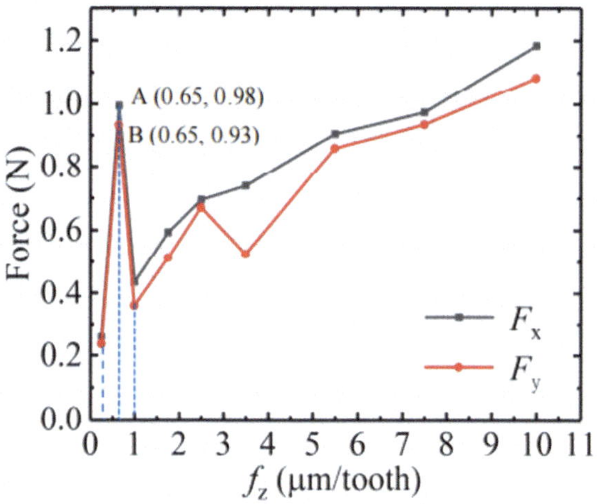


**Fig. 7.4** Milling force $F_x$ and $F_y$ in relation to $f_z$ during micro-milling Al7075

**Fig. 7.5** Experimental setup of micro-milling

The experimental approach employed dry cutting, and the material of workpiece was aluminum alloy (Al7075-T6). Real-time measurements of milling forces in the $F_x$, $F_y$, and $F_z$ directions were obtained using a Kistler-9119AA1 dynamometer. The dynamometer had a sampling frequency of 36 kHz, ensuring an adequate number of data points for capturing force signals. The milling experiment was performed using a two-edge TiAlN-coated flat end milling tool (NS TOOL). The micro-milling tool was made of super hard alloy and had a diameter of 1 mm. The front angle, flank angle, and helix angle of the tool were 12°, 5°, and 30°, respectively.

The size of burrs and surface roughness (Ra) in micro-slot surfaces is significantly influenced by various cutting parameters, including the micro-milling tool, workpiece material etc. Table 7.1 presents the cutting parameter settings for the experiment. N1–N8 are utilized to investigate the impact of edge size, while N9–N24 are employed to study the cutting parameters' influence on the top burr size and Ra values of micro-slot surfaces in both strong and weak size effect regions.

Prior to testing, the workpiece surface underwent finish milling to remove any effects from material preparation and surface defects on cutting depth and quality. The SEM was employed to captured the micro-slot burr's shape and width can be examined, as shown in Fig. 7.6. Figure 7.6a depicts the top burr on the down milling side (DMS), Fig. 7.6b illustrates the machined micro-slot, and Fig. 7.6c shows the top burr on the up milling side (UMS). The Ra value of the micro-slot bottom (MSB) can be determined using Keyence's VK-X series laser scanning confocal microscope.

**Table 7.1** Cutting parameters of experiments

| Test | Cutting speed $v$ (mm/min) | Feed rate per tooth $f_z$ (μm/tooth) | Axial cutting depth $a_p$ (μm) | Tool extend length $l$ (mm) |
|---|---|---|---|---|
| N1 | 5 | 0.25 | 30 | 20 |
| N2 | 20 | 1 | 30 | 20 |
| N3 | 30 | 1.5 | 30 | 20 |
| N4 | 50 | 2.5 | 30 | 20 |
| N5 | 70 | 3.5 | 30 | 20 |
| N6 | 110 | 5.5 | 30 | 20 |
| N7 | 150 | 7.5 | 30 | 20 |
| N8 | 200 | 10 | 30 | 20 |
| N9 | 20 | 1 | 10 | 20 |
| N10 | 20 | 1 | 20 | 20 |
| N11 | 20 | 1 | 40 | 20 |
| N12 | 20 | 1 | 50 | 20 |
| N13 | 125 | 6.25 | 10 | 20 |
| N14 | 125 | 6.25 | 20 | 20 |
| N15 | 125 | 6.25 | 40 | 20 |
| N16 | 125 | 6.25 | 50 | 20 |
| N17 | 16 | 1 | 30 | 20 |
| N18 | 24 | 1 | 30 | 20 |
| N19 | 28 | 1 | 30 | 20 |
| N20 | 32 | 1 | 30 | 20 |
| N21 | 100 | 6.25 | 30 | 20 |
| N22 | 150 | 6.25 | 30 | 20 |
| N23 | 175 | 6.25 | 30 | 20 |
| N24 | 200 | 6.25 | 30 | 20 |

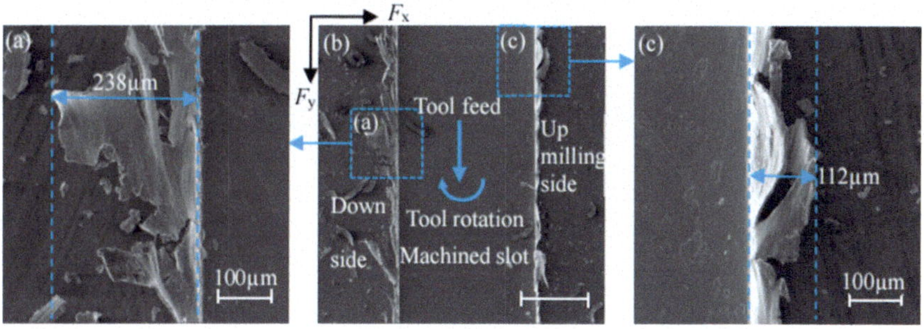


**Fig. 7.6** Captured morphology of micro-slot burr: **a** DMS, **b** micro-slot figure, and **c** UMS

## 7.1.2 *Size Effect in Micro-milling*

### 7.1.2.1 Classifications of Size Effect Band in Micro-milling

The size effect band of micro-milling refers to the impact of the size effect of the micro-tool on the workpiece material, specifically Al7075. This effect influences the shape, size, and surface roughness of the micro-slot burr, and it varies with the parameter $f_z/r_0$.

In the strong size effect zone, a large top burr width and worse surface roughness of the MSB are observed when $f_z/r_0$ is less than 70%, as illustrated in Fig. 7.7. This occurs because when $f_z$ is similar to $r_0$, the amount of cutting material ahead of the tool is minimal, thus magnifying the size effect. In the transition size effect zone ($f_z/r_0$ between 70 and 220%), the top burr is significantly influenced by the edge-size effect, resulting in noticeable changes in its width. However, the surface quality of the MSB remains good, with a relatively small Ra. When $f_z/r_0$ exceeds 220%, $f_z$ is significantly larger than $r_0$. The micro-milling process is less affected by the edge-size effect and resembles traditional milling. As a result, the Ra value of the slot bottom increases. However, the burr primarily contacts the tool's front face, and the workpiece material undergoes more plastic deformation, facilitating easier burr removal and, consequently, a smaller top burr width.

Hence, the optimal size range for burrs can be identified as the weak size effect zone. The strong size effect zone and the transition burr size zone are characterized by larger top burr widths, as shown in Fig. 7.7.

### 7.1.2.2 Optimization of Micro-milling Parameters

The influence of $f_z$ on the morphology of the top burr in the micro-slot was studied by examining micro-milling tests N1–N8 from Table 7.1. SEM images of the micro-slot and burr morphology are shown in Fig. 7.8.

In the strong zone ($f_z/r_0 < 70\%$), the width of top burrs decreases initially and then increases on both the DMS and UMS, as shown in Fig. 7.7a. The top burrs on the

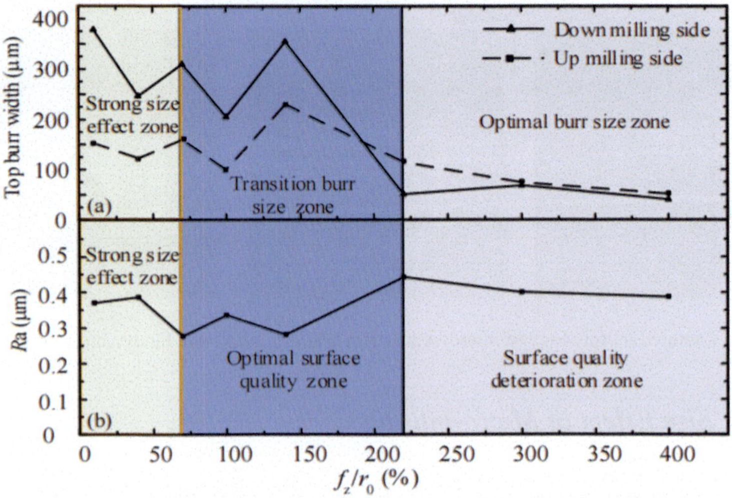


**Fig. 7.7** Variations of the top burrs' width and Ra of the MSB in different size effect band

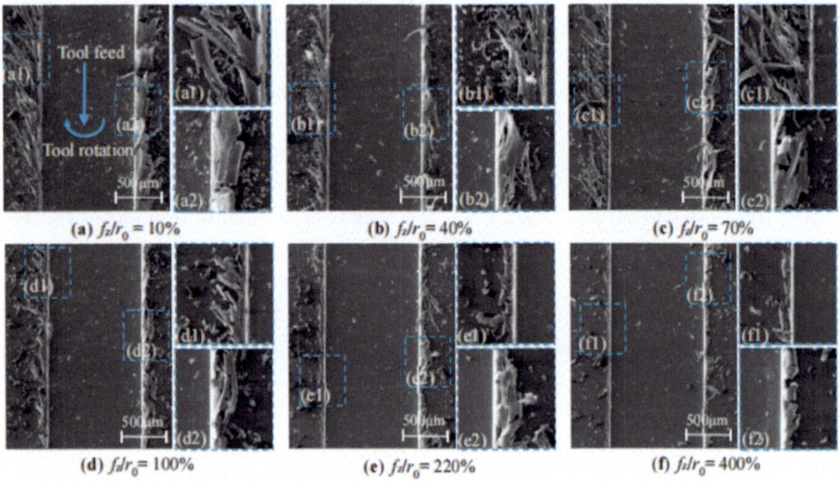


**Fig. 7.8** Variations of micro-slot and burr morphology in the size effect zone: $f_z/r_0$ = **a** 10%; **b** 40%; **c** 70%; **d** 100%; **e** 220%; **f** 400%

DMS exhibit a whisker-shaped morphology (Fig. 7.8a1–c1), while those on the UMS appear block-shaped (Fig. 7.8a2–c2). These observations suggest that the micro-milling process, influenced significantly by the edge-size effect, does not produce continuous chips. Instead, the material in contact with the tool undergoes continuous squeezing and scratching along the micro-slot edge, leading to the formation of strip burrs due to the good plasticity and ductility of the workpiece.

Within the transition size effect zone ($f_z/r_0$ > 70% and $f_z/r_0$ < 220%), the width variations of the top burrs in the micro-slot are more pronounced on both the down milling and UMSs, as depicted in Fig. 7.7a. This phenomenon is attributed to decreased elastic recovery on the machined surface, resulting in a reduction of the ploughing effect area and an increase in the shear effect area. Consequently, the formation of stable chips becomes challenging, leading to the presence of top burrs as the chips fail to be separated promptly at the slot edge. Figure 7.8a–d illustrates that the width of top burrs on the DMS is notably larger than that on the UMS, exhibiting fluctuations as $f_z/r_0$ decreases. At a value of 140% for $f_z/r_0$, Fig. 7.7a shows that the maximum width of top burrs on the DMS is 388 μm, while on the UMS, it is 265 μm. This disparity indicates inconsistent chip shapes and sizes, contributing to uncertain burr formation and significant width variations in the top burrs. Specifically, top burrs on the DMS mainly exhibit a tearing flag shape (Fig. 7.8d1), whereas those on the UMS display a continuous serration shape (Fig. 7.8d2).

The optimal selection of burr size in micro-milling occurs when the $f_z/r_0$ value surpasses 220%, as shown in Fig. 7.7a. Within this range, the impact of the edge-size effect is minimized, leading to fewer small and scarce top burrs during both down and up milling. Figure 7.8e–f indicate that the width of the top burrs decreases with increasing $f_z/r_0$. Specifically, Fig. 7.8f1–f2 demonstrate that the top burrs exhibit predominantly curly and wavy shapes on the down milling and UMSs. At a $f_z/r_0$ value of 220%, Fig. 7.8e1–e2 reveal that the width of the top burr on the DMS becomes smaller than that on the UMS, indicating a separation from the zone influenced by the edge-size effect.

(1) *Influence of axial cutting depth on the burr size effect band*

In Table 7.1, the cutting parameters involving the N9–N16 groups are presented. The influence of axial cutting depth ($a_p$) on the burr size effect band is investigated by analyzing the top burr morphology of the micro-slot under two conditions: $f_z/r_0 =$ 40% (strong size effect zone) and $f_z/r_0 = 250\%$ (weak size effect zone).

Figure 9a, b demonstrate that different $a_p$ values in various size effect zones impact the burr morphology on the two sides of the micro-slot differently. In the DMS, the top burr width ranges from 90 to 179 μm, while the UMS remains around 100 μm, highlighting the significant effect of ap on the DMS due to variation in tool cutting depth. The down milling direction facilitates slitting the top burr and generating long chips as the chip flow-out direction aligns with the workpiece's moving direction.

When $f_z/r_0 = 40\%$, a cutting depth of less than 20 μm results in a DMS top burr width close to 100 μm. As the axial cutting depth increases to 50 μm, the DMS top burr width fluctuates to 179 μm, while the UMS width slightly decreases (Fig. 7.9a). Figures 7.10a–e and 7.11a–e illustrate that both sides present larger patches or striped-shaped top burrs, respectively.

If $f_z/r_0 = 250\%$, it indicates that micro-milling is being performed in the weak size effect zone, where the change in axial cutting depth has minimal impact on the size of the top burr. The width of the top burr remains consistent within a range of 30 μm for both down milling and up milling. In the case of up milling, an initial increase in the size of the top burr is observed with an increase in axial cutting depth,

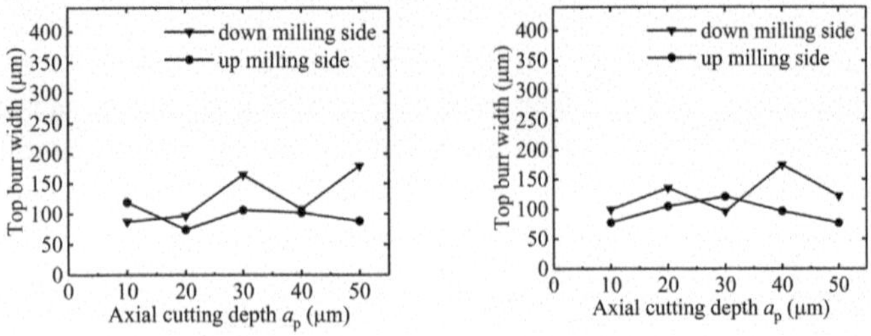


**Fig. 7.9** Various of micro-slot's top burr width with $a_p$ (N9–N16): $f_z/r_0$ = **a** 40%, **b** 250%

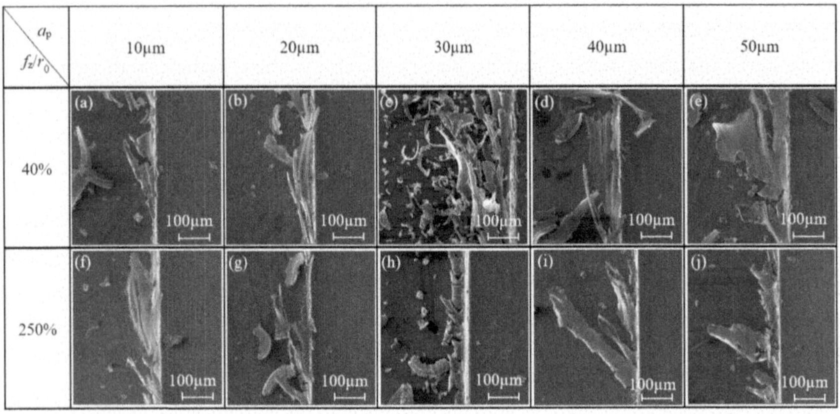


**Fig. 7.10** Changes of top burr morphology in the DMS with respect to $a_p$ in the size effect zone (N9–N16)

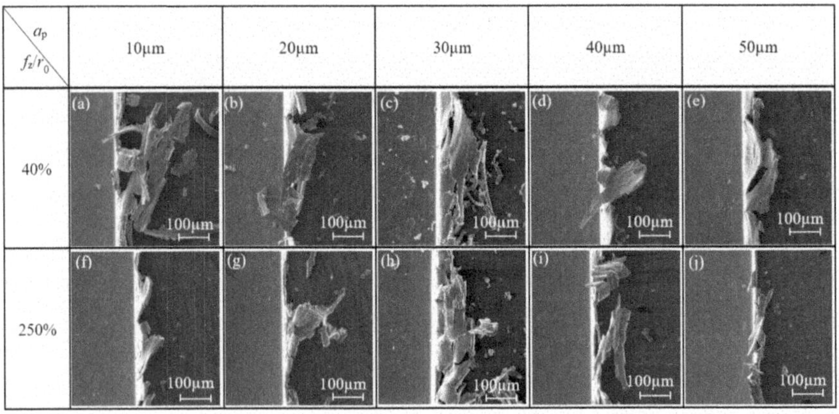


**Fig. 7.11** Various of top burr morphology in the UMS in relation to $a_p$ in the size effect zone (N9–N16)

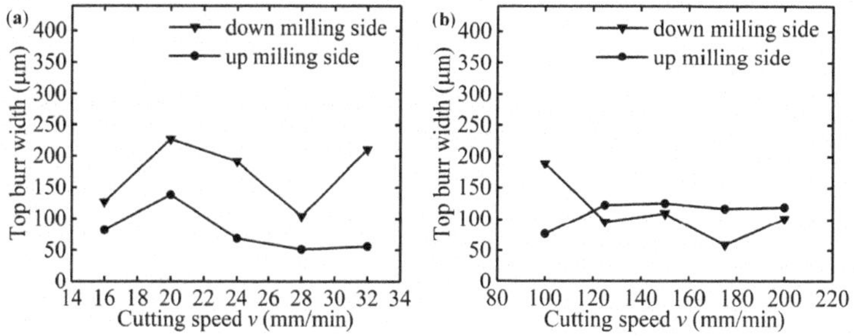


**Fig. 7.12** Various of micro-slot's top burr width with $v$ (N17–N24): $f_z/r_0 =$ **a** 40% and **b** 250%

followed by a subsequent decrease. To separate micro-milling from the strong size effect zone, it is necessary to either increase the axial cutting depth to reduce the top burr or choose a smaller $f_z/r_0$ value, which would limit the range of cutting parameter options available. On the DMS and UMS, the top burr morphology typically shows discontinuous serration or a wavy pattern, as illustrated in Figs. 7.10f–j and 7.11f–j respectively.

(2) *Influence of cutting speed on burr size effect band*

Illustrated in Table 7.1 are the cutting parameters used in the micro-milling straight slot test for the N17–N24 group. This section explores the impact of cutting speed ($v$) on the size of burrs, focusing on the top burr morphology of the micro-slot under the same conditions: $f_z/r_0 = 40\%$ and $f_z/r_0 = 250\%$.

When $f_z/r_0 = 40\%$, it is evident that cutting speed significantly affects the top burr width in the DMS. Figure 7.12a demonstrates that as $v$ increases from 16 to 32 mm/min, the top burr width on the UMS initially increases before gradually decreasing. Meanwhile, the top burr width in the DMS surpasses that in the UMS. Figure 13a–e portray the top burr morphology as a primarily long strip block shape in the DMS and a tearing block shape in the UMS.

In contrast, at $f_z/r_0 = 250\%$, Fig. 7.12b shows that the cutting speed has minimal impact on the width of the top burr in the micro-milling slot. Increasing the cutting speed from 100 to 200 mm/min results in a change of less than 30 μm in the width of the top burr on the UMS. However, on the DMS, the top burr width is smaller than on the UMS when the cutting speed is increased from 12 mm/min. Figure 13f–j demonstrate the transition of the top burr morphology on the DMS from a long and thin strip burr to a small serrated burr, while Fig. 7.14f–j illustrate the change in the top burr morphology on the UMS from a large serrated burr to a small wavy burr.

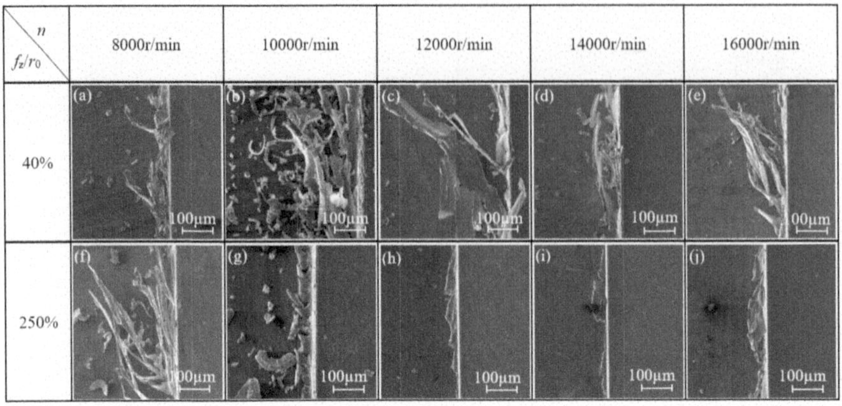


**Fig. 7.13** Different types of top burr shapes on the DMS within the size effect zone (N17–N24)

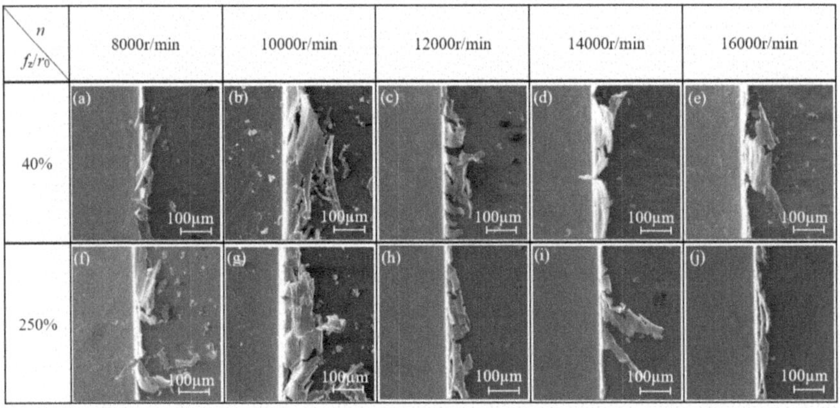


**Fig. 7.14** Different types of top burr shapes on the UMS within the size effect zone (N17–N24)

#### 7.1.2.3 Optimization of Micro-milling Parameters Based on Surface Roughness

An optimal range of cutting parameters for $f_z/r_0$ exists which minimizes the surface roughness of the MSB in micro-milling. Figure 7.7b demonstrates that the quality of the MSB surface deteriorates when $f_z/r_0$ is below 70%, leading to a rapid increase in surface roughness. In this region of significant size effect, where $f_z$ is less than $h_{min}$, the machined material predominantly undergoes elastic recovery. Moreover, the formation of chips during ploughing becomes challenging, and the workpiece material is gradually extruded after multiple tooth feeds, resulting in heightened surface roughness (Fig. 7.15a–c).

On the other hand, when the ratio of $f_z/r_0$ exceeds 220%, micro-milling transitions into a region characterized by a weak size effect, as shown in Fig. 7.7b, where the

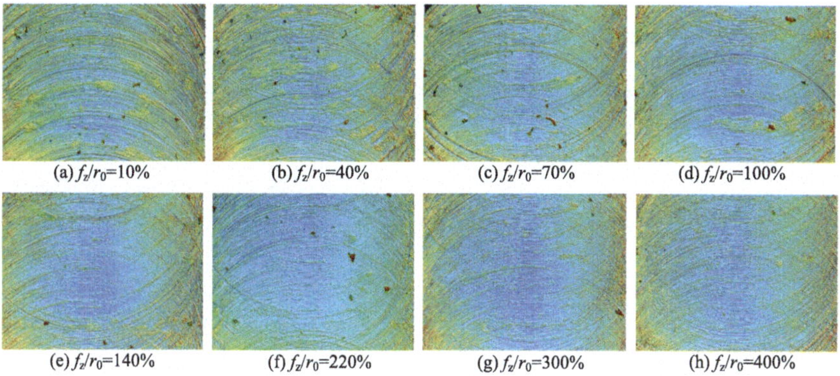


**Fig. 7.15** Surface morphology of the MSB under different $f_z/r_0$ values measured by white light interferometer: $f_z/r_0 =$ **a** 10%; **b** 40%; **c** 70%; **d** 100%; **e** 140%; **f** 220%; **g** 300%; **h** 400%

predominant force is shear force. The influence of tool edge size diminishes, and although surface roughness tends to increase with higher $f_z/r_0$ values, it generally remains stable. Figure 7.16 illustrates the tool marks observed on the machined surface at an $f_z/r_0$ ratio of 400%. The spacing between consecutive tool marks in the feed direction closely matches the feed value between two adjacent teeth in a tool rotation cycle, indicating the micro profile replication by the micro-milling tool, which leads to a high Ra value at the bottom of the micro-slot.

When the $f_z/r_0$ falls within the range of 70–220%, micro-milling undergoes a transition in the edge-size effect zone, indicating a change from ploughing to shear. Chips

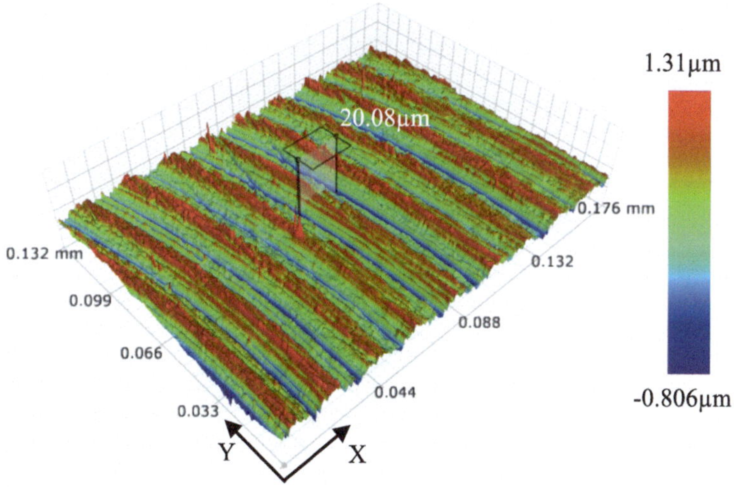


**Fig. 7.16** The MSB surface's three-dimensional morphology captured using the white light interferometer

produced at this point exhibit minimal adhesion to the tools due to $f_z$ being close to $h_{min}$, resulting in a smoother machined surface with fewer defects and reduced surface roughness at the slot bottom, as shown Fig. 7.15d–f. Optimal cutting parameters should be chosen within the $f_z/r_0$ range of 70–220% to achieve a surface roughness level of 0.331–0.44 μm. The optimal $f_z/r_0$ value of 140% was determined from the N1–N8 group test in Table 7.1, with the corresponding surface topography shown in Fig. 7.15e.

(1) *Influence of axial cutting depth on the roughness of MSB*

Figure 7.17 illustrates the influence of axial cutting depth on the surface roughness of the MSB, utilizing cutting parameters N9–N16 from Table 7.1. With $f_z/r_0$ at 40%, increasing the axial cutting depth from 10 to 50 μm results in a gradual rise in surface roughness. In the region of strong size effect, the edge-size impact is notable, causing tool and workpiece material interaction that leads to tool wear. Moreover, the expansion of axial cutting depth enlarges the contact squeeze area, hence increasing the volume of squeezed workpiece material and elevating ploughing resistance. Despite the challenges in chip formation, multiple tool marks are produced on the bottom (Fig. 7.18a–e).

When the $f_z/r_0$ ratio reaches 250%, micro-milling transitions into the fragile size range. Figure 7.17 demonstrates that as the axial cutting depth rises from 10 to 40 μm, the Ra value of the MSB increases from 0.28 to 0.42 μm. This suggests that despite the tool maintaining stable chip formation under the primary shearing force, the resistance to material plastic deformation and chip-tool friction rises. As a result, a residual amount of chips remains at the MSB, with some chips being extruded twice by the tool flank (Fig. 7.18f–l). Once the axial cutting depth reaches 50 μm, there is a slight reduction in surface roughness. This could be attributed to prior experiments causing wear on the milling tool's cutting edge, leading to a larger cutting edge

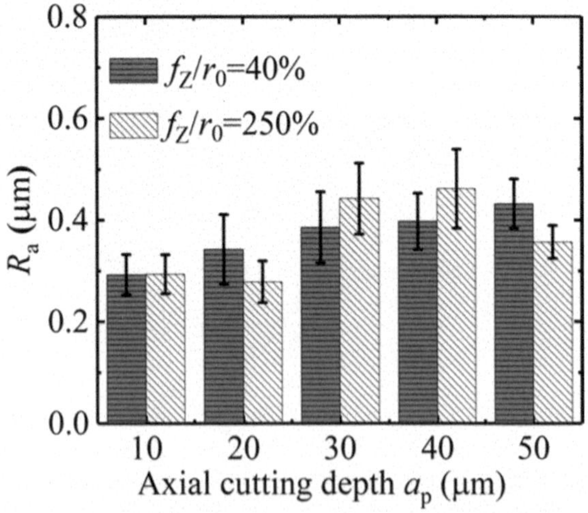


**Fig. 7.17** Surface roughness of the MSB with respect to $a_p$ in the size effect zone (N9–N16)

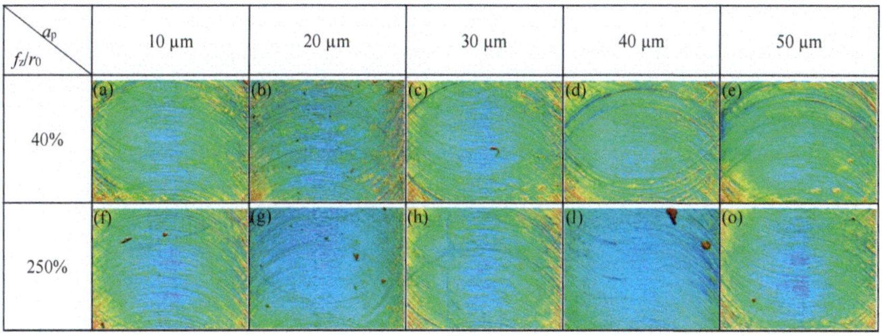


**Fig. 7.18** Surface morphology of the MSB in different $a_p$

arc radius. Consequently, the $f_z/r_0$ ratio decreases, facilitating chip formation and detachment at the micro-slot edge, resulting in the formation of regular and consistent tool marks at the MSB (Fig. 7.18o).

(2) *Influence of cutting speed on the surface roughness*

Figure 7.19a demonstrates the impact of cutting speed on surface roughness when utilizing cutting parameters N17–N24 from Table 7.1 in a micro-milling test. With $f_z/r_0$ at 40%, as the cutting speed gradually increases from 16 to 32 mm/min, the surface roughness of the MSB remains consistently low and even slightly decreases. This observation indicates that in the region strongly affected by size, chip formation is minimal. However, as the cutting speed rises, the duration of contact between the micro-milling tool and the workpiece material decreases, leading to reduced elastic deformation of the material between the tool flank face and the machined surface. This reduction results in decreased squeeze and friction effects, with ploughing cutting forces experiencing minimal deviations. Once the cutting speed reaches 200 mm/min, tool run-out may occur, causing a sudden spike in the surface roughness of the MSB.

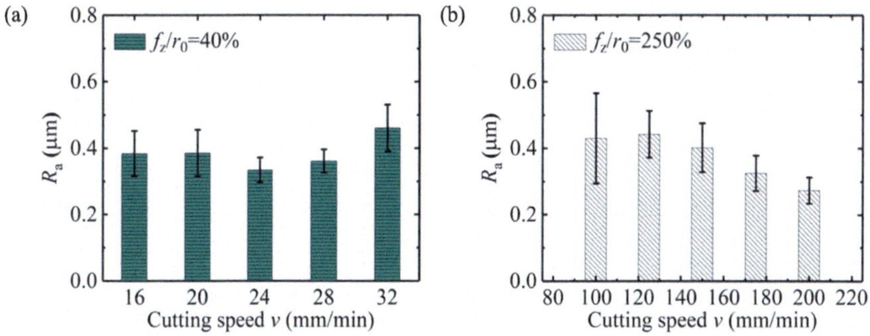


**Fig. 7.19** Surface roughness of the MSB in relation to cutting speed $v$ in the size effect band (N17–N24): $f_z/r_0 =$ **a** 40% and **b** 250%

At a $f_z/r_0$ of 250%, the surface roughness value at the bottom of the micro-slot decreases gradually as the feed rate $v$ increases, as illustrated in Fig. 7.19b. This phenomenon is attributed to micro-milling taking place in the region affected by weak edges, where the workpiece material generates stable chips due to the shear force, leading to consistent tool marks on the MSB surface. Moreover, with the rise in cutting speed, the duration of contact between the tool and the chip decreases, resulting in reduced adhesion of the chip to the tool and facilitating its easier separation from the MSB. Consequently, there are fewer residual chips left on the MSB.

## 7.2 Size Effect of Cutting Depth

Size effects are evident in ultra-precision fly cutting (UPFC) processes as well. UPFC entails the concurrent rotation of a SCD tool with the machine tool spindle, intermittently cutting the workpiece surfaces [11]. UPFC is extensively employed in the fabrication of non-rotationally symmetric surfaces and intricate microstructures such as freeform surfaces, F-theta lenses, pyramid arrays, and tetrahedron arrays, achieving submicron-level form accuracy and nanometric-level surface roughness [12, 13]. In this process, size effects become noticeable when the remaining dimensions of the workpiece reduce to a particular value, significantly affecting its physical behavior and mechanical response due to the influence of grain size [14]. For example, Kim et al. observe that 'size effect' characteristics in metal cutting appear when the UCT is comparable to the size of the tool edge radius [15]. The size effect in metal cutting has several implications, including an unbounded increase in specific cutting energy as the UCT decreases [16–18]. Additionally, this effect results in burr formation on the machined surface and an increase in surface roughness.

In UPFC, the cutting mechanism causes the chip to thin at its edges while the central section remains thicker (ranging from tens to hundreds of microns). This discrepancy in UCT along its length leads to the possibility of the size effect. In particular, the size effect in UPFC manifests in two primary ways: first, there is a higher proportion of surface layer grains compared to inner layer grains, and second, beyond a certain distance from the TIS, the chip thickness becomes smaller than the cutting-edge radius. This section presents a theoretical and experimental study on the generation of SRPs (Subsurface Residual Stresses) in UPFC.

### 7.2.1 *Ultra-precision Fly Cutting Experiments*

Using the Precitech Freeform 750G ultra-precision lathe, equipped with three linear axes and two rotary axes suitable for raster fly cutting, an experiment was conducted. In this setup, the workpiece material was affixed to the B-axis rotatable table using an 'L'-shaped fixture, as depicted in Fig. 7.20. Additionally, a Kistler dynamometer was positioned between the workpiece and the fixture.

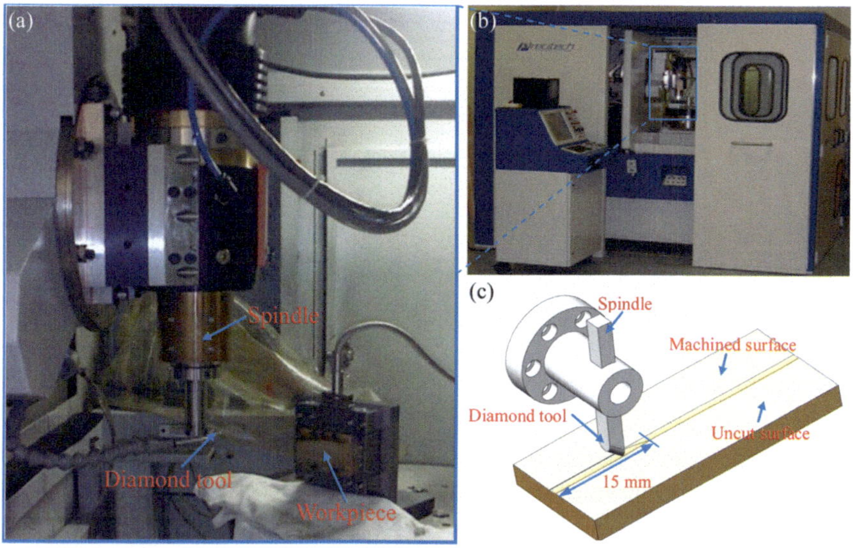


**Fig. 7.20** Experimental setup

Throughout the duration of the experiment, a Contour diamond tool was used for the UPFC procedure. Brass C3602 was chosen to serve as the material for the test. In order to investigate the impact that cutting parameters have on the production of SRPs, a number of different feed rates, cutting depths, and spindle speeds were used. At the end of each scenario, a final down-to-up cut was performed at the very last raster line. This was followed by the execution of one hundred raster lines of horizontal cutting. Detailed information on the cutting parameters that were used in this investigation can be found in Table 7.2. The numbers included in parentheses indicate the value that was initially set for each parameter when the other parameters were changed. The cutting force was monitored using a Kistler 9256C1 sensor during the whole UPFC process. The force signal that was obtained was then analyzed using software developed by Dynoware. As soon as the cutting experiment was over, the machined workpiece was taken out and examined using an Olympus BX60 optical microscope and a Zygo Nexview white light interferometer for the purpose of taking measurements.

### 7.2.2 *Size Effect in Ultra-precision Fly Cutting*

The image in Fig. 7.21 displays the surface appearance achieved when the spindle rotates at a speed of 500 rpm. It emphasizes the presence of a clear feed boundary on the machined surface, dividing the tool marks into two distinct sections. Moreover, it is apparent that the surface quality varies on either side of the feed boundary,

In the process of UPFC, the diamond cutting tool moves continuously as the spindle rotates, intermittently cutting the surface of the workpiece. This discontinuous cutting method causes the tool to approach the material of the workpiece at an angle on the side where the tool enters. As a result, a downward force component ($F_v$ on the TIS in Fig. 7.24b) applies compressive stress to the workpiece material. Conversely, the tool on the outer side cuts into the workpiece material with an upward slant, leading to an upward force component ($F_v$ on the TOS in Fig. 7.24b) that creates a combined tensile-compressive stress on the workpiece material. Thus, the workpiece material experiences alternating stress during rotary fly cutting, which is different from the continuous cutting process seen in SPDT. Additionally, due to the relatively shallow cutting depth compared to the swing distance (radius of the spindle's rotational path), the interaction between the cutting tool and workpiece during cutting is brief, as depicted in Fig. 7.24b. This short contact duration is also evident from the recorded cutting force displayed in Fig. 7.24c, where the cutting phase occupies only about 1% of the entire fly cutting cycle, with the majority of the cycle dedicated to non-cutting movements.

In UPFC, a fresh slope surface is created with each rotary cutting motion. The grains on these fresh slope surfaces in the surface layer have more exposed surfaces and exhibit mechanical properties similar to monocrystalline materials. On the contrary, the grains in the inner layer lack exposed surfaces and show mechanical properties resembling those of polycrystalline materials. As the UCT in UPFC varies from zero to tens of microns before reducing to zero again, and given that the size of the surface layer grain is in the tens of microns scale, the rotary fly cutting process cuts through both the surface layer and inner layer grains. Based on the cutting chips' morphology and the thickness of the surface layer grains, it can be concluded that the surface layer grains make up more than 35% of the total grains, underscoring the significance of their properties. Therefore, the occurrence of SRPs in UPFC is often linked to the size effect.

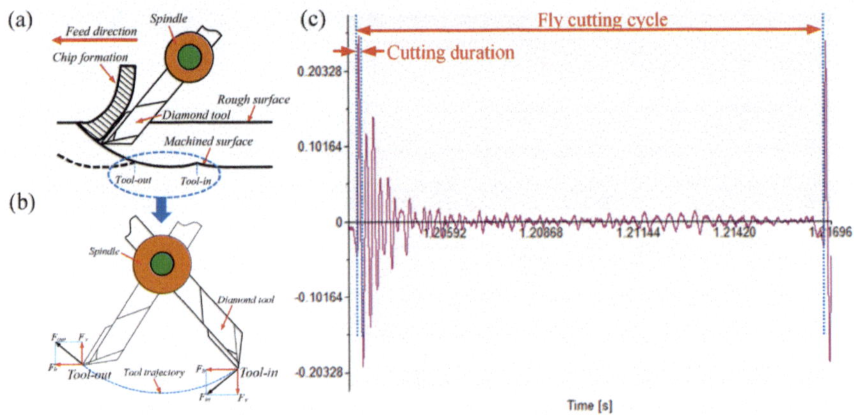


Fig. 7.24 Schematics of fly cutting process and the corresponding cutting force

### 7.2.3 FE Simulation

To capture the size effect, a size factor is introduced into the Johnson Cook model, resulting in the development of a hybrid constitutive model. Subsequently, a finite element simulation is conducted using this newly established hybrid constitutive model. Finally, the effectiveness of the hybrid constitutive model is validated through a comparison between the simulation and experimental results.

#### 7.2.3.1 Johnson Cook Model and Surface Layer Model

The Johnson Cook model is typically represented by a three-term product that accounts for the impact of material properties, strain rate, and deformation temperature ($T_w$) on the flow stress ($\sigma$), as depicted in Eq. (7.1) [26]:

$$\sigma = \left[A + B\varepsilon^n\right]\left[1 + C\ln\frac{\dot{\varepsilon}}{\dot{\varepsilon}_0}\right]\left[1 - \left(\frac{T_w - T_r}{T_m - T_r}\right)^m\right] \tag{7.1}$$

In Eq. (7.1), the first term, typically denoted as $A + B\varepsilon^n$ using the Schmidt approach, streamlines curve fitting by focusing on just two unknown coefficients (A, B) and the hardening coefficient ($n$). The second term introduces the unknown coefficient $C$ to address the impact of strain rate, leveraging split Hopkinson pressure bar (SHPB) experimental data for fitting curves. The third term takes into account the effect of the operating temperature ($T_w$). In contrast to models like the Armstrong model, the Johnson Cook model necessitates only five unknown coefficients that can be established through upsetting and SHPB experiments, effectively tackling a variety of mechanical issues across different strain rates.

This section delves into exploring the phenomenon where the in-tool area of a tool feed impression displays a smooth surface, while the out-tool area exhibits SRPs. To investigate this, the surface layer model is integrated into the Johnson Cook model, which considers a workpiece consisting of two layers: the surface layer and the inner layer. The grains in the surface layer, possessing free surfaces without deformation constraints, are more susceptible to plastic deformation. The fundamental formulation of the surface layer model is as follows [27]:

$$\sigma = \frac{N_s\sigma_s + N_i\sigma_i}{N} = \eta\sigma_s + (1 - \eta)\sigma_i, \left(N = N_s + N_i, \eta = \frac{N_s}{N}\right) \tag{7.2}$$

When conducting compression experiments with standard-sized samples, the influence of surface grains is typically minimal due to the fact that the surface grain number is generally less than 2% of the total grain number.

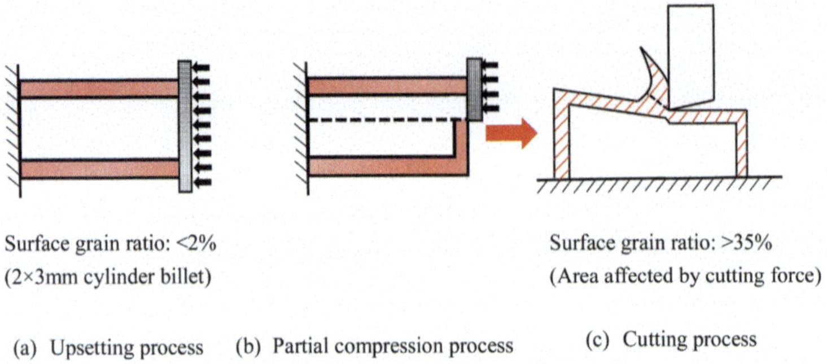


**Fig. 7.25** Surface grain ratio in different deformation processes

In microscale cutting situations, the size effect becomes notable when the ratio of surface grains reaches about 35–55%, depending on the impacted region by the cutting force. As depicted in Fig. 7.25, the red portion represents the surface layer grains. It is observed that the cutting deformation resembles a compression process, where a segment of the material undergoes compression, causing an increase in shear stress. This elevated stress surpasses the yield point, resulting in plastic deformation and sliding along a 45° line in the direction where the plastic strain reaches the fracture strain. Eventually, this leads to the removal of the chip from the workpiece.

Previous studies in this field have provided extensive insights into the size effect and the surface layer model [28]. The Hall–Petch equation [29], Hansen's research [30], crystal plasticity theory [31], and Armstrong's studies [32, 33] contribute to our understanding of shear stress $\tau(\varepsilon)$, which can be described by the lattice friction stress $\tau_0$ ($\varepsilon$) and the dislocation interactions. Based on these findings, the Hall–Petch equation can be expressed as shown in Eq. (7.3) [28].

$$\left.\begin{aligned} &\sigma = M\tau \\ &\sigma = \sigma_0 + kd^{-\frac{1}{2}} \\ &\tau(\varepsilon) = \tau_0(\varepsilon) + k(\varepsilon)d^{-\frac{1}{2}} = \tau_0(\varepsilon) + \alpha\mu b\sqrt{\rho_T} \end{aligned}\right\} \\ \Rightarrow \sigma(\varepsilon) = M\tau(\varepsilon) = M(\tau_0(\varepsilon) + \alpha\mu b\sqrt{\rho_T}) \tag{7.3}$$

The stress contributions from the surface layer and the inner layer grains can be expressed as shown in Eq. (7.4) [28].

$$\begin{cases} \sigma = \sigma_s(\varepsilon) + \sigma_i(\varepsilon) \\ \sigma_s(\varepsilon) = m\tau_0(\varepsilon) = mk_1\varepsilon^{n_1} \\ \sigma_i(\varepsilon) = \sum \sigma_i f_i \\ = \sigma_\alpha f_\alpha + \sigma_\beta f_\beta \\ = (Mk_2\varepsilon^{n_2} + \alpha M\,\mu_\alpha b_\alpha\sqrt{\rho_{T\alpha}(\varepsilon)})f_\alpha \\ \quad +(Mk_2\varepsilon^{n_2} + \alpha M\,\mu_\beta b_\beta\sqrt{\rho_{T\beta}(\varepsilon)})(1-f_\alpha) \\ = (Mk_2\varepsilon^{n_2} + \alpha M\,\mu_\alpha b_\alpha\sqrt{\dfrac{C_1\varepsilon}{b_a d}})f_\alpha + (Mk_2\varepsilon^{n_2} \\ \quad +\alpha M\,\mu_\beta b_\beta\sqrt{\dfrac{C_2\varepsilon}{b_\beta d}})(1-f_\alpha) \end{cases} \tag{7.4}$$

The second equation within Eq. (7.4) suggests that the mechanical properties of the surface layer grains exhibit similarities to those of a single crystal, as indicated by Armstrong's study [26]. Conversely, the inner layer grains are modeled using the polycrystal model described in Eq. (7.3).

In this section, Eq. (7.4) can be further developed as follows:

$$\begin{cases} \sigma_{cuttingarea}(\varepsilon) = \sigma_{se}(\varepsilon) + \sigma_{ie}(\varepsilon) \\ \sigma_{se}(\varepsilon) = \dfrac{\sigma_s(\varepsilon)}{N_s}\cdot N_{se} = \dfrac{mk_1\varepsilon^{n_1}}{N_s}\cdot N_{se} \\ \sigma_{ie}(\varepsilon) = \dfrac{\sigma_\alpha(\varepsilon)}{N_\alpha}\cdot N_{\alpha e} + \dfrac{\sigma_\beta(\varepsilon)}{N_\beta}\cdot N_{\beta e} \\ \sigma_\alpha(\varepsilon) = \left(Mk_2\varepsilon^{n_2} + \alpha M\,\mu_\alpha b_\alpha\sqrt{\dfrac{C_1\varepsilon}{b_a d}}\right)f_\alpha \\ \sigma_\beta(\varepsilon) = \left(Mk_2\varepsilon^{n_2} + \alpha M\,\mu_\beta b_\beta\sqrt{\dfrac{C_2\varepsilon}{b_\beta d}}\right)(1-f_\alpha) \end{cases} \tag{7.5}$$

The initial term of the original Johnson Cook model does not adequately explain the formation of SRPs in the TOS. Equation (7.5) presents several unknown variables, complicating the curve fitting process. Nevertheless, it is posited that the increased ratio of surface grains is the main factor responsible for the abrupt change in surface roughness. Hence, Eq. (7.5) is used to replace the first term of the original Johnson Cook model.

In the Johnson Cook model, the coefficient C is generally determined via SHPB experiments, which illustrate the impact of strain rate on the true stress–strain curve. Nevertheless, comparing SHPB data with static upsetting data of 2 and 1 mm diameters reveals challenges in fitting curves using the Johnson Cook model, as stress increments in micro-scaled billets are substantially higher than those in macro-scaled billets. To address the size effect on the stress–strain curve at high strain rates, the strain rate size effect coefficient $D$ is incorporated into the Johnson Cook model.

As a result, the hybrid constitutive Johnson Cook (HCJC) model is formulated as follows [27]:

$$\sigma = \left[\frac{N_{se}}{N_s} nmk_1\varepsilon^{n_1} + \frac{N_{\alpha e}}{N_\alpha}\left(nMk_1\varepsilon^{n2} + \alpha M\,\mu_\alpha b_\alpha\sqrt{\frac{C_1\varepsilon}{b_a d}}\right)f_a + \frac{N_{\beta e}}{N_\beta}\left(nMk_2\varepsilon^{n2} + \alpha M\,\mu_\beta b_\beta\sqrt{\frac{C_2\varepsilon}{b_\beta d}}\right)(1 - f_a)\right]$$
$$\left[1 + C^* \ln\frac{\dot{\varepsilon}}{\dot{\varepsilon}_0}\right][1 + \eta D]\left[1 - \left(\frac{T_w - T_r}{T_m - T_r}\right)^m\right] \tag{7.6}$$

To implement the aforementioned concept, a general methodology is proposed and depicted in Fig. 7.26. By following this approach, three different grain sizes can be used to obtain three corresponding effective stresses for a given effective strain. By applying a series of effective strains, the relationship between grain size and effective stress can be established, enabling the calculation of $\tau(\varepsilon)$ and $k(\varepsilon)d^{-\frac{1}{2}}$ in Eq. (7.3). The next step involves curve fitting to determine the unknown coefficients in Eq. (7.6). This process brings us closer to the final formulation of the HCJC model.

### 7.2.3.2 Influence of Size Effect on Strain Rate in HCJC Model

In Eq. (7.6), $d = 18.54$ μm; $m = 2$; $M = 3.06$; $\alpha = 0.34$ [34]; $\mu_\alpha = 78{,}500$ MPa (shear modulus for FCC phase); $\mu_\beta = 72{,}000$ MPa (shear modulus for BCC phase); $b_\alpha = \times 3.69 \times 10^{-10} = 2.608 \times 10^{-10}$ m (Burgers vector for FCC phase); $b_\beta = \times 2.94 \times 10^{-10} = 2.546 \times 10^{-10}$ m (Burgers vector for FCC phase). Through the analysis of EDS results and examination of a metallurgical photograph with enhanced clarity, the volume fractions of the $\alpha$, $\beta$, and lead phases are identified as 81, 17, and 2% respectively. Furthermore, the SHPB experiment indicates that the true stress–strain curve at the microscale (1 mm diameter) is considerably higher than at the macroscale. Specifically, the macroscale yield stress at a high strain rate has only increased by 15.6%, whereas at the microscale, it has risen by 57.6%, as demonstrated in Fig. 7.27. These findings underscore the necessity of incorporating a strain-rate size effect coefficient when employing the Johnson Cook model to describe the true stress–strain behavior in both low and high strain rate scenarios.

Under static forming conditions, the strain rate approaches zero and the deformation temperature ($T_w$) corresponds to room temperature. Consequently, Eq. (7.6) can be simplified as follows [27]:

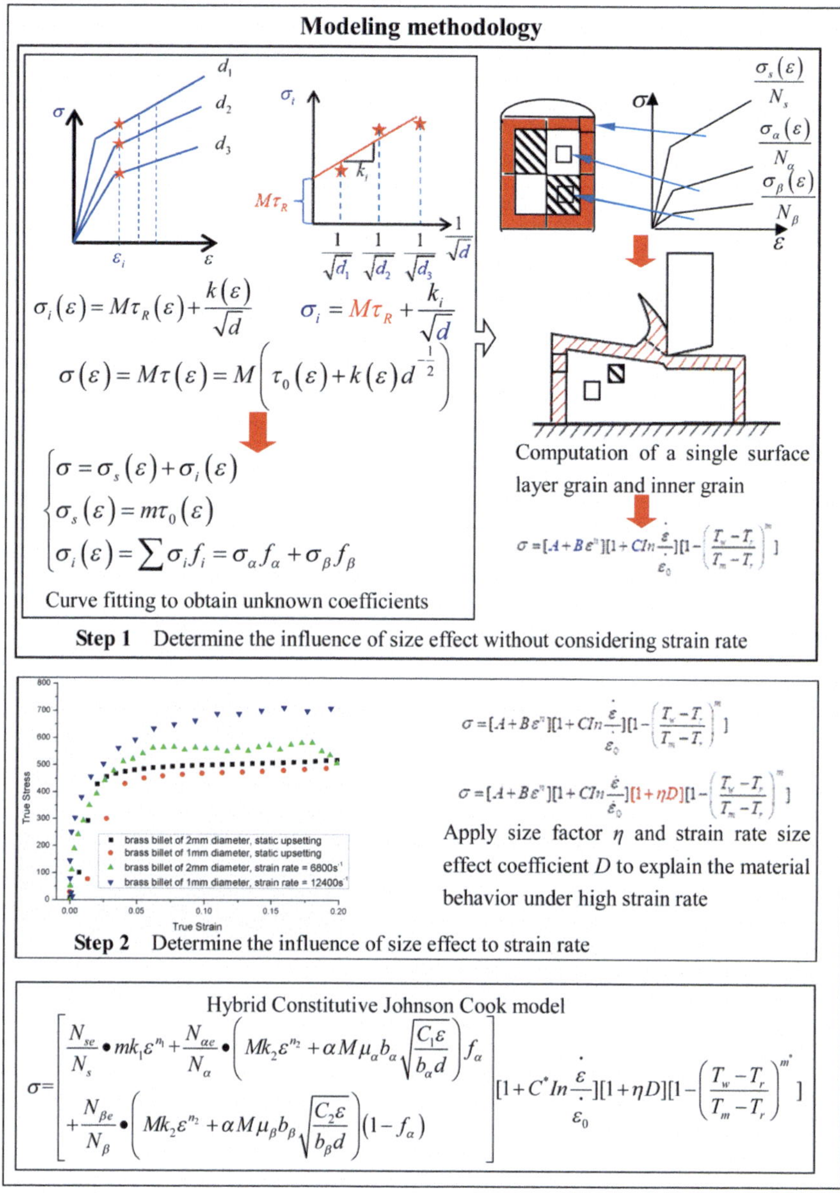


**Fig. 7.26** Modelling methodology of HCJC model

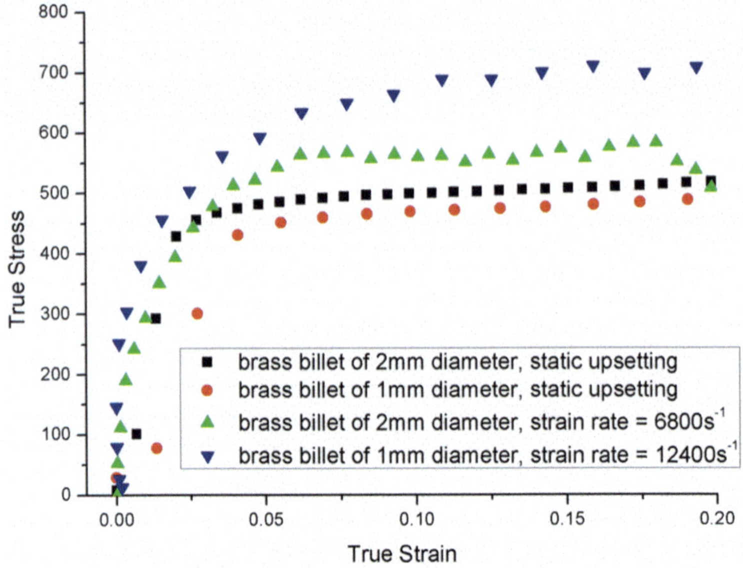


**Fig. 7.27** True stress–strain curve of brass C3602 under low and high strain rates

$$\sigma = \begin{bmatrix} \frac{N_{se}}{N_s} \cdot mk_1\varepsilon^{n_1} + \frac{N_{\alpha e}}{N_\alpha} \cdot \left( Mk_2\varepsilon^{n_2} + \alpha M \mu_\alpha b_\alpha \sqrt{\frac{C_1\varepsilon}{b_\alpha d}} \right) f_\alpha \\ + \frac{N_{\beta e}}{N_\beta} \cdot \left( Mk_2\varepsilon^{n_2} + \alpha M \mu_\beta b_\beta \sqrt{\frac{C_2\varepsilon}{b_\beta d}} \right) (1 - f_\alpha) \end{bmatrix} [1 + C^*][1 + \eta D] \tag{7.7}$$

Figure 7.28 illustrates the curve fitting process for determining the values of $k$, $n$, and $C$. According to Eq. (7.3), for a specific strain (e.g., $\varepsilon = 0.1$), four stresses can be obtained with four different grain sizes. By applying a series of strains, $\tau(\varepsilon)$ and $k(\varepsilon)d^{-\frac{1}{2}}$ can be determined, leading to the determination of the values of $k$, $n$, and $C$. Utilizing the data from the SHPB experiment, the values of $C^* = 0.16$, $D = 46.04$, and $m^* = 1.09$ are obtained. Thus, the final expression of the HCJC model is as follows [27]:

$$\sigma = \left[559.58\varepsilon^{0.17} + 101.75\varepsilon^{0.5}\right]\left[1 + 0.16In\frac{\dot{\varepsilon}}{\dot{\varepsilon}_0}\right]$$
$$[1 + 46.04\eta]\left[1 - \left(\frac{T_w - T_r}{T_m - T_r}\right)^{1.09}\right] \tag{7.8}$$

Upon deriving the final expression of the HCJC model, the strain energy required for damage initiation can be calculated by integrating this model. According to

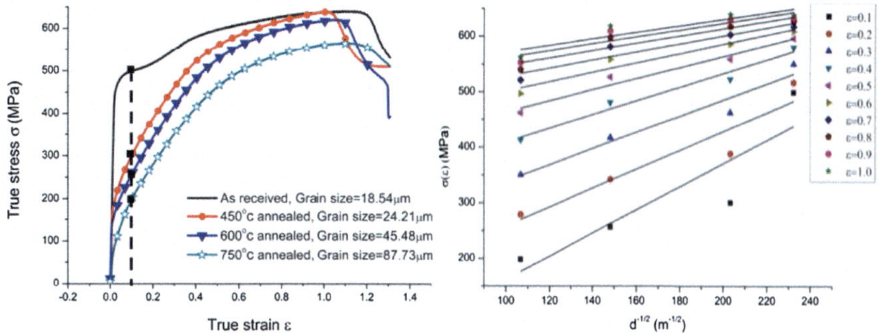


**Fig. 7.28** Determination of the unknown coefficients for HCJC model

Hillerborg's research [35], the fracture energy is determined by integrating the stress-displacement curve of the HCJC model from the onset of damage to the formation of a major crack. The resulting fracture strain and fracture energy can subsequently be applied in FE simulations for predicting fractures.

The comparison between the stress–strain curve derived from Eq. (7.8) and the experimental results reveals that the HCJC model closely matches the experimental data, demonstrating its effectiveness as shown in Fig. 7.29. Using Eq. (7.3), the effective stresses of individual grains in the surface and inner layers are considered. Figure 7.30a illustrates the effective stress–strain relationship for a single grain in the surface layer and the inner layer. The strain energy of a single grain in the surface layer is approximately ten times greater than that of a grain in the inner layer. In UPFC, where surface layer grains constitute over 35% of the total grain count, the contributions of strain energy from grains in both layers are shown in Fig. 7.30. The strain energy of the surface layer grains is roughly six times that of the inner layer grains.

### 7.2.4 SRP Formation and Suppression

Using the derived HCJC model, a two-dimensional FE simulation is conducted in the Abaqus® software to simulate the UPFC process. The rotary fly cutting in UPFC is transformed into an orthogonal cutting simulation based on the chip geometry, as depicted in Fig. 7.31.

Figure 7.32 illustrates the FE simulation of chip formation in UPFC. Initially, at the TIS, a cutting chip forms as the surface layer undergoes shearing by the cutting tool, leading to damage due to shear stress. During the intermediate stage of chip formation, voids appear in the inner layer adjacent to the surface layer before the cutting tool reaches it. The tool subsequently cuts through these voids, creating marks, as depicted in Fig. 7.32. Previous studies [28] and Hillerborg [35] indicate that damage initiation and crack formation during shear deformation occur when the

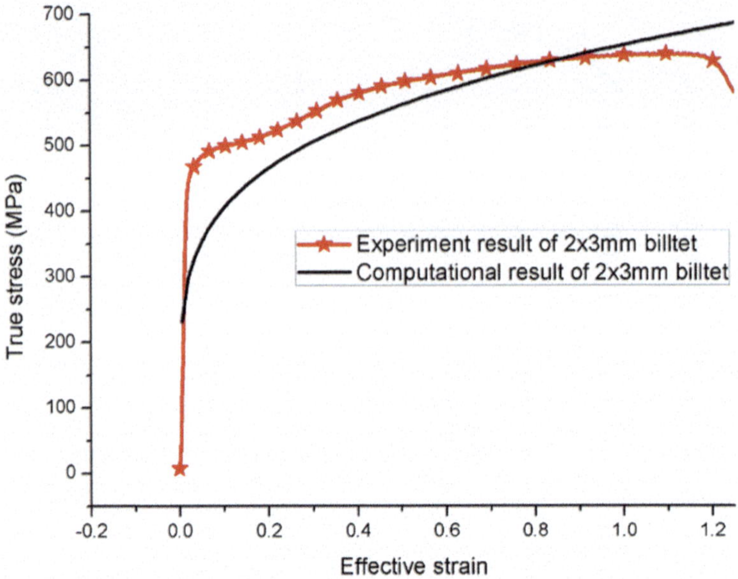


**Fig. 7.29** Experimental and computational results of HCJC model on upsetting of the 2 × 3 mm billet

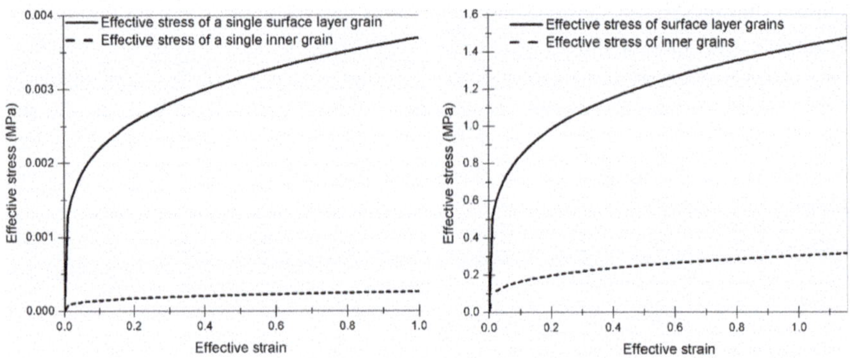


(a) Effective stress of single grain (b) Effective stress of grain layer

**Fig. 7.30** Single grain and grain layer strain energy comparison

absorbed energy surpasses the workpiece's strain energy during cutting. Given that the strain energy of the surface layer is considerably higher than that of the inner layer, the inner grains will reach fracture strain much earlier than the surface grains. This indicates that the inner grains of the chip will be fully damaged and develop significant cracks before being contacted by the cutting tool.

As the cutting progresses, the cutting chip forms completely. The comparison in Fig. 7.33 illustrates the simulated chip formation alongside the observed morphology

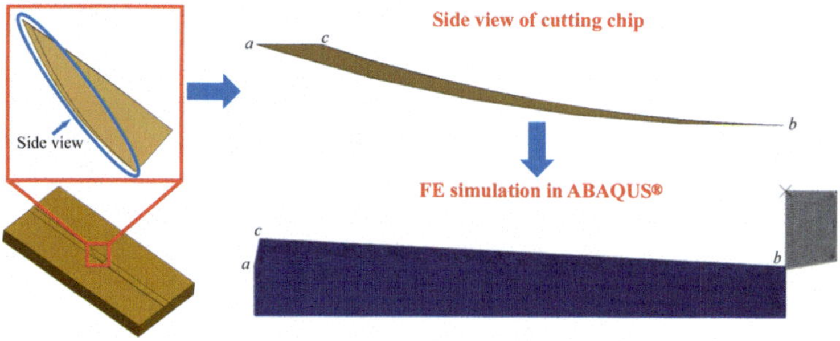


**Fig. 7.31** Cutting kinematics transforms in Abaqus® FE simulation

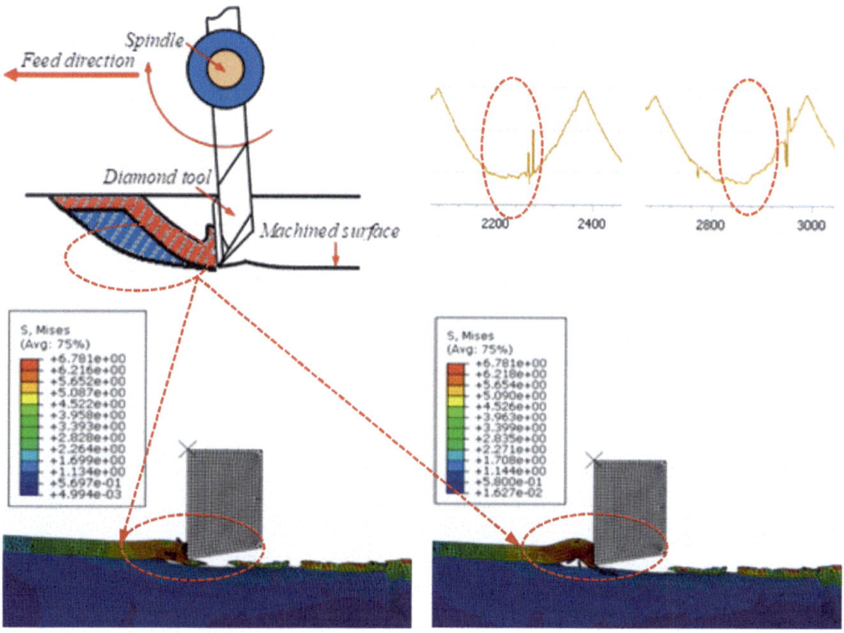


**Fig. 7.32** Tool-in causes void formation of inner layer in two-dimensional FE simulation of UPFC

of the real cutting chip. In Fig. 7.34, the cutting force comparison between the FE simulation and UPFC experimental results is presented. The morphology of the cutting chip and the cutting force show good agreement with the simulated results, confirming the simulation's reliability.

Illustrated in Fig. 7.35, the process of SRP formation is detailed. When the diamond tool enters the surface of the workpiece, both the outer and inner layers of grains experience equal cutting force from the tool. The surface layer near the diamond tool undergoes deformation and damage due to shear stress. Calculations

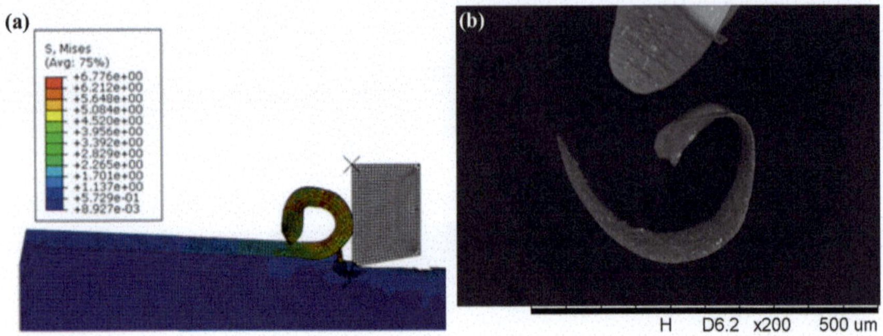


**Fig. 7.33** Comparison of chip morphology between simulated result and measured one

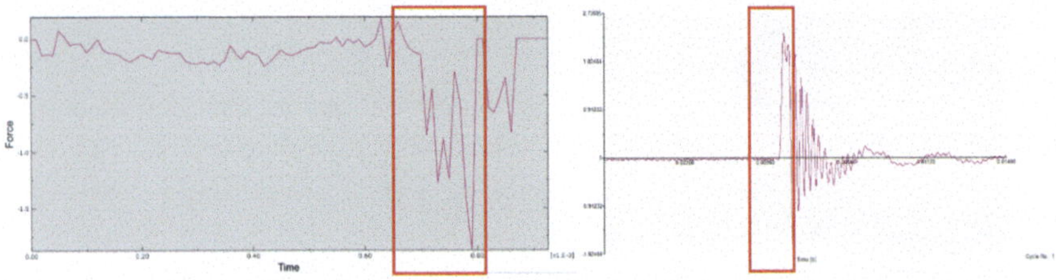


a) Cutting force obtained from FE simulation with Abaqus

b) Cutting force obtained from UPFC experiment

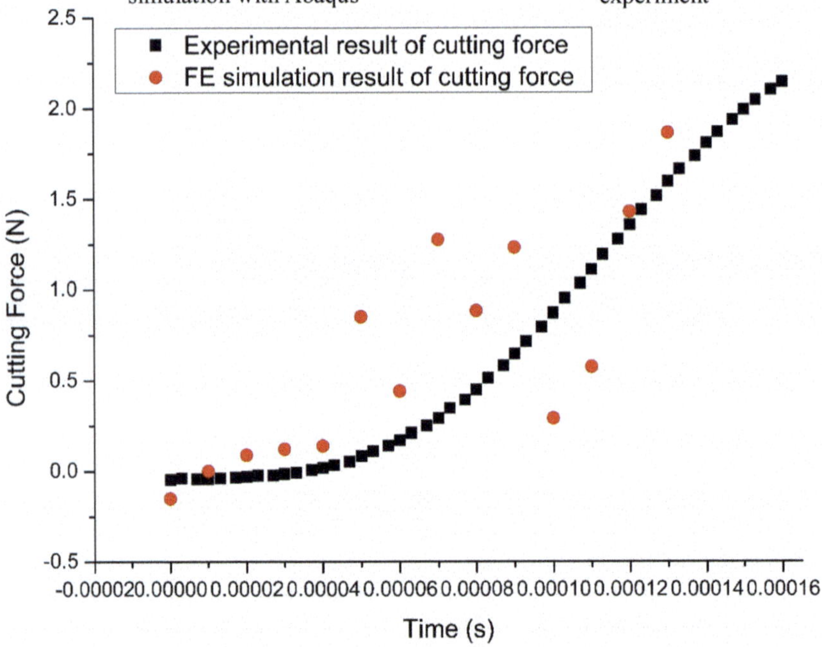


c) Comparison between experimental result and FE simulation

**Fig. 7.34** Cutting force comparison between experiment and FE simulation result

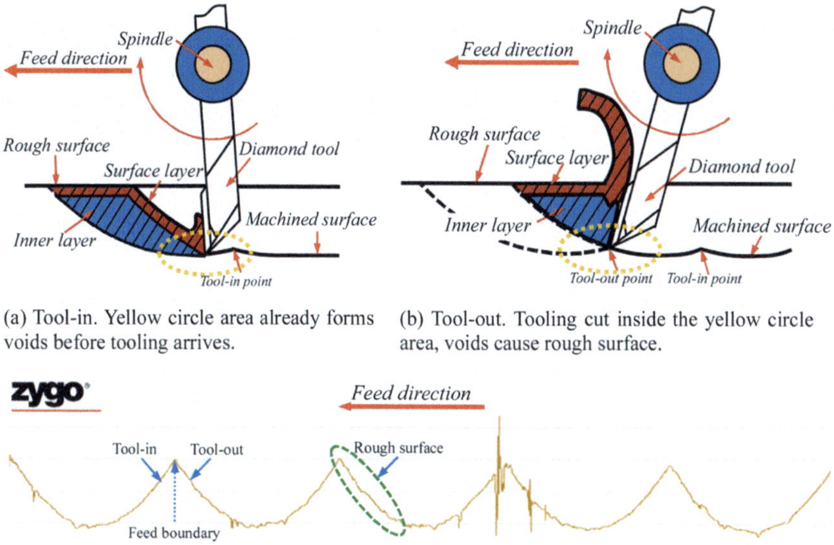


(a) Tool-in. Yellow circle area already forms voids before tooling arrives.

(b) Tool-out. Tooling cut inside the yellow circle area, voids cause rough surface.

(c) Surface profile. Rough surface at tool-out side

**Fig. 7.35** Cutting chip and SRP generation in a fly cutting process

indicate that the inner layer, with a fracture strain of approximately 1.16, will experience damage prior to the surface layer, as it can only withstand around 0.3 MPa for the fracture strain, while the surface layer remains in its elastic phase. Microvoids are created in the inner grain layer when the diamond tool cuts through the surface layer, as indicated by a yellow circle in Fig. 7.35a. These microvoids persist after the tool passes over, resulting in the formation of SRPs as shown in Fig. 7.35c.

The presence of built-up edge at the tool exit area of a cutting tool negatively impacts the surface finish and performance. One effective method to prevent built-up edge formation is by reducing the thickness of the chip, thus minimizing the affected cutting region. Decreasing the chip thickness alters the number of surface grains slightly, but significantly reduces the number of internal grains, resulting in fewer voids. In the context of UPFC, reducing the feed rate per revolution (mm/rev) leads to a decrease in chip thickness, thereby reducing the area under the cutting force. The feed rate per revolution can be determined using the formula:

$$f' = \frac{f}{s_p} \tag{7.9}$$

where $f'$ is the feed per revolution, $f$ represents the feed rate in units of mm/min, and $s_p$ denotes the spindle speed in units of rev/min. From Eq. (7.9), it is evident that increasing the spindle speed or decreasing the feed rate will lower the feed per revolution.

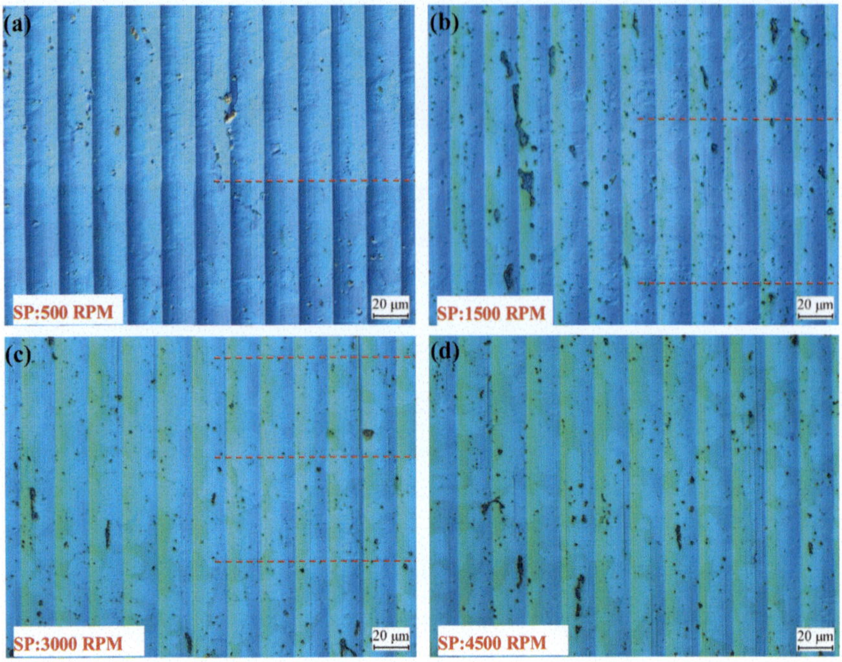


**Fig. 7.36** Surface topography of UPFC under different spindle speeds (measured by Olympus BX60 optical microscope)

Figure 7.36 portrays the surface topography of UPFC at different spindle speeds, namely 500, 1500, 3000, and 4500 rpm. The machined surface topography is visibly influenced by the spindle speed. At 500 rpm, the feed boundary is clearly defined, and SRPs are observed in the TOS, whereas a smooth surface is formed in the TIS. The distribution of SRPs decreases as the spindle speed increases. Notably, at a spindle speed of 4500 rpm, the feed boundary becomes indistinguishable, and no SRPs are observed yet. Additionally, Fig. 7.37 displays the FE simulation result of UPFC processes under different spindle speeds. The surface roughness at 500 rpm spindle speed is found to be worse compared to that achieved at 4500 rpm.

Similar observations can be made regarding different feed rates, as presented in Fig. 7.38. The SRPs in the TOS are easily identifiable at a feed rate of 900 mm/min. However, as the feed rate decreases, the length of the tool feed imprint in rotary cutting becomes shorter, making it more challenging to identify the SRPs.

Remarkably, the cutting depth does not affect SRP formation since, under fixed cutting parameters, altering the cutting depth solely influences the chip length without impacting the reduction in chip thickness.

This set of experimental results confirms that increasing the spindle speed and decreasing the feed rate effectively suppresses SRP formation, leading to improved surface finish. Conversely, the cutting depth exhibits little influence on SRP formation.

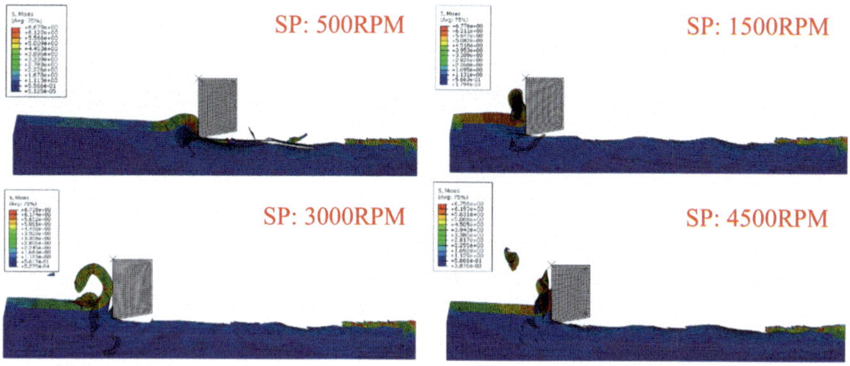


**Fig. 7.37** Simulation result of UPFC processes under different spindle speeds

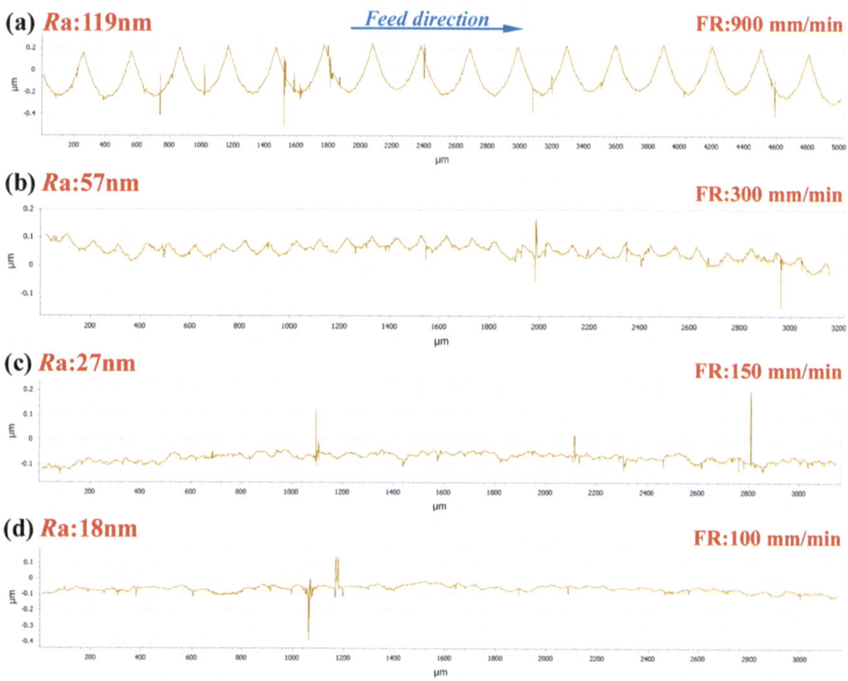


**Fig. 7.38** Sectional profile of tool feed imprint in a white light interferometer measurement at difference feed rates (Zygo Nexview)

## References


1. Thepsonthi T (2014) Modeling and optimization of microend milling process for micromanufacturing. PhD Thesis, Rutgers Univ, New Brunswick, NJ, USA
2. Bissacco G, Hansen HN, Chiffre LD (2005) Micromilling of hardened tool steel for mould making. Appl J Mater Process Technol 167:201–207
3. Biermann D, Kahnis P (2010) Analysis and simulation of size effects in micromilling. J Prod Eng 4:25–34
4. De Oliveira FBO, Rodrigues AR, Coelho RT, de Souza AF (2015) Size effect and minimum chip thickness in micromilling. Int J Mach Tool Manuf 89:39–54
5. Lai X, Li H, Li C, Lin Z, Ni J (2008) Modelling and analysis of micro scale milling considering size effect, micro cutter edge radius and minimum chip thickness. Int J Mach Tool Manuf 48:1–14
6. Dépincé P, Hascoët JY (2006) Active integer of tool deflection effects in end milling. Part 1. Prediction of milled surfaces. Int J Mach Tool Manuf 46:937–944
7. Aramcharoen A, Mativenga PT (2009) Size effect and tool geometry in micromilling of tool steel. Precis Eng 33:402–407
8. Liang YC, Yang K, Bai QS, Chen JX, Wang B (2009) Modeling and experimental analysis of microburr formation considering tool edge radius and tool-tip breakage in microend milling. J Vac Sci Technol B Microelectron Nanometer Struct Process Meas Phenom 27:1531–1535
9. Zhang T, Liu Z, Xu C (2013) Influence of size effect on burr formation in micro cutting. Int J Adv Manuf Technol 68:1911–1917
10. Tamin MN, Sudin I, Mon TT (2008) Thermal-mechanical responses of Ti-6Al-4V during orthogonal cutting process. In: Proceedings of defect diffusion forum. Balatonfured, Hungary, pp 673–678
11. Zhang G, To S, Xiao G (2014) A novel spindle inclination error identification and compensation method in ultraprecision raster milling. Int J Mach Tools Manuf 78(3):8–17
12. Wu Y, Peng W, Liu Y (2013) A novel fabrication method for micro optical waveguide mold based on fly-cutting technology. Opt Int J Light Electron Opt 124(9):867–869
13. Zhu Z, To S, Zhang S (2015) Theoretical and experimental investigation on the novel end-fly-cutting-servo diamond machining of hierarchical micro-nanostructures. Int J Mach Tools Manuf 94:15–25
14. Deng JH, Fu MW, Chan WL (2011) Size effect on material surf deformation behavior in microforming process. Mater Sci Eng A 528(13–14):4799–4806
15. Kim KW, Lee WY, Sin HC (1999) A finite element analysis for the characteristics of temperature and stress in micro-machining considering the size effect. Int J Mach Tools Manuf 39(9):1507–1524
16. Chen N, Chen M, Wu C, Pei X, Qian J, Reynaerts D (2017) Research in minimum undeformed chip thickness and size effect in micro end-milling of potassium dihydrogen phosphate crystal. Int J Mech Sci 134:387–398
17. Weng J, Zhuang K, Zhu D, Guo S, Han D (2018) An analytical model for the prediction of force distribution of round insert considering edge effect and size effect. Int J Mech Sci 138–139:86–98
18. Liu K, Melkote SN (2007) Finite element anal of the influence of tool edge radius on size effect in orthogonal microcutting process. Int J Mech Sci 49(5):650–660
19. Dautzenberg JH, Hijink JAW, van der Wolf ACH (1982) The minimum energy principle applied to the cutting process of various workpiece materials and tool rake angles. CIRP Annal 31:91–96
20. Malekian M, Mostofa MG, Park SS, Jun MBG (2012) Modeling of minimum uncut chip thickness in micro machining of aluminum. J Mater Process Technol 212:553–559
21. Li C, Lai X, Li H, Ni J (2007) Modeling of threedimensional cutting forces in microendmilling. J Micromech Microeng 17:671–678
22. Yuan ZJ, Zhou M, Dong S (1996) Effect of diamond tool sharpness on minimum cutting thickness and cutting surf integrity in ultraprecis machining. J Mater Process Technol 62:327–330

23. Liu X, DeVor RE, Kapoor SG (2005) An analytical modelling for the prediction of minimum chip thickness in micromachining. J Manuf Sci Eng 128:474–481
24. Afazov SM, Ratchev SM, Segal J (2010) Modelling and simulation of micro-milling cutting forces. J Mater Process Tech 210:2154–2162
25. Liu ZB, Wang XB (2010) An experimental study on nonlinear characteristics of vibration signal in microendmilling process. Acta Armament 31:84–87
26. Johnson GR, Cook WHA (1983) A constitutive model and data for materials subjected to large strains, high strain rates, and high temperatures. In: Proceedings of 7th International Symposium on Ballistics, pp 541–547
27. Zhang GQ, Ran JQ, To S, Wu XY, Huang P (2020) Size effect on surface generation of multiphase alloys in ultra-precision fly cutting. J Manuf Process 60:23–36
28. Ran JQ, Fu MW (2014) A hybrid model for analysis of ductile fracture in micro-scaled plastic deformation of multiphase alloys. Int J Plast 61:1–16
29. Petch NJ (1953) The cleavage strength of polycrystals. J Iron Steel Inst 174:25–28
30. Hansen N (1985) Polycrystalline strengthening. Metall Trans A 16:2167–2190
31. Argon AS (2008) Strengthening mechanisms in crystal plasticity. Contemp Phys 54(6):306–307
32. Armstrong R, Codd I, Douthwaite RM, Petch NJ (1962) The plastic deformation of polycrystalline aggregates. Philos Mag 7(73):45–58
33. Armstrong RW (1982) The yield and flow stress dependence on polycrystal grain size, yield, flow and fracture of polycrystals, pp 1–31.
34. Rodríguez R, Gutiérrez I (2003) Correlation between nanoindentation and tensile properties: influence of the indentation size effect. Mater Sci Eng A 361(1):377–384
35. Hillerborg A, Modéer M, Petersson PE (1976) Anal of crack formation and crack growth in concrete by means of fracture mechanics and finite elements. Cem Concr Res 6(6):773–781

# Chapter 8
# Synthetic Diamond and Application in Precision Machining

**Abstract** Synthetic diamonds possess excellent properties, including unparalleled hardness, thermal conductivity, and wear resistance, making them ideal for use as cutting tools in precision machining. Therefore, the quality and properties of synthetic diamonds are essential for their application in this field. This chapter explores the synthesis of diamonds, examining the processes and properties that make synthetic diamonds suitable for precision machining applications. Additionally, this chapter delves into the cutting performance of diamond tools, covering aspects from selection and fabrication to their actual use in ultra-precision machining.

**Keywords** Synthetic diamond · Diamond orientations · Fabrication of diamond tool · Cutting performance

## 8.1 Diamond Synthesis Process

### *8.1.1 Classification of Diamonds*

Diamond, renowned for its crystalline clarity and vibrant hues, is a covalently bonded crystal consisting of carbon atoms [1]. In this structure, each atom forms four bonds with its neighboring atoms [2]. This closely packed arrangement of atoms endows diamond with a multitude of exceptional properties. Notably, diamond exhibits remarkable semiconductor characteristics, including an ultra-wide forbidden bandwidth, high electron mobility, high breakdown electric field, and exceptional thermal conductivity. Due to its ultra-hardness and diverse properties, diamond serves as an ideal material for manufacturing cutting tools, abrasives, high-frequency devices, and high-power electronic components.

As depicted in Fig. 8.1, diamonds can be categorized based on whether nitrogen and boron impurities are present or absent within their crystal structure [3]. Diamonds that contain impurity nitrogen atoms are categorized as Type I diamonds. Among these, diamonds with nitrogen atoms in the form of solitary atoms (replacing carbon atoms) are classified as Type Ib diamonds. On the other hand, Type IaA diamonds

 
G. Zhang and J. Wang, *Material-Oriented Cutting Processes in Precision Machining*, Springer Tracts in Mechanical Engineering,
https://doi.org/10.1007/978-981-96-2504-8_8

feature nitrogen atoms arranged as double nitrogen atom pairs. Diamonds devoid of significant nitrogen impurities fall under the designation of Type II diamonds. Type II diamonds can be further categorized as Type IIa, which lacks boron, and Type IIb, which incorporates boron atoms.

Natural diamonds form deep beneath the earth's surface, approximately 150 km into the mantle [4]. Through geological processes spanning hundreds of millions of years, diamonds are brought to the surface by events like volcanoes erupting. In 1796, Tennant demonstrated the carbon composition of diamonds by burning them and producing $CO_2$. This discovery prompted further exploration into the synthesis of diamond. Consequently, researchers began attempts to synthesize diamonds using graphite. As the understanding of diamond materials deepened, it became apparent that artificially synthesized and modified diamonds could possess desired properties compared to natural diamonds.

Diamond synthesis encompasses the process of preparing synthetic materials that possess structures and properties akin to those of natural diamonds. Generally, diamond synthesis can be classified primarily into two categories: High-Pressure High-Temperature (HPHT) and CVD [5]. HPHT diamond synthesis enables the production of large SCD, which can also serve as substrates for obtaining high-purity CVD diamonds with fewer defects. As a result, HPHT synthesis finds wide

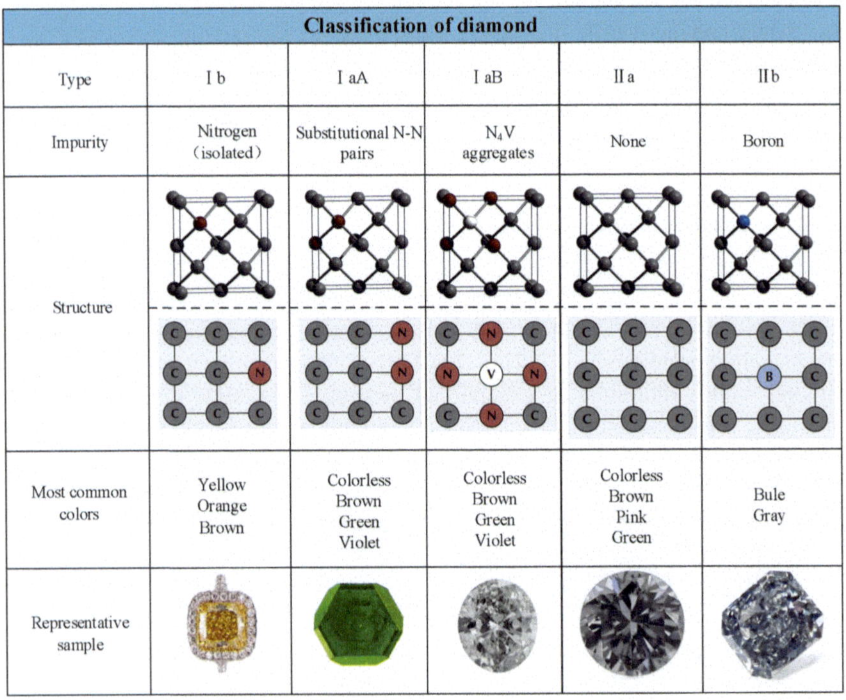


**Fig. 8.1** Classification of diamonds [3]

application in the manufacturing of abrasives and precision cutting tools. Moreover, it holds promise for use in optical components, next-generation electronic devices, quantum computing etc. [6–9].

### 8.1.2 Diamond Synthesis Method

The temperature gradient growth method (TGGM) is one of the primary techniques currently used for synthesizing high-quality, large diamond single crystals [10]. The use of TGGM to synthesize diamond single crystals under HPHT conditions requires stringent stability in both pressure and temperature. This is especially critical for the synthesis of the type IIa. Any fluctuations in pressure or temperature during the synthesis process can lead to a deterioration in crystal quality, resulting in defects such as inclusions, twinning, and pits.

Consequently, the precision control of the synthesizing equipment is of paramount importance. Moreover, the synthesis of good diamond single crystals typically spans from several hours to hundreds of hours, demanding that the experimental apparatus provide long-duration, high-precision control. These extreme requirements necessitate specialized high-pressure equipment to maintain the appropriate conditions for diamond synthesis.

The four primary types of equipment used for HTHP diamond synthesis (> 5 GPa) include: (1) Belt, (2) Toroid, (3) BARS, and (4) Cubic presses. Figure 8.2 illustrates the relationships among maintenance costs, operational difficulty, and reaction volumes.

Producing high-quality single-crystal diamonds remains challenging due to crystal defects that can cause issues like current leakage, reduced breakdown voltage, and hinder light propagation and the use of the nitrogen-vacancy center [10, 11]. Even the purest Type IIa diamonds, created through HPHT or CVD methods, still show some defects in their structures [12–15]. Therefore, eliminating these defects is crucial for synthesizing high-purity diamonds.

In the pursuit of high-quality diamonds via HPHT, catalysts such as Fe, Co, Ni, or their alloys are commonly employed [16, 17]. While Ni and Co can be easily incorporated into the diamond lattice through substitution [18, 19], the incorporation of Fe is considerably more challenging. Consequently, Fe-based catalyst alloys have shown significant potential for synthesizing high-purity diamonds. However, it is often observed that larger diamonds synthesized using these materials still exhibit filiform defects [20, 21]. Because of this, it is very important to understand how filiform flaws form in diamonds and take effective steps to get rid of or greatly reduce their occurrence. This will improve the preparation of diamond tools and make them better at cutting.

In this section, Type Ib diamonds were synthesized using $Fe_{64}Ni_{36}$ as a catalyst under 5.5 GPa pressure and temperatures ranging from 1250 to 1300 °C. The conventional rapid cooling method led to the presence of thread-like defects in the synthesized diamonds due to uneven temperature distribution during synthesis, resulting

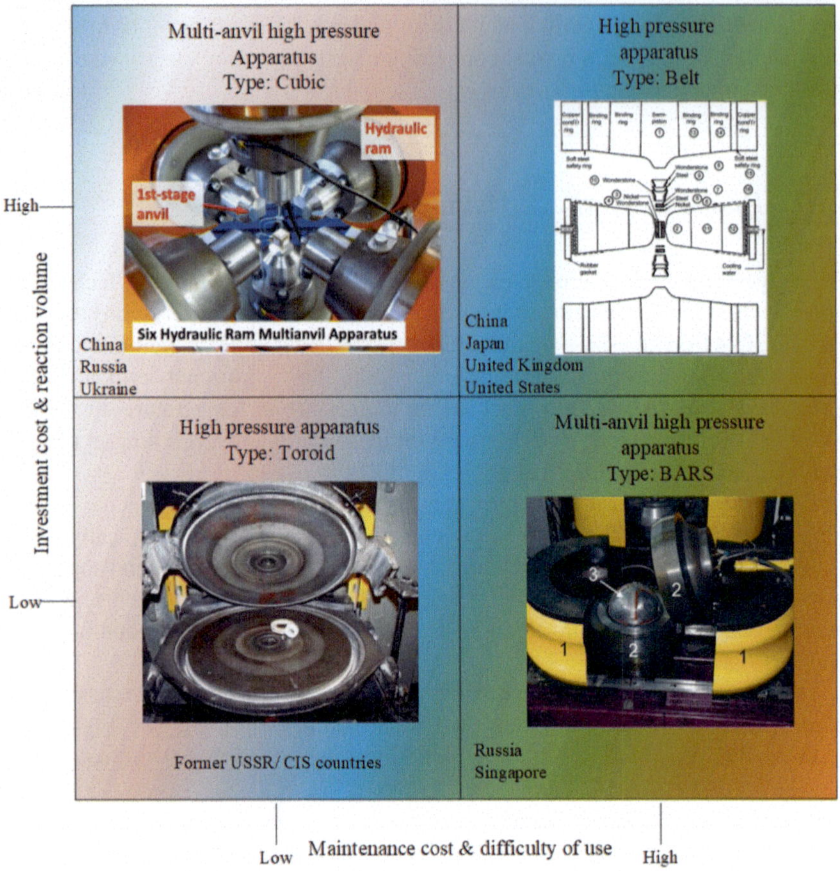


**Fig. 8.2** Comparison of the costs, reaction volumes, and ease of use associated with currently available HPHT equipment in the market [3]

in thermoelastic stresses [22]. Analyzing thermoelastic stress is vital for predicting synthetic material quality. Finite element simulations in this study showed a relatively even distribution of von Mises stress in the diamonds after synthesis, indicating a potential link between thermoelastic stress and filiform defect formation. The application of conditional gradient cooling technology (GCT) successfully eliminated these defects.

### *8.1.3 Diamond Synthesis Experiments*

Diamonds were synthesized using the TGGM under HPHT conditions within an SPD-6 $\times$ 1200 cubic anvil high-pressure apparatus (CHPA). In the later stages of diamond synthesis, both traditional rapid cooling and gradient cooling technologies were employed for the cooling process of the synthesis assembly.

The diamond synthesis assembly, as shown in Fig. 8.3, utilized traditional rapid cooling for 3 min, and gradient cooling for 60, 180, and 300 min. A $Fe_{64}Ni_{36}$ invar alloy acted as the catalyst between the carbon source and the seed bed, initiating diamond growth from a type Ib seed diamond with {100} crystal planes measuring 0.7 mm. Pressure calibration during synthesis was achieved using Bi, Tl, and Ba for their pressure-induced phase transitions. Temperature near the seed diamond was monitored with a Pt-30%Rh/Pt-6%Rh thermocouple. Following diamond synthesis, the catalyst alloy surrounding the synthetic diamond was dissolved in hot dilute nitric acid. Filiform defects in the diamonds were observed using optical microscopy (OM). Transient analysis for thermoelastic stress simulation within the synthetic diamond-seed bed-catalyst alloy system was conducted using finite element modeling [23–26]. The boundary conditions governing the interface between the growth diamond, seed bed, and catalyst alloy were postulated to be consistently bonded, hence ensuring the absence of any relative motion between. For the purpose of transient study, a finite element model was used to simulate the thermoelastic stress experienced by a synthetic diamond-seed bedcatalyst alloy. The Raman spectroscopy experiment was conducted using a Renishaw model Via Raman microscope equipped with backscattering geometry. The 532 nm line of an argon-ion laser was used for this purpose. The Vertex80 V Fourier-transform infrared (FTIR) spectrometer was used to analyze the infrared absorption spectra. The spectrometer spans a spectral range of 400–4000 $cm^{-1}$ and operates in transmittance mode with a spectral resolution of 2 $cm^{-1}$.

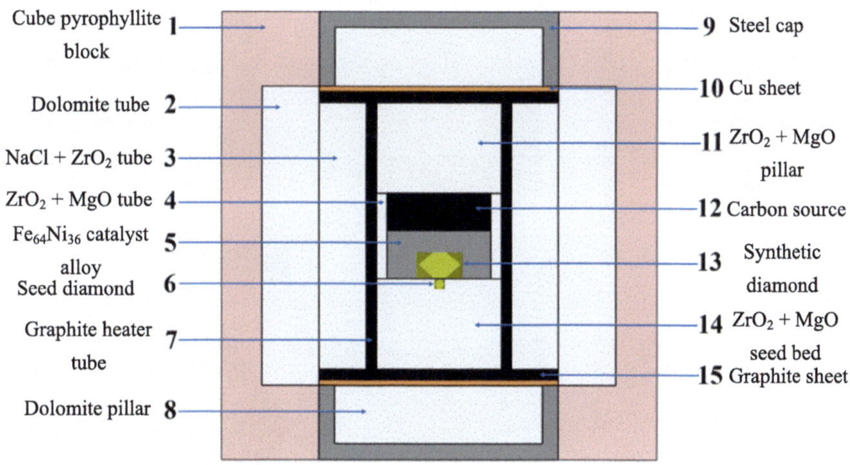


**Fig. 8.3** Schematic assembly diagram of diamond synthesis

### 8.1.4 Synthetic Diamond Properties

Figure 8.2 consists of OM images showcasing diamonds synthesized using traditional rapid cooling technology at 1250, 1270, and 1300 °C (Fig. 8.4a–c, respectively). These are top-down images (Fig. 8.4a–c) and side views (Fig. 8.4a′–c′) of the synthetic diamonds captured under transmitted light mode. The synthetic diamonds exhibit various shapes, including cubic, cub-octahedral, and octahedral, as depicted in Fig. 8.4. Filiform defects are observed throughout the synthesized diamonds, appearing perpendicular to the {100} crystal planes. The presence and extent of filiform defects may be influenced by the distinct mechanical properties exhibited by different crystal planes. Research has indicated that synthetic diamonds exhibit higher hardness on the {111} plane compared to the {100} planes [27]. Furthermore, the orientation of filiform defects in diamonds aligns with the direction of load applied during indentation hardness tests [28]. During the later stages of synthesis assembly cooling, the thermal stress experienced by the diamond may undergo a similar destructive process as the load applied during the diamond hardness test. In this scenario, higher hardness corresponds to higher fracture strength. Consequently, it can be inferred that the {111} planes of the diamond are capable of withstanding greater thermal stress compared to the {100} planes. This suggests that the filiform defects in diamonds primarily occur perpendicular to the {100} planes. The summarized results and specifics of diamond synthesis experiments employing traditional rapid cooling technology are presented in Table 8.1.

**Fig. 8.4** Optical microscopy images of diamonds synthesized using the traditional rapid cooling method at various temperatures: **a** and **a′** 1250 °C; **b** and **b′** 1270 °C; **c**, **c′** 1300 °C

**Table 8.1** Experimental details of diamond synthesis parameters

| Run | Cooling time (min) | Crystal morphology | Defect depict | Synthesis temperature (°C) | Synthesis time (h) |
|---|---|---|---|---|---|
| E-1 | 3 | Cubic | Numerous | 1250 | 40 |
| E-2 | 3 | Cubic-octahedron | Numerous | 1270 | 40 |
| E-3 | 3 | Octahedron | Numerous | 1300 | 40 |

Studies conducted previously have indicated that filiform imperfections in diamonds could stem from the misalignment of diamond crystals during the formation of cubic diamonds [20, 21]. The parameters necessary for the growth of cubic diamonds bear similarities to those governing the transition from graphite to diamond [29]. It is possible that there was a slight interruption in the growth process before it quickly resumed, leading to the misorientation within the cubic diamond. However, it is noteworthy that filiform defects were also observed in diamonds with cub-octahedral and octahedral shapes. These growth conditions were far from the graphite-diamond transition curve, particularly in the case of octahedral diamonds. Therefore, while misorientation or inclusions may contribute to the appearance of defects, they are unlikely to be the main cause of filiform defects.

Figure 8.5 shows the temperature variation in the synthesis chamber (Fig. 8.5a), the catalyst alloy (Fig. 8.5b), and the synthetic diamond (Fig. 8.5c) at the end of the diamond synthesis process. Figure 8.5 shows that there is a 50 °C change in temperature along the axis between the carbon source and the seed bed. There is a temperature change of about 20 °C at the point where the seed bed meets the catalyst metal. Different temperature levels and the fact that the synthetic diamond, catalyst metal, and seed bed all have different expansion factors cause the thermoelastic stress to be spread out unevenly within the diamond.

To analyze and predict the presence of filiform defects, finite element simulations were conducted to characterize the thermoelastic stress within diamond crystals. The von Mises stress ($\sigma_{\mathrm{Mises}}$) was a key parameter used to assess stress during

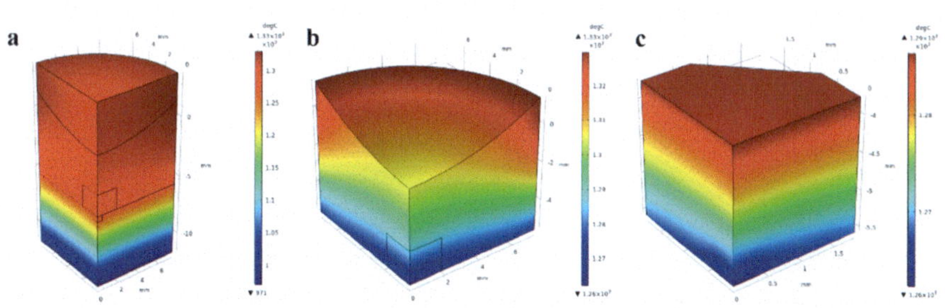


**Fig. 8.5** Temperature distribution in various regions within the diamond synthesis assembly: **a** within the synthesis chamber zone; **b** within the catalyst alloy region; **c** within the synthetic diamond zone

crystal growth and was computed based on different stress components. In cylindrical coordinates, the von Mises stress can be calculated using the following equation [30]:

$$\sigma_{Mises} = \sqrt{\frac{(\sigma_{\gamma\gamma} - \sigma_{zz})^2 + (\sigma_{\varphi\varphi} - \sigma_{\gamma\gamma})^2 + (\sigma_{\varphi\varphi} - \sigma_{zz})^2 + 6\sigma_{\gamma z}^2}{2}} \tag{8.1}$$

To streamline the computational process of the finite element simulations, a simplified approximation was employed, dividing the synthesis assembly into four equivalent parts and analyzing one region at a time. After performing the finite element simulations, the von Mises scalar was visualized using color graphics. Following is Fig. 8.6 of the von Mises stress patterns and numbers in a simple finite element diamond model at the end of the diamond synthesis process. From the results displayed in Fig. 8.6, it is evident that the stress within the diamond exhibits significant variation, with the maximum and minimum values of the von Mises stress being 10.01 and 0.63 GPa, respectively.

After the completion of diamond synthesis, it was necessary to lower the temperature of the synthesis assembly to around 150 °C for the safe unloading of high pressure. During the cooling process, the thermoelastic stresses within the diamond were relieved. However, with traditional rapid cooling technology, which only took a few minutes, it was not possible to completely alleviate the thermoelastic tension since there was not enough time, resulting in the formation of filiform defects. Moreover, this rapid cooling process also generated high levels of thermoelastic stress within the diamond due to the catalyst alloy and seed bed. Thus, this section propose that employing the GCT with a longer cooling time might effectively reduce the occurrence of filiform defects in diamond.

In order to test this hypothesis, experiments were conducted in this section involving diamond synthesis using GCT and varying cooling durations of 60, 180, and 300 min. Figure 8.7 illustrates the gradient cooling rates observed in experiments E-1 through E-6. The cooling rate for conventional rapid cooling technology was around 373.33 °C/min (Fig. 8.7a). On the other hand, the cooling rates in experiments E-4 to E-6 were reduced to around 18.67, 6.22, and 3.73 °C/min, respectively

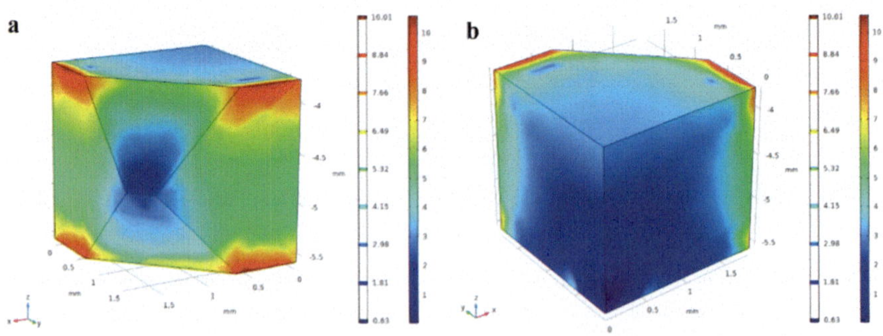

**Fig. 8.6** Distribution of Von Mises stress in the simplified finite element diamond model: **a** Diamond model seen from the outside; **b** Diamond model seen from the inside

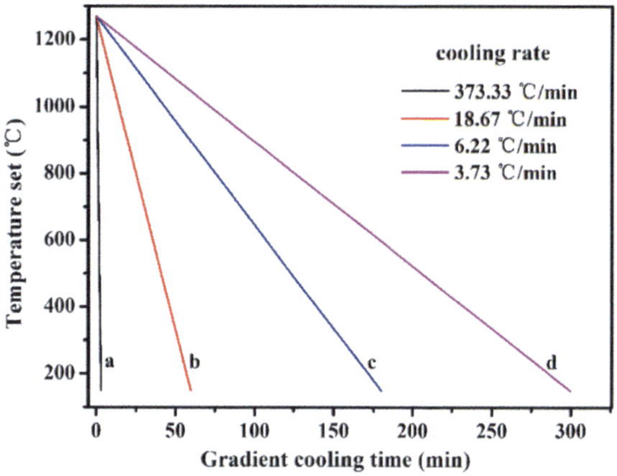


**Fig. 8.7** The gradient cooling rates for the experiments E-1 to E-6: **a** 3 min for experiments E-1 to E-3, **b** 60 min for experiment E-4, **c** 180 min for experiment E-5, and **d** 300 min for experiment E-6

(Fig. 8.7b–d). Figure 8.6 displays the OM images of diamonds created with the GCT. Figure 8a–c show top-view optical microscope images, while images 8(a′-c′) display side-view images of synthetic diamonds under transmitted light. With a gradient cooling time of 60 min, there was a noticeable decrease in filiform defects in Fig. 8.8a in comparison to Fig. 8.4b. With a cooling time of 180 min, only a few small defects were observed in the diamond shown in Fig. 8.8b, b′. In contrast to Fig. 8.4, the majority of filiform defects were oriented at a right angle to the {100} planes. After a cooling time of 300 min, the elimination of filiform defects led to the formation of a yellow transparent diamond crystal, as shown in Fig. 8.8c, c′. Table 8.2 provides a summary of the additional details and results obtained from the experiments on gradient cooling diamond synthesis.

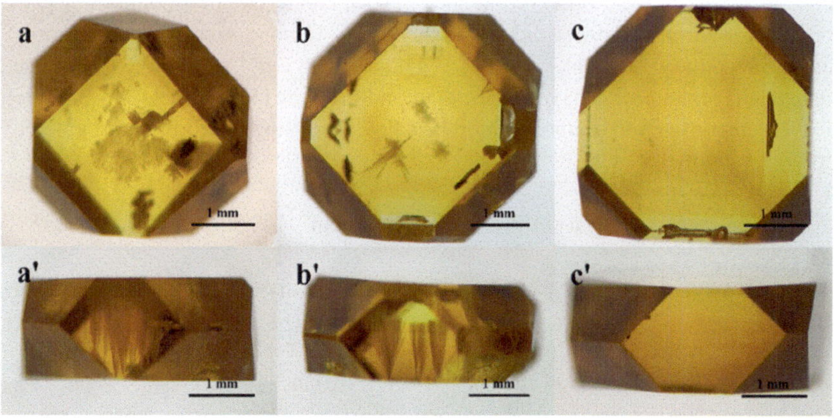


**Fig. 8.8** Optical images of the diamond produced using GCT with varying gradient cooling durations: **a** and **a′** after 60 min in the E-4 experiment; **b** and **b′** after 180 min in the E-5 experiment; **c** and **c′** after 300 min in the E-6 experiment

**Table 8.2** Experimental details of GCT parameters

| Run | Gradient cooling time (min) | Crystal morphology | Defects observed | Synthesis temperature (°C) | Synthesis time (h) |
|---|---|---|---|---|---|
| E-4 | 60 | Cubic-octahedron | Modicum | 1270 | 40 |
| E-5 | 180 | Cubic-octahedron | Barely | 1270 | 40 |
| E-6 | 300 | Cubic-octahedron | None | 1270 | 40 |

Raman spectroscopy was employed to identify residual stresses, inclusions, and evaluate the crystallinity of diamond [31]. The Raman spectra from the undamaged regions of diamond samples in experiments E-2, E-4 to E-6 are presented in Fig. 8.9a–d respectively. Correspondingly, Fig. 8.9a′–c′ show the defective areas of diamond samples from experiments E-2, E-4, and E-5. Optical microscopy images indicating the regions analyzed for Raman spectra are positioned in the top right corner of each panel in Fig. 8.9. The characteristic Raman peak of diamond was observed at 1331.5 $cm^{-1}$ in Fig. 8.9, with minor variations between 1331.41 and 1331.55 $cm^{-1}$, and the half-peak width ranged from 3.72 to 4.16 $cm^{-1}$.

By observing Fig. 8.9, it is evident that the Raman peak intensity was higher in the intact area than in the defect area, and the half-peak width was narrower in the intact region. These findings indicate that the compressive stress was greater in the defect area and the crystalline quality was superior in the intact region. Among the experiments, E-6 produced diamond with the highest crystallinity and the lowest compressive stresses within its lattice.

FTIR spectroscopy was utilized for identifying impurity defects in the diamond lattice and determining their concentration. The FTIR spectra of diamond samples are presented in Fig. 8.10a–d. Within these spectra, the intrinsic vibration absorption peaks of diamond were observed at 1970, 2030, and 2160 $cm^{-1}$. Importantly, the presence of filiform defects did not affect the C–C bond vibrations in diamond. Additionally, the FTIR peaks at 1130, 1344, and 2688 $cm^{-1}$ were linked to isolated nitrogen (C-center nitrogen) within the diamond structure [32–35]. The concentration of C-center nitrogen in diamond could be calculated using the formula provided by [36–38]:

$$N_{c(ppm)} = \mu(1130\,\text{cm}^{-1})/\mu(2120\,\text{cm}^{-1}) \times 5.5 \times 25 \tag{8.2}$$

where $\mu$ *represents* the corrected coefficients and can be determined by following formula:

$$\begin{aligned} \mu(1130\,\text{cm}^{-1}) &= [A(1290\,\text{cm}^{-1}) - A(1370\,\text{cm}^{-1})]/0.31 \\ \mu(2120\,\text{cm}^{-1}) &= [[40 \times A(2030\,\text{cm}^{-1}) + 87 \times A(2160\,\text{cm}^{-1})]/127 \\ &\quad - A(2120\,\text{cm}^{-1})] \end{aligned} \tag{8.3}$$

where $A$ represents the absorption intensity of the relevant peaks in the FTIR spectra.

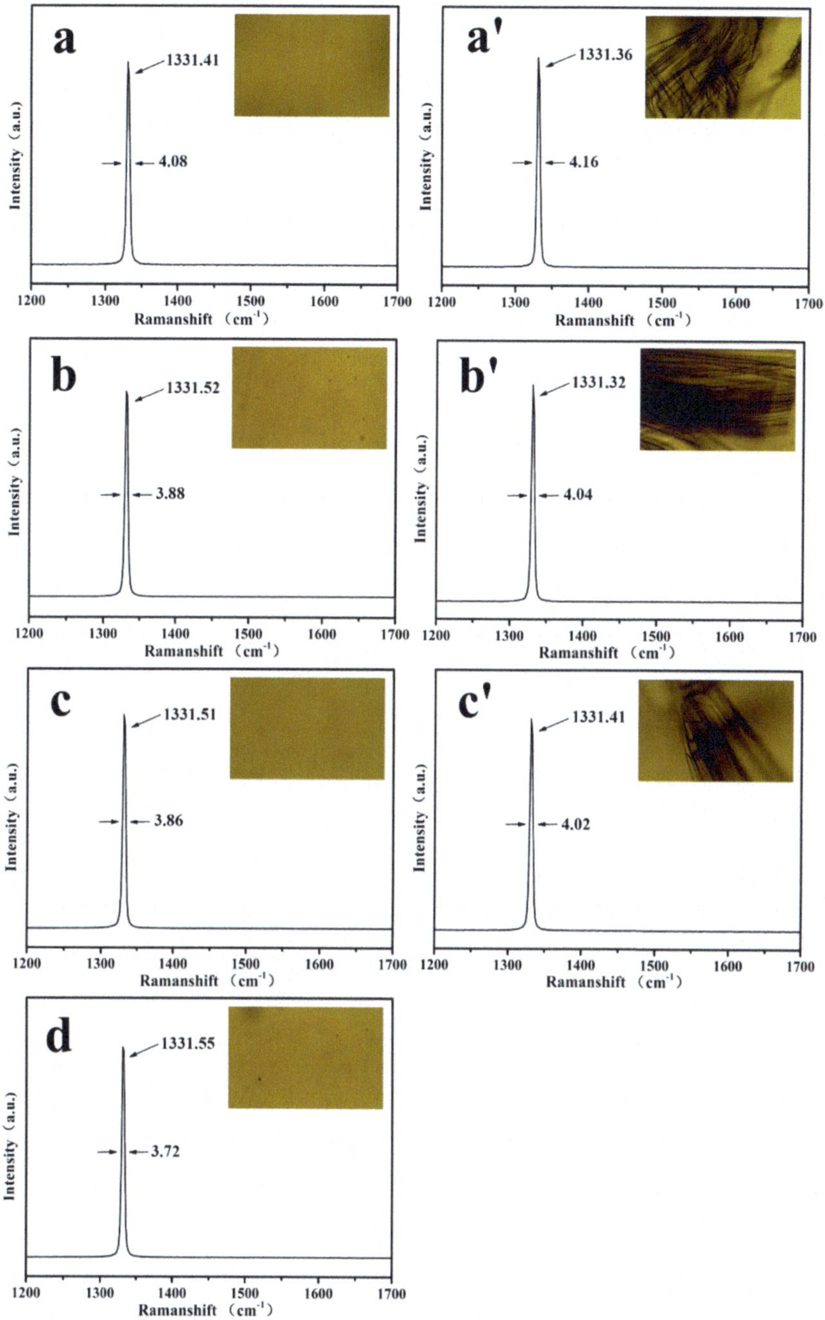


**Fig. 8.9** The Raman spectra of the intact areas of diamonds synthesized from experiments E-2, E-4 to E-6 as **a**–**d**, and the defect areas of diamonds from experiments E-2, E-4, and E-5 as **a′**–**c′**

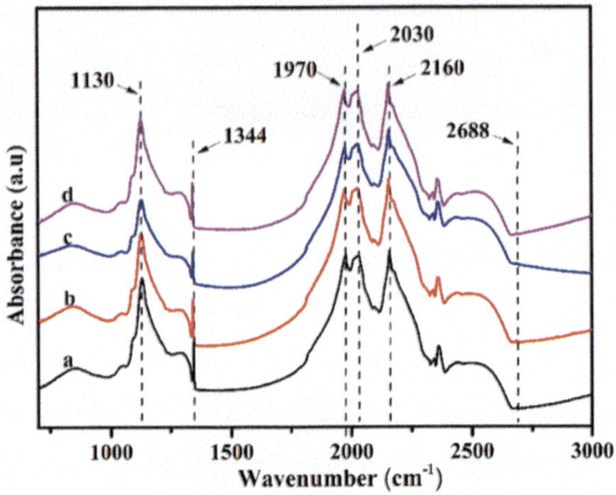


**Fig. 8.10** FTIR spectra of synthesized diamond: **a–d** were obtained from experiments E-2, E-4 to E-6, respectively

Nitrogen concentrations in the diamonds from experiments E-2, E-4 to E-6 were estimated at 264, 260, 237, and 246 ppm with an uncertainty of around 10%. Despite varying concentrations, all values fell within the margin of error, indicating that filiform defects had no significant impact on nitrogen levels in the diamonds.

The chemical composition and nitrogen content in diamonds from experiments E-2, E-4 to E-6 remained unaffected by the filiform defects as shown in both the Raman and FTIR spectra. This suggests that the filiform defects primarily caused physical damage to the diamond, with a higher likelihood of occurrence in diamonds grown using TGM compared to those grown using FGM. In FGM diamond growth, temperature fluctuations near the diamond were minimal, and the diamond was encapsulated by a thin metal film of only 50 μm, wrapped in graphite [39]. In contrast, during TGM diamond synthesis (as illustrated in Fig. 8.3), the catalyst alloy had a thickness ranging from several millimeters to tens of millimeters. Moreover, the thermal expansion coefficient of the catalyst alloy was considerably higher than that of graphite. Consequently, the thermoelastic stress generated in TGM diamonds during synthesis and cooling was significantly greater than that in FGM diamonds. Based on our experiments and results, it can be inferred that the rapid relief of thermoelastic stress in diamonds and the high thermoelastic stress induced during the rapid cooling of the synthesis assembly were the primary factors contributing to filiform defect formation. Given that thermoelastic stresses are inherent to HPHT growth conditions in diamonds, the utilization of GCT can help mitigate the damage caused by thermoelastic stress during diamond growth.

## 8.2 Diamond Tool Preparation and Cutting Performance

### 8.2.1 Diamond Tools Preparation

Ultra-precision machining (UPM) is a cost-effective and efficient method that allows for the direct fabrication of high-quality parts with surface roughness at the nanometer scale and form error at the sub-micrometer level [40, 41]. This technique has found wide applications in various fields, including optics, biotechnology, medicine, electronics, and communications [42, 43]. Diamond, due to its exceptional physicochemical properties, is considered an ideal tool material for UPM.

Diamonds can be classified into different types based on the presence of impurities, such as nitrogen (N) and boron (B). These types include Ia (containing aggregated N impurities), Ib (containing single substitutional N impurities), IIa (the purest form with less than 1 ppm N impurity), and IIb (containing B impurity and few N impurities) [44, 45]. The complex formation conditions of natural diamonds contribute to variations in their wear resistance and cutting performance as cutting tool materials. This variation can be attributed to factors such as residual stress, cracking, dislocations, and vacancies within the diamond lattice [46–48]. Therefore, relying solely on the color and visible purity of natural diamonds for cutting tools may yield highly variable experimental results in terms of cutting performance.

In contrast, HPHT synthetic diamonds offer relative control over the type and concentration of impurities and defects within the diamond lattice. Consequently, the experimental data and processing stability achieved in UPM using synthetic diamond tools are generally more reliable than those obtained by using natural diamonds cutting tools. Current trends indicate that HPHT synthetic diamonds, specifically type Ib, are predominantly used in UPM, with limited research available on other types.

Different types of diamonds exhibit distinct physicochemical properties. For example, type IIa diamonds possess the highest thermal conductivity and hardness compared to other types of diamonds [49]. Type IIb diamonds, on the other hand, serve as p-type semiconductors and exhibit exceptional thermal stability [50–52]. Moreover, the use of boron-doped diamond coatings in tools results in prolonged tool lifetimes due to the excellent properties of boron-doped diamond [53]. It is evident that different types of diamonds offer varying physicochemical characteristics. Type IIb diamonds demonstrate remarkably high resistance to crater wear [54]. Nevertheless, there is a lack of measurement and characterization data regarding the diamond tools and workpieces. To date, there is no existing literature on the use of HPHT synthetic type IIa diamonds in UPM. Additionally, chipping of diamond tool edges frequently occurs during UPM as a result of excessive compressive stresses, microcracks, and vacancies within the diamond lattice. Synthetic diamonds can mitigate defects in the lattice by tailoring the synthetic processes and assembly.

In this section, various synthetic diamonds were prepared in a similar synthesis system to minimize experimental errors caused by other diamond defects. UPM experiments revealed that copper workpieces machined using type IIb diamond tools exhibited the lowest surface roughness. This investigation is believed to contribute to a better understanding of the mechanisms behind diamond tool wear and the improvement of tool life in UPM.

It is well known that SCD displays anisotropic properties, with distinct characteristics such as hardness and mechanical strength on each crystalline plane [55]. The high hardness and anisotropic grinding removal rate of diamond present significant challenges in the production of diamond tools. Therefore, the search for cost-effective methods to achieve the finest cutting edge radius and ensure high-quality edge integration has become a pressing issue.

Through the use of a micro-chipping mechanism, the utilization of abrasive diamond powder and grit has been shown to be an extremely efficient approach for polishing diamond surfaces. Various ways for polishing diamonds may be classified into two categories: non-contact and contact. Non-contact methods primarily include plasma etching [56] and laser polishing [57]. Contact methods consist of ion beam polishing [58], chemical polishing [59], thermo-chemical polishing [60, 61], dynamic friction polishing [62], and mechanical lapping [63]. Essentially, these methods enhance mechanical polishing for specific applications. Among these, mechanical lapping is frequently preferred. According to Zong's research, lapping experiments used (110)-(100) and (100)-(100) orientations as tool rake and flank faces, respectively. The experimental results indicated that the tool rake face surface roughness for (110)-(100) orientations is 0.8 nm (Ra) with a lapped cutting edge radius of 35–50 nm. For (100)-(100) orientations, the surface roughness is 0.7 nm (Ra) with a lapped cutting edge radius of 40 nm. These findings demonstrate a significant improvement in the lapping quality of diamond tools [64]. Numerous lapping theories have been proposed to explain the mechanisms of the lapped surface layer of diamond [65–68]. Zong et al. [69, 70] introduced the removal mechanisms of the 'soft' and 'hard' crystal orientations in diamond tool fabrication, which provided the rules of diamond tool fabrication for industry, as shown in Fig. 8.11.

### 8.2.2 *Cutting Performance in UPM*

The ultra-precision cutting tools, made from the synthesized diamonds, were polished using similar processes. Each diamond tool had a rake angle of 0°, a clearance angle of 10°, and a tapered flank face.

Figure 8.12a–c display optical images depicting three varieties of diamond tools: type Ib, type IIa, and type IIb. Using a Moore Nanotech 450UPL ultra-precision lathe, experiments in ultra-precision machining (UPM) were conducted. The experimental setup for UPM is illustrated in Fig. 8.12d. Workpiece materials were securely gripped by a chuck system on the spindle, while the diamond tool was affixed to an adjustable tool holder. The spindle speed was fixed at 2000 rpm, with a cutting depth of 3 μm and a feed rate of 5 μm/r. In this study, TU1 Cu, a high-purity material with an oxygen level and impurities content of no less than 99.97%, was employed as the workpiece material.

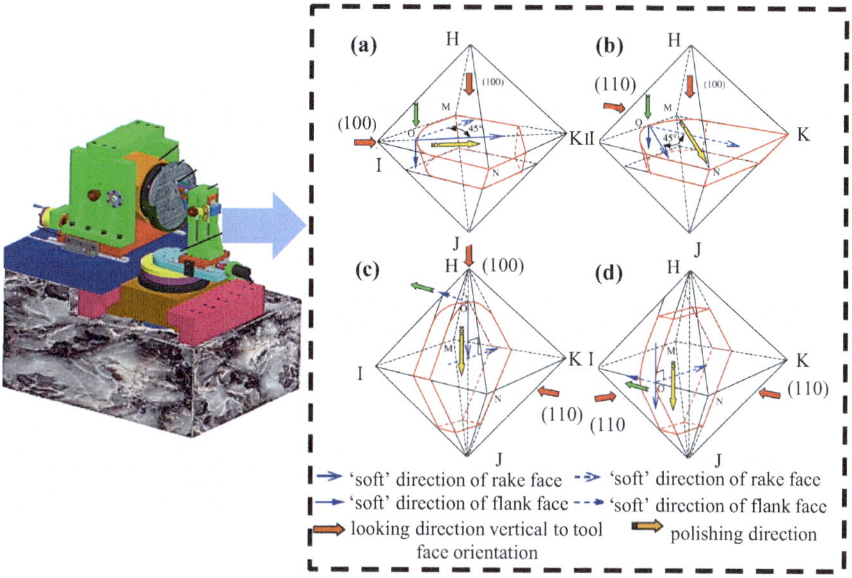


**Fig. 8.11** Different polishing orientations of diamond tools on rake and flank faces: **a** R(100) F(100), **b** R(100) F(110), **c** R(110) F(100), and **d** R(110) F(110). Here, R denotes the rake face, and F represents the flank face [69]

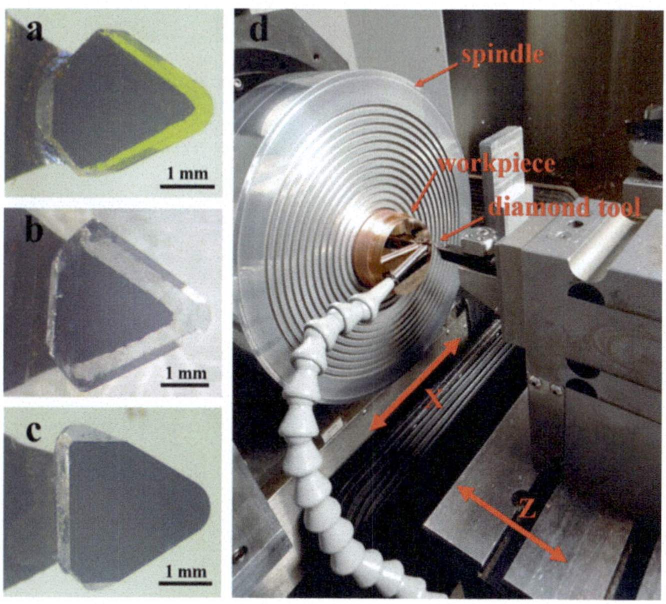


**Fig. 8.12** Micrographs of the diamond tools using in experiments: **a** type Ib, **b** type IIa, **c** type IIb and **d** experimental setup

Optical micrographs in Fig. 8.12 revealed that type Ib diamond tools are yellow and transparent, type IIa tools are colorless and transparent, and type IIb tools are black and opaque, with no visible impurities or defects observed. Experiments N-4 to N-6 were conducted using the side edges of tools No. 1 to No. 3, which have a 45° angle between the main and side edges. Additional details on the UPM experiments and diamond tool parameters can be found in Table 8.3.

Raman spectroscopy is an exceptionally precise method capable of identifying various features of diamond tools, such as the $sp^2$ graphite structure, inclusions as defects, and residual stresses. As illustrated by Fig. 8.13a–f, the Raman peaks of the diamond tool edges before the UPM experiments are prominently located at 1332.20 $cm^{-1}$. These peaks display a strong and narrow peak with a flat background, as depicted in Fig. 8.13. In a pure diamond with a perfect cubic structure, Raman peaks are typically observed at 1332.5 $cm^{-1}$ [71, 72]. Variations in the wavenumbers indicate the presence of residual stresses within the diamond lattice.

It is important to mention that diamond tool edge chipping is a common issue in UPM, even when utilizing highly machinable copper work materials [73]. One of

**Table 8.3** Experimental details and diamond tool's parameters

| Run | Tool no | Diamond type | Rake face/ Flank face | Tool radius (μm) | Cutting distance (km) | Spindle speed (rpm) | Feed rate (μm/r) | Cutting depth (μm) |
|---|---|---|---|---|---|---|---|---|
| N-1 | 1 | Ib | (100)/(110) | 490.8 | 50 | 2000 | 5 | 3 |
| N-2 | 2 | IIa | (100)/(110) | 487.4 | 50 | 2000 | 5 | 3 |
| N-3 | 3 | IIb | (100)/(110) | 482.3 | 50 | 2000 | 5 | 3 |
| N-4 | 1 | Ib | (100)/(100) | 490.8 | 50 | 2000 | 5 | 3 |
| N-5 | 2 | IIa | (100)/(100) | 487.4 | 50 | 2000 | 5 | 3 |
| N-6 | 3 | IIb | (100)/(100) | 482.3 | 50 | 2000 | 5 | 3 |

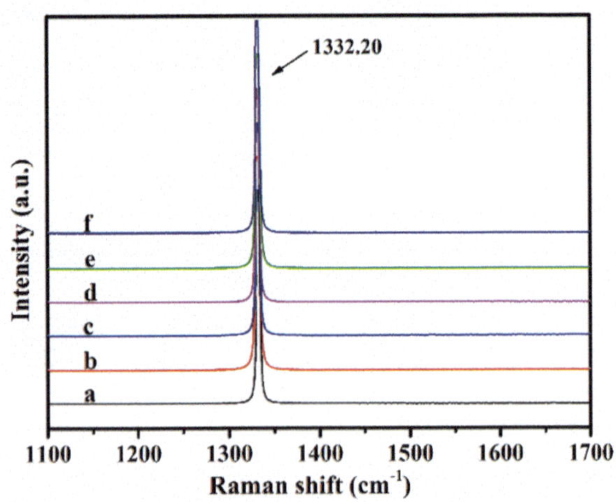


**Fig. 8.13** The Raman peaks of the diamond tools before the UPM experiments: **a–c** the main edge of type Ib, type IIa, and type IIb diamond tools and **d–f** the side edge of type Ib, type IIa, and type IIb diamond tools

the primary reasons for this problem is the presence of significant internal stresses within the diamond lattice. On the other hand, as can be shown in Fig. 8.13, the Raman peaks of the diamond tools that were used in this UPM experiment are very similar to the ideal diamond Raman peak, which is located at 1332.5 cm-1. There is evidence that the diamond tools used in this UPM experiment have a high-quality $sp^3$ structure and exhibit negligible residual stresses inside the diamond lattice [74]. This evidence is provided by the Raman spectra.

In order to assess the chemical composition, molecular structure, and the amount of nitrogen (N) impurities that are present inside the diamond lattice, FTIR methods are consistently used. The infrared spectra of the diamond tools that were used in this experiment are shown in Fig. 8.14. It is possible that singly substituted N impurities are responsible for the peaks that were found at 1130 and 1344 $cm^{-1}$ in Fig. 8.14a, b, respectively [75, 76]. An assessment of N impurity concentration in the diamond can be conducted using FTIR spectra [77]. For the type Ib diamond tool, the N concentrations at the main and side edges were approximately 190 ppm. Conversely, the absence of peaks at 1130 and 1344 $cm^{-1}$ in Fig. 8.14c–d suggests a substituted N concentration of less than 1 ppm in the diamond tool [78].

Because the type IIb diamond tool had a dark look, FTIR tests in both the reflectance and transmission modes did not reveal any signals that could be detected. In order to further investigate the chemical bonding states and boron concentration in the diamond lattice, XPS measurements were carried out for the purpose of further analysis. As shown in Fig. 8.15, the B 1 s spectra had four peaks at 189.0, 190.6, 191.5, and 192.3 eV. These peaks were located in the spectrum. The peaks that occurred at 189.0 and 190.6 eV were most likely related with BC2O bonding, while the peaks that occurred at 191.5 and 192.3 eV were ascribed to h-BN bonding and BCO2 bonding, respectively [79–83]. The XPS survey revealed a B concentration of around 4400 ppm.

Surface roughness is a crucial parameter in workpiece quality assessment, particularly for precision components, where lower roughness implies higher surface quality

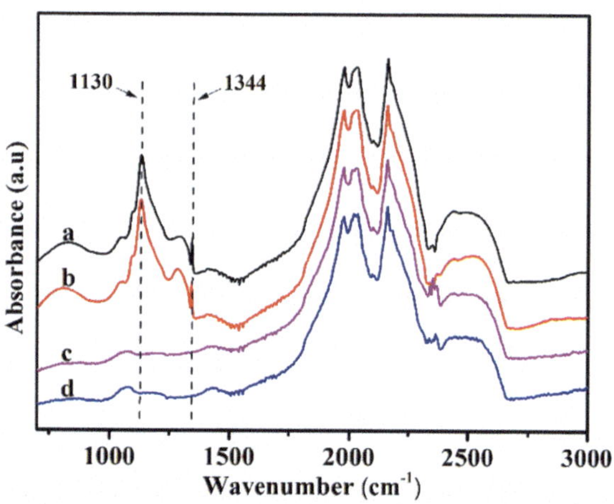


**Fig. 8.14** Recorded infrared absorption spectra for diamond tools of type Ib and type IIa: **a** main edge of type Ib tool, **b** side edge of type Ib tool, **c** main edge of type IIa tool, **d** side edge of type IIa tool

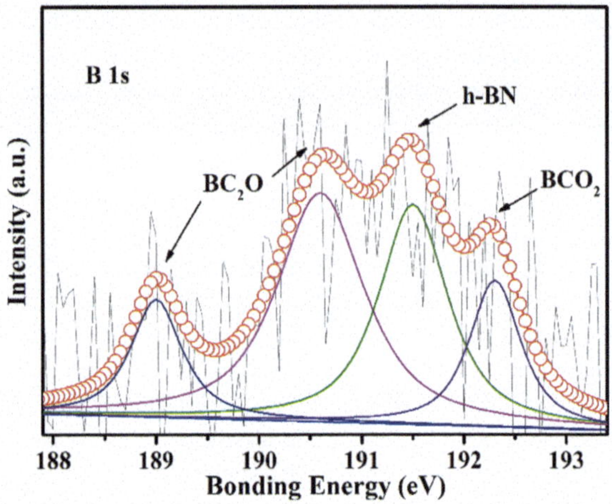


**Fig. 8.15** XPS measurement of the type IIb diamond tool

[84]. The variation in surface roughness during the machining process provides essential information regarding the reliability of the diamond tool. After UPM experiments, the surface roughness values ranged from 6.07 to 26.06 nm. Figure 8.16a–f illustrate the obtained surface roughness measurement results for the N-1 to N-6 experiments. The workpiece's surface roughness achieved using the type IIb diamond tool with an oriented R(100)F(110) exhibited lower values compared to the other types of diamond tools with an oriented R(100)F(110) after 10 km of UPM experiments, as evidenced in Fig. 8.16a–c. It is essential to emphasize that, after the completion of ten kilometers of UPM testing, the workpiece displayed the least amount of surface roughness when it was orientated with the type IIb diamond tool in the R(100)F(100) orientation that was used. It is interesting to note that after 10 km of ultra-precision milling (UPM), all surface roughness measurements of the workpiece obtained with the oriented R(100)F(110) diamond tool were greater than those collected with the oriented R(100)F(100) diamond tool. These practical results are in agreement with the findings of recent theoretical research, which shows that the directed R(100)F(110) diamond tool provides a larger surface roughness in comparison to other oriented tools [85]. The fact that the parallelism between the high tensile strength of the flank face and the cutting direction is stronger for the oriented R(100)F(100) diamond tool than it is for the oriented R(100)F(110) diamond tool is one of the possible explanations for this outcome.

The 3D surface shapes of the block in the N-1 to N-6 experiments after 50 km of UPM can be seen in Fig. 8.17. The surface roughness values for N-1 to N-6 experiments after 50 km of UPM were 17.98, 20.81, 16.01, 10.99, 10.50, and 8.87 nm, respectively. This showed that the type IIb diamond tool with the oriented R(100)F(100) had the best UPM performance and the longest durability.

SEM was employed to examine the wear morphology of the diamond tools. Figure 8.18 presents the SEM micrographs of the diamond tools after undergoing 50 km of UPM experiments. Specifically, Fig. 8.18a–c display the SEM micrographs

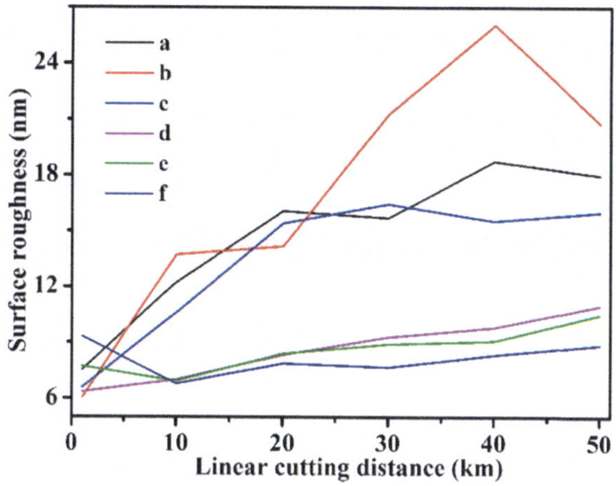


**Fig. 8.16** The measured surface roughness in experiments N-1 to N-6 as the linear cutting distance increased: **a–f** corresponded to experiments N-1, N-2, N-3, N-4, N-5, and N-6, respectively

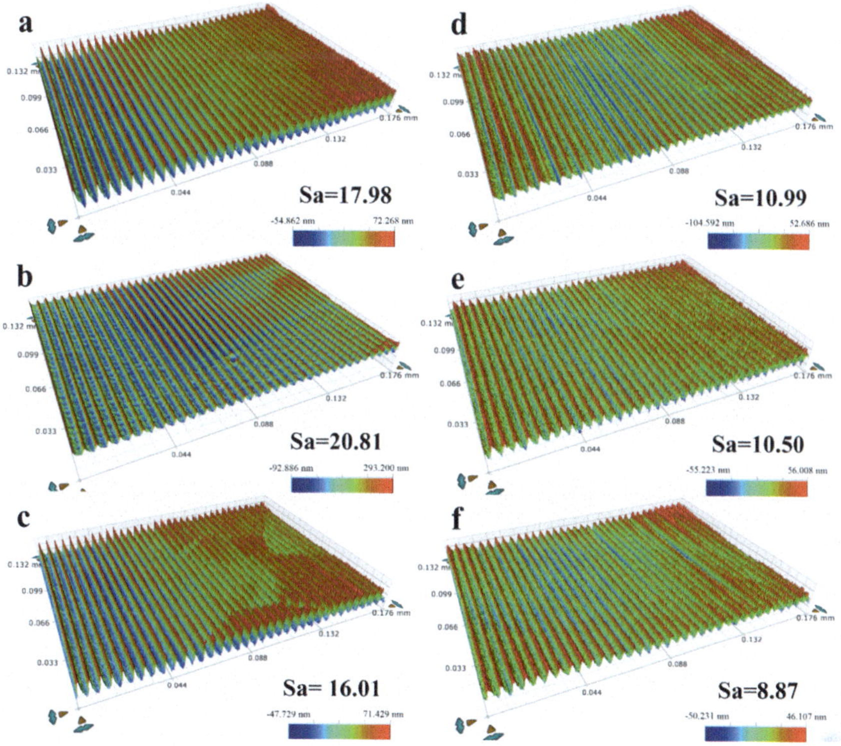


**Fig. 8.17** The surface morphology of the workpiece after 50 km UPM. **a–f** were obtained after the N-1, N-2, N-3, N-4, N-5, and N-6 experiments, respectively

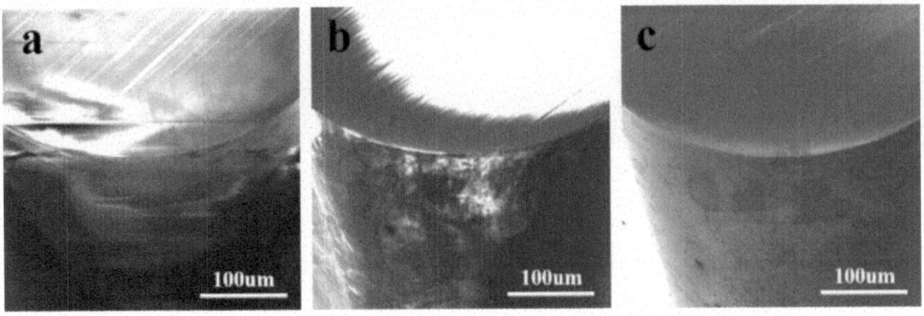


**Fig. 8.18** SEM micrographs of diamond tools following 50 km UPM experiments, with images showing the condition of **a** Type Ib, **b** Type IIa, and **c** Type IIb diamond tool

of the type Ib, type IIa, and type IIb diamond tools, respectively. Upon observation, no prominent mechanical or physical wear indications were apparent on the surface of the diamond tools.

The variation in workpiece surface roughness following UPM experiments with different diamond tools can be attributed to the properties of the raw diamond materials. Previous studies have shown that type IIa diamond exhibits the highest thermal conductivity among various diamond types. Additionally, the hardness of the diamond is influenced by the concentration of nitrogen (N) and boron (B) within the diamond lattice. Type IIa diamond typically exhibits higher hardness than other types of diamond. However, despite possessing high thermal conductivity and hardness, type IIa diamond does not necessarily offer the best cutting performance, suggesting that other diamond properties play a role in determining cutting performance. It is suggested that chemical wear should be considered as the main factor influencing workpiece surface roughness variation.

Chemical reactions between carbon (C) and oxygen ($O_2$) can be represented as follows: $C + O_2 \rightarrow CO_2$ or $2C + O_2 \rightarrow 2CO$. These reactions occur spontaneously as the change in Gibbs free energy ($\Delta G$) is negative, indicating thermodynamic feasibility [86]. Diamond, being an allotrope of carbon, reacts with oxygen to form diamond oxide. The Gibbs free energy change for diamond oxide formation with 1 mol of oxygen is − 100 kJ/mol. Hence, when a diamond tool interacts with copper (Cu) in UPM experiments, the carbon atoms on the diamond surface may undergo oxidation and be eliminated. Notably, the UPM experiments in this study were carried out in the presence of air, requiring an examination of the oxidation resistance of the diamond tool. Figure 8.19 displays the thermogravimetric (TG/DTG) curves obtained for type Ib, type IIa, and type IIb diamond samples. The estimated onset temperatures for type Ib, type IIa, and type IIb diamonds were determined to be 1000, 1015, and 1096 °C, respectively. At these onset temperatures, weight reductions of 8.3, 7.8, and 7.9% were observed. The peak weight loss rates for type Ib, type IIa, and type IIb diamonds were recorded at 1139 °C (0.42%/min), 1170 °C (0.39%/min), and 1247 °C (0.41%/min), respectively. These findings suggest that type IIb diamond displays superior thermal stability in comparison to type Ib and type IIa diamonds. The presence of boron in type IIb diamond leads to the formation of a $B_2O_3$ surface

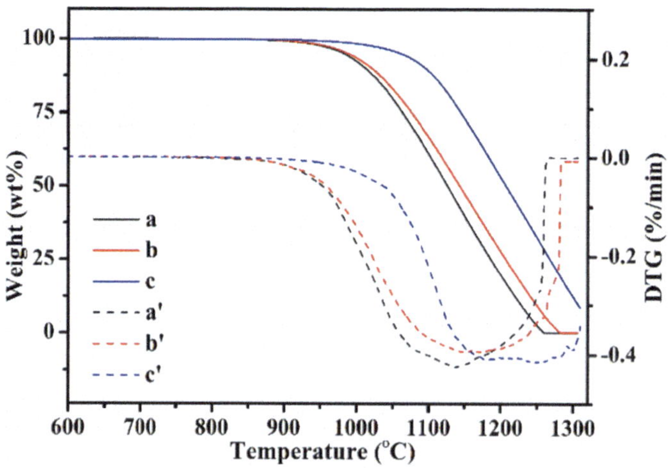


**Fig. 8.19** TG and DTG curves of type Ib, type IIa, and type IIb diamond samples by heating them at a rate of 10 °C/min up to 1300 °C under air conditions. The TG curves for type Ib, type IIa, and type IIb diamond samples are labeled as **a**, **b**, and **c** respectively, while the corresponding DTG curves are labeled as **a′**, **b′**, and **c′** respectively

layer, which enhances its thermal stability [87]. It is plausible to suggest that the $B_2O_3$ oxide film on the type IIb diamond tool, when used for UPM on copper workpieces, contributes to lower friction coefficient, reduced frictional heat, and consequently, minimized thermal degradation. This, in turn, results in a decreased likelihood of chemical reaction and graphitization wear.

Moreover, it has been demonstrated that diamond synthesis using a Cu catalyst can be achieved under HPHT conditions of 7.0 GPa and 1500–1800 °C [88]. This finding corroborates the catalytic role of Cu in diamond crystallization and growth. Cu also acts as a catalyst for the reverse transformation of diamond into graphite under metastable diamond conditions during UPM experiments. The Gibbs free energy change for oxide formation is higher for Cu compared to carbon, indicating the greater stability of $CO_2$ and CO oxides relative to $Cu_2O$ and CuO oxides. Consequently, carbon atoms can deoxidize $Cu_2O$ and CuO, resulting in the production of $CO_2$ and CO [89]. Therefore, when the diamond tool comes into contact with Cu oxides during the UPM process, the surface carbon atoms on the tool will be selectively removed. It is worth mentioning that type IIb diamond tools exhibit the additional advantage of weakening graphite and reducing wear through deoxygenation due to the presence of boron in the diamond lattice. Consequently, based on their superior machining performance and enhanced tool durability, type IIb diamond tools are recommended as the preferred choice for UPM applications.

In general, SCD tools can typically achieve a stable cutting distance of over 10 km when machining non-ferrous metal materials, such as brass and aluminum alloys. The slow and uniform wear of the tools makes their wear cost acceptable in comparison to the value of the machined workpieces. However, the application of single-crystal

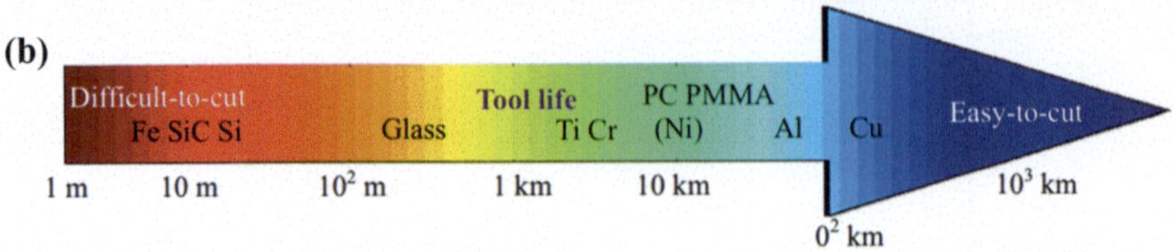


**Fig. 8.20** Schematic map of diamond tool lifespan for various workpiece materials [91]

diamond tools remains limited; for instance, they are not yet capable of free-form machining hard and brittle materials as well as ferrous metals without relying on a variety of auxiliary technologies. For methods of machining hard and brittle materials using SPDT, readers can consult Chap. 1 of this book, while methods for SPDT of ferrous metals are discussed in Chap. 5. Figure 8.20 summarizes the cutting distances of single-crystal diamond tools when turning various materials [91].

## References

1. Green BL, Collins AT, Breeding CM (2022) Diamond spectroscopy, defect centers, color, and treatments. Rev Mineral Geochem 88(1):637–688
2. Zhang J, Wang J, Zhang G, Huo Z, Huang Z, Wu L (2023) A review of diamond synthesis, modification technology, and cutting tool application in ultra-precision machining. Mater Des 112577
3. Zhang JB, Wang JP, Zhang QG, Huo ZX, Huang ZJ, Wu LJ (2024) A review of diamond synthesis, modification technology, and cutting tool application in ultra-precision machining. Mater Des 237:112577
4. Boyd F, Gurney JJ (1986) Diamonds and the African lithosphere. Science 232:472–477
5. Gracio J, Fan Q, Madaleno J (2010) Diamond growth by chemical vapour deposition. J Phys D Appl Phys 374017
6. Pezzagna S, Rogalla D, Wildanger D, Meijer J, Zaitsev A (2011) Creation and nature of optical centres in diamond for single-photon emission—overview and critical remarks. New J Phys 13:035024
7. Maze JR, Cappellaro P, Childress L, Dutt MVG, Hodges JS, Hong S, Jiang L, Stanwix PL, Taylor JM, Togan E, Zibrov AS, Hemmer P, Yacoby A, Walsworth RL, Lukin MD (2009) Nanoscale magnetic sensing using spin qubits in diamond. Proc SPIE 7225:722509
8. Childress L, Hanson R (2013) Diamond NV centers for quantum computing and quantum networks. MRS Bull 38:134–138
9. Palyanov YN, Kupriyanov IN, Borzdov YM, Khokhryakov AF, Surovtsev NV (2016) High-pressure synthesis and characterization of ge-doped single crystal diamond. Cryst Growth Des 16:3510–3518
10. Masuya S, Hanada K, Oshima T, Sumiya H, Kasu M (2017) Formation of stacking fault and dislocation behavior during the high-temperature annealing of single-crystal HPHT diamond. Diamond Relat. Mater. 75:155–160
11. Masuya S, Kasu M (2018) Dislocations in chemical vapor deposition (111) single crystal diamond observed by synchrotron X-ray topography and their relation with etch pits. Diamond Relat Mater 90:40–46
12. Tallaire A, Mille V, Brinza O, Tran Thi TN, Brom JM, Loguinov Y, Katrusha A, Koliadin A, Achard J (2017) Thick CVD diamond films grown on high-quality type IIa HPHT diamond substrates from New Diamond Technology. Diamond Relat Mater 77:146–152

13. Sumiya H, Tamasaku K (2012) Large defect-free synthetic type IIa diamond crystals synthesized via high pressure and high temperature. Jpn J Appl Phys 51:090102
14. Tatsumi N, Tamasaku K, Ito T, Sumiya H (2017) Behavior of crystal defects in synthetic type-IIa single-crystalline diamond at high temperatures under normal pressure. J Cryst Growth 458:27–30
15. Naamoun M, Tallaire A, Doppelt P, Gicquel A, Legros M, Barjon J, Achard J (2015) Reduction of dislocation densities in single crystal CVD diamond by using self-assembled metallic masks. Diamond Relat Mater 58:62–68
16. D'Haenens-Johansson UFS, Katrusha A, Moe KS, Johnson P, Wang WY (2015) Large colorless HPHT-grown synthetic gem diamonds from new diamond technology, Russia. Gems Gemol 51:260–279
17. Guo MM, Li SS, Hu MH, Su TC, Wang JZ, Gao GJ, You Y, Nie Y (2020) The growth characteristics of type IIa large single crystal diamond with Ti/Cu as nitrogen getter in FeNi-C system. Chin Phys B 29:018101
18. Jia XP, Hayakawa S, Ugajin K, Wakatsuki N, Sugiyama E, Takahashi T, Ohnishi A, Gohshi Y, Wakatsuki M (1998) Iron and chromium as impurities in artificial diamonds. Koatsuryoku no Kagaku to Gijutsu 7:998–1000
19. Jia XP, Hayakawa S, Li W, Gohshi Y, Wakatsuki M (1999) Cobalt impurities in synthetic diamond. Diamond Relat Mater 8:1895–1899
20. Shiryaev AA, Zolotov DA, Suprun OM, Ivakhnenko SA, Averin AA, Buzmakov AV, Lysakovskyi VV, Dyachkova IG, Asadchikov VE (2018) Unusual types of extended defects in synthetic high pressure-high temperature diamonds. CrystEngComm 20:7700–7705
21. Suprun EM, Kalenchuk VA, Ivakhnenko SA, Kutsai AM, Lysakovskii VV, Zanevskii OA, Dufu V, Shenlin V (2016) Defectand-impurity state of type Ib diamond single crystals of cubic habit. J Superhard Mater 38:377–381
22. Li YD, Jia XP, Yan BM, Chen N, Fang C, Li Y, Sun SS, Ma HA (2016) Influence of carbon convection field on high quality large single crystal diamonds morphology under high pressure and high temperature. RSC Adv 6:40330–40335
23. Li R, Xu BJ, Zhang QC, Gu X, Zheng GL, Ma HA, Jia XP (2016) Finite-element analysis on pressure transfer mechanism in large-volume cubic press. High Pressure Res 36:575–584
24. Li R, Ding MM, Shi TF (2018) Finite element design for the HPHT synthesis of diamond. J Cryst Growth 491:111–115
25. Li R, Wang YF, Liu J, Jia XP, Ma HA (2020) Design an effective solution for commercial production of 18/20 mesh diamond grits using large volume cubic high pressure apparatus. Diamond Relat Mater 101:107593
26. Han QG, Li MZ, Jia XP, Ma HA, Li YF (2011) Modeling of effective design of high pressure anvils used for large scale commercial production of gem quality large single crystal diamond. Diamond Relat Mater 20:969–973
27. Blank V, Popov M, Pivovarov G, Lvova N, Terentev S (1999) Mechanical properties of different types of diamond. Diamond Relat Mater 8:1531–1535
28. Yan CS, Mao HK, Li W, Qian J, Zhao YS, Hemley RJ (2004) Ultrahard diamond single crystals from chemical vapor deposition. Phys Status Solidi A 201:R25–R27
29. Kennedy CS, Kennedy GC (1976) The equilibrium boundary between graphite and diamond. J Geophys Res 81:2467–2470
30. Kurz M, Pusztai A, Muller G (1999) Development of a new powerful computer code Crys++ especially designed for fast simulation of bulk crystal growth processes. J Cryst Growth 198(199):101–106
31. Chen N, Ma HA, Yan BM, Chen LC, Chen LX, Guo LS, Miao XY, Fang C, Jia XP (2018) Characterization of various centers in synthetic Type Ib diamond under HPHT annealing. Cryst Growth Des 18:3870–3876
32. Fang C, Shen WX, Zhang YW, Mu PY, Zhang ZF, Jia XP (2019) Si doping effects on the growth of large single-crystal diamond in a Ni-based metal catalyst system under high pressure and high temperature. Cryst Growth Des 19:3955–3961

33. Guo LS, Ma HA, Chen LC, Chen N, Miao XY, Wang Y, Wang ZK, Yang ZQ, Fang C, Jia XP (2019) Synthesis of large diamond crystals with C3H8N4O2 as organic additive under HPHT conditions. Int J Refract Hard Met 79:47–52
34. Hainschwang T, Fritsch E, Massi L, Rondeau B, Notari F (2012) The C center isolated nitrogen-related infrared absorption at 2688 cm-1: perfect harmony in diamond. J Appl Spectrosc 79:737–743
35. Wang JK, Li SS, Hu MH, Su TC, Gao GJ, Guo MM, You Y, Nie Y (2020) C3H6N6 doping effect of synthetic diamond under high pressure and high temperature. Int J Refract Hard Met 87:105150
36. Chen N, Ma HA, Chen LX, Yan BM, Fang C, Liu XB, Li YD, Guo LS, Chen LC, Jia XP (2018) Effects of S on the synthesis of type Ib diamond under high pressure and high temperature. Int. J. Refract. Hard Met. 71:141–146
37. Fang C, Jia XP, Chen N, Li YD, Guo LS, Chen LC, Ma HA, Liu XB (2016) HPHT synthesis of N-H co-doped diamond single crystals. J Cryst Growth 436:34–39
38. Fang S, Ma HA, Cai ZH, Wang CX, Fang C, Zhao ZD, Lu ZY, Wang YK, Chen LC, Jia XP (2020) Study on the crack phenomenon of heavy FeS-doped Ib diamond crystals with 111 surface as growth surface in Fe-Ni-C system. CrystEngComm 22:602–609
39. Sung JC, Sung M, Sung E (2006) Diamond growth on an array of seeds: the revolution of diamond production. Thin Solid Films 498:212–219
40. Zhang SJ, To S, Wang SJ, Zhu ZW (2015) A review of surface roughness generation in ultra-precision machining. Int J Mach Tools Manuf 91:76–95
41. Zhang G, Dai Y, To S, Wu X, Lou Y (2019) Tool interference at workpiece centre in single-point diamond turning. Int J Mech Sci 151:1–12
42. Zhang SJ, To S, Zhang GQ (2017) Diamond tool wear in ultra-precision machining. Int J Adv Manuf Technol 88:613–641
43. Hashim U, Nadia E, Salleh S (2009) Nanotechnology development status in Malaysia: industrialization strategy and practices. Int J Nanoelectronics 1:119–134
44. Chen N, Ma HA, Yan BM, Chen LC, Chen LX, Guo LS et al (2018) Characterization of various centers in synthetic Type Ib diamond under HPHT annealing. Cryst Growth Des 18:3870–3876
45. Chen N, Ma HA, Fang C, Li YD, Liu XB, Zhou Z, Jia XP (2017) Synthesis and characterization of IIa-type S-doped diamond in FeNi catalyst under high pressure and high temperature conditions. Int J Refract Metals Hard Mater 66:122–126
46. Wang ZY, Dong LH, Wang DS, Dong YH (2012) Study of HPHT single crystal diamond as precision cutting tool material. Precis Eng 36:162–167
47. Fridrichova J, Bacik P, Skoda R, Antal P (2015) Use of spectroscopic methods for determination of diamond origin and treatment. Acta Geol. Slovaca 7:11–18
48. Bangert U, Barnes R, Gass MH, Bleloch AL, Godfrey IS (2009) Vacancy clusters, dislocations and brown colouration in diamond. J Phys Condens Matter 21:364208
49. Olson JR, Pohl RO, Vandersande JW, Zoltan A, Anthony TR, Banholzer WF (1993) Thermal conductivity of diamond between 170 and 1200 K and the isotope effect. Phys Rev B 47:14850
50. Herrmann M, Matthey B, Gestrich T (2019) Boron-doped diamond with improved oxidation resistance. Diamond Relat Mater 92:47–52
51. Mavrin BN, Denisov VN, Popova DM, Skryleva EA, Kuznetsov MS, Nosukhin SA et al (2008) Boron distribution in the subsurface region of heavily doped IIb type diamond. Phys Lett A 372:3914–3918
52. Gaillou E, Post JE, Rost D, Butler JE (2012) Boron in natural type IIb blue diamonds: chemical and spectroscopic measurements. Am Mineral 97:1–18
53. Ramasubramanian K, Arunachalam N, Rao MR (2017) Investigation on tribological behaviour of boron doped diamond coated cemented tungsten carbide for cutting tool applications. Surf Coat Technol 332:332–340
54. Oomen JM, Eisses J (1992) Wear of monocrystalline diamond tools during ultraprecision machining of nonferrous metals. Precis Eng 14:206–218
55. J Field (2012) The mechanical and strength properties of diamond. Rep Prog Phys 75:126505

56. Buchkremer-Hermanns H, Long C, Weiss H (1996) ECR plasma polishing of CVD diamond films. Diam Relat Mat 845–849
57. Gloor S, Lüthy W, Weber HP, Pimenov SM, Ralchenko VG, Konov VI et al (1999) UV laser polishing of thick diamond films for IR windows. Appl Surf Sci 138–139:135–139
58. Grogan DF, Zhao T, Bovard BG, Macleod HA (1992) Planarizing technique for ion-beam polishing of diamond films. Appl Opt 31:1483–1487
59. Wang CY, Zhang FL, Kuang TC, Chen CL (2006) Chemical/mechanical polishing of diamond films assisted by molten mixture of LiNO3 and KNO3. Thin Solid Films 698–702
60. Tokura H, Yang C-F, Yoshikawa M (1992) Study on the polishing of chemically vapour deposited diamond film. Thin Solid Films 212:49–55
61. Sun Y, Wang S, Tian S, Wang Y (2006) Polishing of diamond thick films by Ce at lower temperatures. Diam Relat Mat 15:1412–1417
62. Chen Y, Nguyen T, Zhang LC (2009) Polishing of polycrystalline diamond by the technique of dynamic friction—Part 5: quantitative analysis of material removal. Int J Mach Tools Manuf 49:515–520
63. Liu H, Zong W, Cheng X (2021) Behaviors of carbon atoms induced by friction in mechanical polishing of diamond. Comput Mater Sci 186:110069
64. Zong WJ, Cheng K, Li D, Sun T, Liang YC (2007) The ultimate sharpness of single-crystal diamond cutting tools—Part I: theoretical analyses and predictions. Int J Mach Tools Manuf 852–863
65. Couto M, van Enekevort WJP, Wichman B, Seal M (1992) Scanning tunneling microscopy of polished diamond surfaces. Appl Surf Sci 62:263–268
66. Grillo S, Field J (1997) The polishing of diamond. J Phys D Appl Phys 202
67. Brezoczky B, Seki H (1990) Triboattraction: friction under negative load. Langmuir 6:1141–1145
68. Zong WJ, Wu D, Li ZQ (2016) Strength dependent design method for the crystal orientation of diamond Berkovich indenter. Mater Des 89:1057–1070
69. Zong WJ, Li ZQ, Sun T, Cheng K, Li D, Dong S (2010) The basic issues in design and fabrication of diamond-cutting tools for ultra-precision and nanometric machining. Int J Mach Tools Manuf 411–419
70. Zong WJ, Sun T, Li D, Cheng K (2009) Design criterion for crystal orientation of diamond cutting tool. Diam Relat Mat 18:642–650
71. Kim JS, Park YM, Bae MK, Kim CW, Kim DW, Shin DC, Kim TG (2018) Cutting performance of tungsten carbide tools coated with diamond thin films after etching for various times. Mod Phys Lett B 32:1850236
72. Chen N, Zhang G, Li R, Xu G, Wang F, Ma H, Jia X (2020) Defect and stress reduction in high-pressure and high-temperature synthetic diamonds using gradient cooling technology. Cryst Growth Des 20:3358–3364
73. Tanaka H, Shimada S, Higuchi M, Yamaguchi T, Kaneeda T, Obata K (2005) Mechanism of cutting edge chip and its suppression in diamond turning of copper. CIRP Ann 54:51–54
74. Prawer S, Nemanich RJ (2004) Raman spectroscopy of diamond and doped diamond. Philos Trans R Soc A 362:2537–2565
75. Li Y, Jia X, Song M, Ma HA, Zhou Z, Fang C et al (2015) Effects of hydrogen on diamond single crystal synthesized under high pressure and high temperature. Mod Phys Lett B 29:1550162
76. Zhou Z, Wang F, Huang H, Jiang H, Zheng Y, Peng J, Yue L (2020) Synthesis and characterization of diamond with KBH4 additive under high pressure and high temperature conditions. Mod Phys Lett B 34:2050190
77. Chen N, Ma HA, Chen LC, Yan BM, Fang C, Liu XB et al (2018) Effects of S on the synthesis of type Ib diamond under high pressure and high temperature. Int J Refract Met Hard Mater 71:141–146
78. Guo MM, Li SS, Hu MH, Su TC, Wang JZ, Gao GJ et al (2020) Growth characteristics of type IIa large single crystal diamond with Ti/Cu as nitrogen getter in FeNi-C system. Chin Phys B 29:018101

79. Kim YK, Park JH, Lee JW (2018) Facile nano-templated $CO_2$ conversion into highly interconnected hierarchical porous carbon for high-performance supercapacitor electrodes. Carbon 126:215–224
80. Liu X, Wang Y, Dong L, Chen X, ZN, Zhang Y, Zang J (2016) One-step synthesis of shell/core structural boron and nitrogen co-doped graphitic carbon/nanodiamond as efficient electrocatalyst for the oxygen reduction reaction in alkaline media. Electrochim Acta 194:161–167
81. Lee S, Heo Y, Bratescu MA, Ueno T, Saito N (2017) Solution plasma synthesis of a boron-carbon-nitrogen catalyst with a controllable bond structure. Phys Chem Chem Phys 19:15264–15272
82. Seo JH, Wu H, Mikael S, Mi H, Blanchard JP, Venkataramanan G et al (2016) Thermal diffusion boron doping of single-crystal natural diamond. J Appl Phys 119:205304
83. Prakash A, Sundaram KB (2017) Optical and XPS studies of BCN thin films by co-sputtering of B4C and BN targets. Appl Surf Sci 396:484–491
84. Yip WS, To S (2018) Sustainable manufacturing of ultra-precision machining of titanium alloys using a magnetic field and its sustainability assessment. Sustain Mater Technol 16:38–46
85. Zong WJ, Sun T, Li D, Cheng K (2009) Design criterion for crystal orientation of diamond cutting tool. Diamond Relat Mater 18:642–650
86. Shimada S, Inamura T, Higuchi M, Tanaka H, Ikawa N (2000) Suppression of tool wear in diamond turning of copper under reduced oxygen atmosphere. CIRP Ann 49:21–24
87. Sumiya H, Ikeda K, Arimoto K, Harano K (2016) High wear-resistance characteristic of boron-doped nano-polycrystalline diamond on optical glass. Diamond Relat Mater 70:7–11
88. Kupriyanov IN, Khokhryakov AF, Borzdov YM, Palyanov YN (2016) HPHT growth and characterization of diamond from a copper-carbon system. Diamond Relat Mater 69:198–206
89. Furushiro N, Higuchi M, Yamaguchi T, Shimada S, Obata K (2009) Polishing of single point diamond tool based on thermo-chemical reaction with copper. Precis Eng 33:486–491
90. Yuan Z, He J, Yao Y (1992) The optimum crystal plane of natural diamond tool for precision machining. CIRP Ann 41:605–608
91. Zhang S, To S, Zhang G (2017) Diamond tool wear in ultra-precision machining. Int J Adv Manuf Technol 88:613–641

The manufacturer's authorised representative in the EU is Springer Nature Customer Service Centre GmbH, Europaplatz 3, 69115 Heidelberg, Germany. If you have any concerns regarding our products, please contact ProductSafety@springernature.com

Printed and bound by CPI Group (UK) Ltd, Croydon, CR0 4YY

30/09/2026

02225302-0004